MATHEMATICS FOR MANAGEMENT

MATHEMATICS

FOR MANAGEMENT

GARY J. BRONSON
College of Business Administration
Fairleigh Dickinson University, Madison

RICHARD BRONSON
Department of Mathematics and Computer Science
Fairleigh Dickinson University, Teaneck

iep A Dun-Donnelley Publisher, New York

Copyright © 1977 by Thomas Y. Crowell Company, Inc.
All Rights Reserved
Except for use in a review, the reproduction or utilization of this work in any form or by any electronic, mechanical, or other means, now known or hereafter invented, including photocopying and recording, and in any information storage and retrieval system is forbidden without the written permission of the publisher.
Published simultaneously in Canada by Fitzhenry & Whiteside, Ltd., Toronto.

Library of Congress Cataloging in Publication Data

Bronson, Gary J.
  Mathematics for management.

  (The IEP series in mathematics)
  Includes index.
  1. Business mathematics. I. Bronson, Richard, joint author. II. Title.
HF5691.B76 1977   513'.93   76-30746
ISBN 0-7002-2503-X

IEP—A Dun-Donnelley Publisher
666 Fifth Avenue
New York, New York 10019

Typography by The Bronx

Manufactured in the United States of America

To
Uncle Max,
Rochelle,
and the twins

CONTENTS

PREFACE	xi
Part 1 EQUATIONS	**1**
1. Elementary Concepts	**3**
1.1 Signed Numbers	3
1.2 One Equation, One Unknown	7
1.3 Roundoff	12
1.4 Exponents	14
1.5 The Quadratic Formula	21
1.6 Scientific Notation	24
1.7 Logarithms	25
1.8 Computing with Logarithms	30
1.9 Cartesian Coordinate System	33
1.10 Graphs	39
1.11 Sigma Notation	47
2. Equations and Curves	**53**
2.1 Linear Equations and Straight Lines	54
2.2 Properties of Straight Lines	64
2.3 Break-Even Analysis	72
2.4 Quadratic Equations and Curves	76
2.5 Polynomial Equations and Curves	82
2.6 Exponential Equations and Curves	84

3. **The Value of Money** — 87
   - 3.1 Compound Interest — 87
   - 3.2 Time Value of Money — 92
   - 3.3 Present Value of a Cash Flow — 98
   - 3.4 Future Value of an Annuity — 106
   - 3.5 Effective Interest — 114

4. **Matrix Operations** — 117
   - 4.1 Matrices — 117
   - 4.2 Elementary Operations — 120
   - 4.3 Matrix Multiplication — 126
   - 4.4 Simultaneous Linear Equations — 132
   - 4.5 Elementary Row Operations — 138
   - 4.6 Solutions of Simultaneous Linear Equations — 144
   - 4.7 Matrix Inversion — 149

5. **Linear Programming** — 155
   - 5.1 Inequalities — 155
   - 5.2 Systems of Linear Inequalities — 161
   - 5.3 Optimizing a Functional: A Geometric Approach — 168
   - 5.4 Linear Programming Problems — 175
   - 5.5 Slack Variables and Standard Form — 187
   - 5.6 Maximizing a Functional: The Simplex Method — 191
   - 5.7 Minimizing a Functional: The Simplex Method — 199

Part 2 CALCULUS — 205

6. **The Derivative** — 207
   - 6.1 Concept of a Function — 207
   - 6.2 Mathematical Functions — 213
   - 6.3 Average Rate of Change — 218
   - 6.4 Instantaneous Rates of Change — 224
   - 6.5 The Derivative — 236
   - 6.6 Additional Rules — 242
   - 6.7 Higher-Order Derivatives — 246

7. **Applications of the Derivative** — 249
   - 7.1 Optimization through Differentiation — 250
   - 7.2 Modeling — 259
   - 7.3 Profit, Revenue, and Cost — 261
   - 7.4 Inventory Control* — 267
   - 7.5 Econometrics* — 274

* Advanced or optional material.

8.	**The Integral**	**279**
	8.1 Areas	279
	8.2 Antiderivatives	288
	8.3 The Definite Integral	294
	8.4 Applications of the Integral	299
	8.5 Substitution of Variables	306

Part 3 PROBABILITY — 311

9.	**Sets**	**313**
	9.1 Sets	313
	9.2 The Algebra of Sets	316
	9.3 Venn Diagrams	321
	9.4 Sample Spaces	329
10.	**Permutations and Combinations**	**333**
	10.1 Trees	333
	10.2 Fundamental Theorem of Counting	340
	10.3 Permutations	344
	10.4 Combinations	348
	10.5 Counting Complex Processes*	353
11.	**Probability**	**361**
	11.1 Equal Probability—Simple Processes	362
	11.2 Equal Probability—Complex Processes*	367
	11.3 The Rules of Probability	373
	11.4 Conditional Probability	381
	11.5 Independent Events	388
	11.6 Bernoulli Trials	393
	11.7 Expected Value	397

Appendix A.	**Curve Fitting**	**401**
	A.1 Constant Curve Fit	403
	A.2 Linear Least-Squares Fit	408
	A.3 Quadratic Least-Squares Fit	417
	A.4 Exponential Least-Squares Fit	423
	A.5 Selecting an Appropriate Curve	427
Appendix B.	**Mathematics of Finance**	**429**
	B.1 Finite Geometric Sums	429
	B.2 Installment Loans and Interest Charges	433
	B.3 Mortgages and Amortization	436

* Advanced or optional material.

Appendix C.	**Tables**	**441**
C.1	Common logarithms	442
C.2	Exponential functions	444
C.3	Values of $(1+i)^n$	445
C.4	Values of $(1+i)^{-n}$	449
C.5	Values of $a_{\overline{n}|i}$	453
C.6	Values of $s_{\overline{n}|i}$	457

Solutions to Selected Odd-Numbered Problems   **461**

Index   **487**

PREFACE

Over the past twenty years quantitative methods have become essential to business and business decision making. The arithmetic operations appropriate to the small neighborhood stores of the past are ill-suited to the complex problems facing the large corporations of the present. Economic forecasting, allocation of resources, capital budgeting, and determining of product effectiveness are but a few such problems.

The aim of this book is to introduce commercially applicable quantitative methods to business students. We understand that the average business major knows that mathematics is a powerful tool for solving problems, but feels uneasy, sometimes fearful, about the subject matter. Our primary concern in writing this text is to present the material in a clear, understandable, and nonintimidating manner. By numerous examples we relate each new mathematical concept to a commercial problem so that the student never loses sight of the ultimate goal: to develop mathematical tools to solve business problems. We are not interested in mathematics as an end unto itself.

The book assumes student familiarity with algebraic concepts but not facility in using them. Chapter 1, Elementary Concepts, covers algebraic topics that will be used in the rest of the text. Students with strong mathematical backgrounds are encouraged to skim this chapter and begin the text with Chapter 2. Generally, however, we have found that most students are sufficiently ill at ease with algebraic concepts, and that class time should be devoted to Chapter 1.

The remainder of Part One deals with mathematical equations and their application to a variety of business concerns. Chapter 2 covers the traditional material on polynomial equations and their graphs, while Chapter 4 deals with systems of equations in a matrix setting. Chapters 3 and 5 are devoted to the two most important applications of equations, elementary finance and linear programming.

Part Two introduces and develops the rudiments of differential and integral calculus. The derivative is preceded by a detailed discussion of functions and rates of change, and an intuitive development of limits. The methodology of calculating derivatives and integrals is followed by realistic applications in the business world.

Our goal in Part Two is to bring the student to an appreciation and awareness of calculus as a powerful mathematical approach for analyzing and modeling commercial systems. The business student will probably not use calculus techniques, especially integration, directly, but he or she needs enough awareness of its potential to be able, as a decision maker, to use and understand the abilities of a technical staff. The applications in Chapters 7 and 8 were developed with this philosophy in mind. They provide a setting for introducing mathematical models and exposing students to realistic applications of calculus. Although we do not expect the reader of the book to become an expert in modeling, we do hope that these applications develop an understanding of how calculus can be used. Asterisks in the table of contents and the text indicate the more advanced or optional sections.

Part Three treats the elementary concepts of probability and serves as the bridge between an introductory course in mathematical methods and the usual follow-up course in business statistics.

This book may serve as a text for either a one- or two-semester course. Part One by itself is sufficient for the basis of a one-semester course. Topics in Appendixes A and B can be added if time permits. Parts Two and Three provide more than enough material for a second semester.

Interest rates, cash flows, and annuities are purposely treated near the beginning of this book (Chapter 3) both as a useful application of equations and as motivation. The subject matter is relatively easy to grasp and since most students find it interesting, it provides an opportunity for early success in using mathematical methods. The material in Appendix A on least-squares analysis is included to answer the usual question asked by students as to the origin of the equations they have been using.

Finally, we are indebted to a number of people for helping to make this book a reality. First, our appreciation goes to our students, who used most of the material in this book in prepublication form. We also owe our thanks to Edith Dolin and Jeffrey Kingsley for their comments and suggestions on various portions of the book, to Willard Preston for his assistance in proofreading the entire manuscript, and to Gwendolyn Markus for her expert typing of the manuscript. We are also indebted to Alan Turner and Jean Woy for their encouragement and help in this endeavor, and to Marge Lakin and Carol J. Dean for the final editing of the manuscript. Finally, and most importantly, we owe our deep appreciation and thanks to our wives, Rochelle and Evelyn.

The IEP SERIES in MATHEMATICS
under the consulting editorship of

RICHARD D. ANDERSON
Louisiana State University

ALEX ROSENBERG
Cornell University

MATHEMATICS FOR MANAGEMENT

EQUATIONS

Part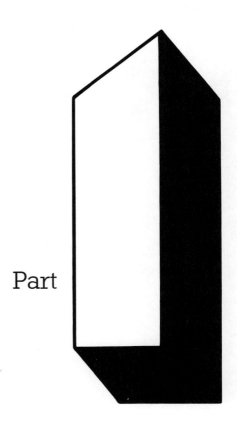

Chapter

Elementary Concepts

This chapter is a review of the topics in algebra used in the rest of the book. Readers who already have a working knowledge of this material are advised to skim over the chapter and go directly to Chapter 2. Others are advised to spend as much time as is necessary to master this material before proceeding further.

1.1 Signed Numbers

Many people have no difficulty performing the basic arithmetic operations (addition, subtraction, multiplication, and division) on positive numbers, but find similar operations on negative numbers mind-boggling. Perhaps the operations would make more sense if they were related to profit and loss in a commercial setting.

In business, positive numbers represent profits, and negative numbers represent losses. In particular, $-\$5.00$ denotes a loss of \$5.00, $-\$10.25$ denotes a loss of \$10.25, and $+\$3.00$ denotes a profit or gain of \$3.00. By convention, a number without a sign is taken as positive. Therefore, $3 = +3$, $5 = +5$, and $9.75 = +9.75$.

In the context of profits and losses, the addition sign is read "followed by." Then, $-3 + 5$ is a \$3 loss *followed by* a \$5 gain; the result is a net gain of \$2, so $-3 + 5 = 2$. For some, the process is clarified when viewed in the context of betting at a horse race. If a person loses $3 on the first race and then wins $5 on the second race, he or she will have a net gain of $2 at the end. Again, $-3 + 5 = 2$.

To calculate $3 + (-7)$, we reason similarly. A $3 profit followed by a $7 loss results in a net loss of $4. Alternatively, if a person wins $3 on the first race but loses $7 on the second race, he or she will then be behind $4. Either way, $3 + (-7) = -4$.

The same reasoning is valid when adding two negative numbers. The quantity $(-10) + (-8)$ denotes a $10 loss followed by an $8 loss. Or it can be viewed as a person losing $10 on the first bet and then losing $8 on the second bet. The end result is the same, a total loss of $18. Therefore $-10 + (-8) = -18$.

Viewed as profits and losses, the following results should be straightforward:

$5 + 7 = 12$
$-9 + 3 = -6$
$2 + (-4) = -2$
$8 + (-5) = 3$
$-7 + (-8) = -15$.

The multiplication of signed numbers is a two-step operation. The first step is to multiply the numbers disregarding any negative signs (treat all numbers as positive). The second step is to determine the appropriate sign for the result. Here the following rules apply. When both numbers have the same sign (either both positive or both negative), the result is positive. When both numbers have different signs (one positive and one negative), the result is negative.

To calculate -5 times $+2$, we first multiply 5 by 2 (disregarding all negative signs) and obtain 10. To determine the appropriate sign for -5 times 2, we note that both numbers have different signs (a negative 5 and a positive 2), so the result is negative. Accordingly -5 times $+2$ is -10.

Mathematically, two sets of parentheses next to each other denote multiplication. We write $(3)(7)$ for 3 times 7, $(8)(-4)$ for 8 times -4, and $(-3)(-6)$ for -3 times -6. In particular, $(-5)(2)$ denotes -5 times 2, which was just found to be $(-5)(2) = -10$.

To calculate $(8)(-\frac{1}{2})$, we first disregard the negative sign and multiply the positive numbers, obtaining $(8)(\frac{1}{2}) = 4$. Since $+8$ and $-\frac{1}{2}$ have different signs, their product is negative. Accordingly, $(8)(-\frac{1}{2}) = -4$. To calculate $(-3)(-6)$, we first calculate $(3)(6) = 18$. Since -3 and -6 have the *same* sign (both negative), we conclude that $(-3)(-6) = 18$. Similarly,

$(3)(-4) = -12$
$(-3)(4) = -12$
$(-3)(-4) = 12$
$(5.1)(-0.2) = -1.02$
$(-\frac{1}{3})(-\frac{6}{7}) = \frac{2}{7}$.

1.1 Signed Numbers

Division follows the same pattern as multiplication. First divide the two numbers disregarding signs and assuming all numbers to be positive. Then determine the appropriate sign exactly as in multiplication. When both numbers have the same sign, their quotient is positive; when both numbers have different signs, their quotient is negative.

To calculate, $-6 \div 3$ or, equivalently, $-6/3$, we first divide $+6$ by $+3$, obtaining 2. Since -6 and $+3$ have different signs, their quotient is negative. Thus $-6/3 = -2$. To calculate $-12/-3$, we first find $12/3 = 4$. Since both -12 and -3 have the same sign, their quotient is positive and $-12/-3 = +4$. Similarly,

$$8/-4 = -2$$
$$4/-8 = -0.5$$
$$-11/-5 = 2.2$$
$$2.1/-3.2 = -0.65625.$$

The most difficult operation for many is subtraction. Fortunately, subtraction can be converted to addition and the standard addition rules will then apply. The key step is to introduce a plus sign before the subtraction sign and then incorporate the minus sign into the second number. We write $8 - 10$ as $8 + (-10)$; we write $-7 - 11$ as $-7 + (-11)$; we write $-3 - (-8)$ as $-3 + [-(-8)]$. Each subtraction then becomes an addition. If as a result there are two negative signs next to each other, such as $-(-8)$, simplify by recalling that a negative times a negative is a positive. That is, $-(-8) = +8$.*

To calculate $-7 - 11$, we rewrite the expression as $-7 + (-11)$. Now we have a \$7 loss followed by an \$11 loss, which results in a net loss of \$18. That is, $-7 - 11 = -7 + (-11) = -18$. Similarly,

$$8 - 10 = 8 + (-10) = -2$$
$$-3 - (-8) = -3 + [-(-8)] = -3 + 8 = 5$$
$$-9 - (-2) = -9 + [-(-2)] = -9 + 2 = -7$$
$$7 - 4 = 7 + (-4) = 3$$
$$-3 - 4 = -3 + (-4) = -7.$$

Occasionally, one is faced with a series of operations such as $(-8)(-3 + 5)$. The procedure is to combine two numbers at a time. Since $(-8)(-3 + 5)$ is -8 times the quantity $-3 + 5$, we first calculate the sum $-3 + 5$, which equals 2. Then

$$(-8)(-3 + 5) = (-8)(2) = -16.$$

---

* Every negative sign can be considered a -1. Therefore $-(-8) = (-1)(-8) = +8$ from the multiplication rules just developed.

A more complicated expression is $7[-3 - 2(8 - 10)]$. We first compute $8 - 10$ and then multiply this result by 2. Here

$$7[-3 - 2(8 - 10)] = 7[-3 - 2(-2)] = 7[-3 - (-4)].$$

The quantity $-3 - (-4)$ can be rewritten as $-3 + [-(-4)]$ which is $-3 + 4 = 1$. Finally,

$$7[-3 - 2(8 - 10)] = 7[-3 - (-4)] = 7[1] = 7.$$

Similarly,

$$\begin{aligned}5[3(2 - 8) - 4(5 - 6)] &= 5[3(-6) - 4(5 - 6)] \\ &= 5[-18 - 4(5 - 6)] \\ &= 5[-18 - 4(-1)] \\ &= 5[-18 - (-4)] \\ &= 5[-18 + (-(-4))] \\ &= 5[-18 + 4] \\ &= 5[-14] \\ &= -70\end{aligned}$$

and

$$\begin{aligned}[5 + (-3)][2 - 7] &= [2][2 - 7] \\ &= 2[2 + (-7)] \\ &= 2(-5) \\ &= -10.\end{aligned}$$

The general rule is to calculate the expression within the innermost parentheses or brackets first and then systematically work toward the outermost parentheses or brackets.

All signed numbers obey certain properties of arithmetic. For example, the order in which addition is performed is immaterial. In particular, $5 + 7 = 7 + 5$, $-2 + 8 = 8 + (-2)$, $5 + [8 + (-3)] = [5 + 8] + (-3)$, and $-6 + [-7 + 4] = [-6 + (-7)] + 4$. The only requirement is that each number in a sum be added once. If we let a, b, and c denote any signed numbers, either positive or negative, we can state these rules formally as

$a + b = b + a$ (commutative law for addition)
$(a + b) + c = a + (b + c)$ (associative law for addition).

Similar rules hold for multiplication; the order in which multiplication is performed is irrelevant. Clearly, $(5)(7) = (7)(5)$, $(-2)(8) = (8)(-2)$, $(5)[(8)(-3)] = [(5)(8)](-3)$, and $(-6)[(-7)(4)] = [(-6)(-7)](4)$. Again, the only requirement is that each number in a product be multiplied once. Formally,

$(a)(b) = (b)(a)$ (commutative law for multiplication)
$(a)[(b)(c)] = [(a)(b)](c)$ (associative law for multiplication).

The left side of the last equation indicates that first b and c are multiplied together and the result then multiplied by a. The right side indicates that first a and b are multiplied together and the result then multiplied by c. The equality indicates both procedures yield the same result. Specifically, with $a = -3$, $b = -2$, and $c = 8$, $(-3)[(-2)(8)] = (-3)(-16) = 48$ and $[(-3)(-2)](8) = (6)(8) = 48$, which are indeed equal.

Other rules are useful if the operations of addition and multiplication are mixed. They are

$(a)[b + c] = (a)(b) + (a)(c)$ (left distributive law)
$[b + c](a) = (b)(a) + (c)(a)$ (right distributive law).

If, as an example, $a = -3$, $b = -2$, and $c = 8$, the left side of the equation for the left distributive law becomes $(-3)[-2 + 8] = (-3)(6) = -18$, while the right side becomes $(-3)(-2) + (-3)(8) = 6 + (-24) = -18$.

Note that each side of the equations for the distributive laws is arithmetically different. On the left side, two numbers are first added and the result then multiplied by a. On the right side, two pairs of numbers are first multiplied and the results then added. The final results on both sides, however, are equal.

Exercises

Evaluate the following expressions.

1. $3 + (-6)$
2. $-4 + 7$
3. $19.7 + (-18.1)$
4. $-6.2 + (-8.1)$
5. $-9 + (-\frac{1}{2})$
6. $-4.1 + 7.1$
7. $9(18)$
8. $9(-8)$
9. $(-9)(18)$
10. $(-9)(-18)$
11. $(2)(-\frac{1}{4})$
12. $(-5)(-\frac{1}{6})$
13. $(-6.1)(2.3)$
14. $(-8)(-1.4)$
15. $(-8)/(-2)$
16. $8/(-2)$
17. $-8/2$
18. $-2/8$
19. $4/(-5)$
20. $(-5)/(-4)$
21. $-2.2/4$
22. $8 - 4$
23. $4 - 8$
24. $-4 - 8$
25. $-4 - (-8)$
26. $-8 - 4$
27. $-8 - (-4)$
28. $2.1 - 5.6$
29. $-5.6 - 2.1$
30. $\frac{1}{10} - \frac{1}{5}$
31. $2[5 + (-3)]$
32. $-2[1 + (-6)]$
33. $-4(1 - 3) + 2(2 - 5)$
34. $6[2(-1 + 7) - 3]$
35. $(1.6)(1.9 - 2.1) - 6.3$
36. $4[(-1)(2 - 9) + 7(3 - 4)]$
37. $\dfrac{8[1 - (-8)] - 2[7 - 1]}{2}$
38. $\dfrac{[(8 - 12)/4] + 4(9 - 3)}{5 - 9}$
39. $\dfrac{(5 - 11)(8 - 14) + 4(2 + 3)}{7[2(1 + 30 - 3(2 - 5)]}$
40. $\dfrac{8[-3(4 - 1) - 6(4 - 8)]}{2[-8 - (-7)]}$

1.2 One Equation, One Unknown

One major use of arithmetic operations on signed numbers is solving a single equation for an unknown number. For example, finding x if $-2x = 10$, or finding y if $2 - y = 4$. The unknown number usually is written as a letter.

The problem is to find a numerical value (or values) for the unknown that will satisfy the given equation.

Two notational conventions are universally followed when writing equations with unknowns. First, parentheses are omitted for the product of a known number and an unknown quantity. In particular, $8y = (8)(y)$, $-3x = (-3)(x)$, and $\frac{1}{2}p = (\frac{1}{2})(p)$. Second, if the product involves a 1, the 1 is not written but simply understood. Accordingly, $y = 1y$, $x = 1x$, and $p = 1p$. Similarly, $-y = (-1)y$, $-x = (-1)x$, and $-p = (-1)p$.

A numerical value for an unknown is a *solution* if that value, when substituted for the unknown, makes the equality valid. To determine whether $x = 4$ is a solution of $-2x = 10$, substitute $x = 4$ into the equation. Then $-2x = (-2)(4) = -8$ which does not equal 10. The equation $(-2)(4) = 10$ is not valid; hence $x = 4$ is not a solution.

Example 1 Determine whether or not $x = 2$ is a solution of

$$\frac{5x + 3(x - 7)}{2x + 4} = -3.$$

Solution Substituting $x = 2$ into this equation, the left side becomes

$$\frac{(5)(2) + 3(2 - 7)}{2(2) + 4} = \frac{10 + 3(-5)}{4 + 4} = \frac{10 + (-15)}{8} = -\frac{5}{8}.$$

Since this does *not* equal the right side, 2, the proposed value of x is not a solution.

Example 2 Determine whether or not $p = \frac{5}{9}$ is a solution of $7 - p = 2 + 8p$.

Solution When $p = \frac{5}{9}$ is substituted into this equation, the left side becomes $7 - \frac{5}{9} = 7 + (-\frac{5}{9}) = \frac{63}{9} + (-\frac{5}{9}) = \frac{58}{9}$, and the right side becomes $2 + 8(\frac{5}{9}) = 2 + \frac{40}{9} = \frac{18}{9} + \frac{40}{9} = \frac{58}{9}$. Since these values *are* equal, $p = \frac{5}{9}$ is a solution.

One method for solving an equation for an unknown is trial and error. Guess a solution and then substitute it into the equation to see if it is valid. If not, guess again and continue guessing solutions until the correct one is found. Clearly this method is time-consuming. It could take many guesses before the correct value is found. It also could take forever; one may never guess the correct value.

A more systematic procedure is to use the arithmetic operations developed in Section 1.1 to isolate the unknown on one side of the equation. The correct value for this unknown will then be on the other side of the equation. The key is remembering that the unknown is really a number and should be treated as such. Remember that

$1x = x$

and

$0x = 0$.

Furthermore, the associative laws and distributive laws remain valid. For example,

$5 + (7 + y) = (5 + 7) + y = 12 + y$ (associative law for addition)
$(-2)(8y) = [(-2)(8)]y = -16y$ (associative law for multiplication)
$7p + 5p = (7 + 5)p = 12p$ (right distributive law)
$5a + (-5a) = [5 + (-5)]a = 0a = 0$ (right distributive law).

Suppose we have the quantity $x + 5$ and we want to isolate x. By adding -5 to this quantity, we obtain $(x + 5) + (-5)$. One application of the associative law for addition yields $(x + 5) + (-5) = x + [5 + (-5)] = x + 0 = x$, and the unknown is isolated. If we have the quantity $-7y$, we can isolate y by multiplying by $-\frac{1}{7}$ or by dividing by -7. With multiplication, one application of the associative law yields $(-\frac{1}{7})[-7y] = [(-\frac{1}{7})(-7)]y = 1y = y$. In general, additive factors such as $+5$ in $x + 5$ can be removed by adding their negatives, and multiplicative factors such as -7 in $-7y$ can be removed by multiplying by their reciprocals. In doing so, however, we must remember that, *whenever any arithmetic operation is performed on one side of an equation, an identical operation must be performed on the other side of the equation*. There are *no* exceptions. Any quantity can be added to, subtracted from, multiplied by, or divided into one side of an equation as long as the same operation is performed on the other side.

Example 3 Solve $-2x = 10$ for x.

Solution We try to isolate x on one side of the equation. To remove the multiplicative factor -2 from the left side of the equation, we multiply *both* sides of the given equation by $-\frac{1}{2}$ (or, alternatively, divide by -2). Then,

$(-\frac{1}{2})(-2x) = (-\frac{1}{2})(10)$
$[(-\frac{1}{2})(-2)]x = -5$ (associative law of multiplication)
$1x = -5$
$x = -5$.

Example 4 Solve $x + 7 = 5$ for x.

Solution We try to isolate x on one side of the equation. To remove the additive factor $+7$ from x, we add -7 to both sides of the equation. Then,

$(x + 7) + (-7) = 5 + (-7)$
$x + [7 + (-7)] = -2$ (associative law for addition)
$x + 0 = -2$
$x = -2$.

Example 5 Solve $2 - y = 4$ for y.

Solution We first remove $+2$ from the left side, thereby leaving only y terms on that side of the equation. This is done by adding -2 to both sides. Accordingly,

$$-2 + (2 - y) = -2 + 4$$
$$(-2 + 2) - y = 2 \quad \text{(associative law for addition)}$$
$$0 - y = 2$$
$$-y = 2.$$

We do not have y yet, but we are close. If we multiply the last equation by -1, we obtain

$$(-1)(-y) = (-1)(2)$$
$$y = -2.$$

Example 6 Solve $7 - p = 2 + 8p$.

Solution We begin by grouping all the p terms on the same side of the equation. One way is to add $-8p$ to both sides. Then,

$$(7 - p) + (-8p) = (2 + 8p) + (-8p)$$
$$[7 + (-p)] + (-8p) = 2 + [8p + (-8p)] \quad \text{(associative law for addition)}$$
$$7 + [(-1p) + (-8p)] = 2 + 0 \quad \text{(associative law for addition)}$$
$$7 + [(-1 + -8)p] = 2 \quad \text{(distributive law)}$$
$$7 + (-9p) = 2.$$

Next we isolate the p terms on the left side by adding -7 to both sides of the equation.

$$(-7) + [7 + (-9p)] = -7 + 2$$
$$(-7 + 7) + (-9p) = -5 \quad \text{(associative law for addition)}$$
$$0 + (-9p) = -5$$
$$-9p = -5.$$

Finally, we multiply both sides of this equation by $-\frac{1}{9}$ to obtain p by itself. Thus

$$(-\tfrac{1}{9})(-9p) = (-\tfrac{1}{9})(-5)$$
$$[(-\tfrac{1}{9})(-9)]p = \tfrac{5}{9} \quad \text{(associative law for multiplication)}$$
$$1p = \tfrac{5}{9}$$
$$p = \tfrac{5}{9}.$$

Example 7 Solve $3(x - 7) = \dfrac{5x + 9}{4}$ for x.

Solution To eliminate fractions, at least initially, we multiply both sides of the equation by 4. Then,

$$4[3(x - 7)] = 4\left[\frac{5x + 9}{4}\right]$$

$$12(x - 7) = 5x + 9.$$

Now, $12x - 84 = 5x + 9$ (distributive law)
$7x - 84 = 9$ (adding $-5x$ to both sides)
$7x = 93$ (adding 84 to both sides)
$x = \frac{93}{7}$ (multiplying both sides by $\frac{1}{7}$).

Example 8 Solve $\dfrac{5x + 3(x - 7)}{2x + 4} = -3.$

Solution To eliminate fractions, at least initially, we multiply both sides of the equation by $2x + 4$. Note that, since x is a number (unknown at the moment), so too is $2x + 4$; therefore we are multiplying by a number. Here

$$\left[\frac{5x + 3(x - 7)}{2x + 4}\right](2x + 4) = (-3)(2x + 4)$$

$$5x + 3(x - 7) = -3(2x + 4).$$

Then, $5x + (3x - 21) = -6x - 12$ (distributive law)
$8x - 21 = -6x - 12$ (adding $5x$ to $3x$)
$14x - 21 = -12$ (adding $6x$ to both sides)
$14x = 9$ (adding 21 to both sides)
$x = \frac{9}{14}$ (multiplying both sides by $\frac{1}{14}$).

Exercises

In Exercises 1 through 6 determine whether or not the proposed values of the unknowns are solutions of the given systems.

1. $2x + 3 = 1; x = -1.$
2. $y + 4 = 2y; y = 1.$
3. $2(p + 7) = 3p + 4; p = 1.$
4. $x + 3 = 2(x + 1) + 1; x = 0.$
5. $\dfrac{(s + 3)(s - 2)}{2s + 1} = s + 7; s = 1.$
6. $\dfrac{(2t + 3)(t - 1) + 1}{2(t + 3) + 1} = 3t - 4; t = 2.$

7. Determine whether or not $x = 1$ is a solution of the following equation if it is known that $y = 2$ and $z = 0$:

$$\frac{x(y-1) + yz}{y(x-z)} = \frac{x}{y}.$$

In Exercises 8 through 26 solve the given equations for the unknown numbers.

8. $x + 7 = 2$.
9. $7 = 2 + x$.
10. $y - 8 = -2$.
11. $8x = -16$.
12. $8p = 15$.
13. $-4p = 16$.
14. $s + 10 = 2s$.
15. $t - 10 = 4 - t$.
16. $2t + 1 = t - 5$.
17. $2x = 3(x + 1)$.
18. $5y - 1 = 4(y - 2)$.
19. $8(p - 2) = 7(2p + 1)$.
20. $7p + 1 = 7(p - 1) + 3p$.
21. $2(a + 7) - 4 = 3(a - 1) + 2a$.
22. $\dfrac{(x-1) + 5(x-4)}{8} = 2x + 1$.
23. $\dfrac{2(y-1) + 4}{y} = 8$.
24. $\dfrac{t-4}{5} = \dfrac{3t+1}{8}$.
25. $\dfrac{3(2t-6) + 4(t-8)}{7(6+t) - 8(t-4)} = -3$.
26. $\dfrac{8t + 9(1-t) + 7}{3} = \dfrac{2(t-8) - 5(t+1)}{2}$.

1.3 Roundoff

Expressing numbers in decimal form is necessary in most commercial transactions. One reason involves money. All financial figures are given in decimal form; the numbers to the left of the decimal point represent dollars, and the numbers to the right of the decimal point represent cents. Quoting the cost of an item as $1.20 is clearer than quoting six-fifths of a dollar. A second reason for using decimals is mathematical. It is easier to add 0.2 and 1.5 than it is to add $\frac{1}{5}$ and $1\frac{1}{2}$. A third reason, and an increasingly important one, is that computers accept and print numbers only in decimal form.

1.3 Roundoff

Most signed numbers in decimal form are nonending. For example, $\frac{1}{3} = 0.3333333\ldots$, $\sqrt{2} = 1.4142135\ldots$, and $\pi = 3.1415926\ldots$. Nonending decimals are inconvenient arithmetically and useless financially. No one pays $0.3333333… for goods or charges $1.4142135… for services. Instead one approximates these quantities by finite decimals. One-third of a dollar becomes 33¢, and the square root of $2.00 becomes $1.41. Converting nonending decimals to finitely long decimals is called *roundoff* or *rounding*.

The most common form of roundoff is called arithmetic rounding or just rounding for short. Here one first decides how many digits are to be retained and then changes all digits to the right of those being kept to zero. Zero digits to the right of a decimal point that are not themselves followed by a nonzero number are simply disregarded. Before changing any digits to zero, however, one checks the first digit not being kept. If it is greater than or equal to 5, the previous digit (which is the last one being kept) is increased by 1. If the first digit not being kept is less than 5, no change is made in the previous digit.

As an example, consider the number $\pi = 3.1415926\ldots$. To round this number to three decimal places, first look at the digit four places to the right of the decimal point. It is 5, so we increase the previous digit by 1 and write $\pi = 3.142$ rounded to three decimal places. To round the same number to two decimal places first look at the digit three places to the right of the decimal point. It is 1, which is less than 5. Thus $\pi = 3.14$ rounded to two decimal places.

Rounding is equally applicable to finite decimals. Anytime one uses the above procedure to reduce the number of digits in a number, one is rounding. In particular, $81.314 = 81.31$ rounded to two decimal places, $8595.62 = 8596$ rounded to units, and $0.0051724 = 0.0052$ rounded to four decimal places.

A second form of rounding is called rounding up. Here the last digit being kept in a number is automatically increased by 1 if *any one* of the discarded digits is not 0. For example, $8.1403 = 8.15$ rounded up to two decimal places (note that one of the discarded digits is a $3 \neq 0$), $1.38112 = 1.382$ rounded up to three decimal places, and $1.900 = 1.9$ rounded up to one decimal place. In the last example, we did not increase the 9, since all the discarded digits were 0.

Rounding up is used by banks for mortgage payments and merchants for determining prices. If a mortgage payment is $241.5723 exactly, a bank will charge $241.58. If the retail price of a product is $1.2905, a merchant will charge $1.30. Rounding up is employed when the user does not wish to absorb the losses incurred by regular rounding.

Still a third form of rounding is rounding down or truncation. Here the last number being kept is never changed, regardless of the magnitude of the numbers being discarded. Thus $89.318 = 89.31$ rounded down to two decimals, and $13.75 = 13$ rounded down to units.

Rounding down is often used in reporting the number of finished goods produced. At the end of a day, a company may have produced $82\frac{3}{4}$ cars, but reports a production of only 82 cars.

Exercises

In Exercises 1 through 5 round the given numbers to two decimal places and then to three decimal places.
1. $\frac{2}{3}$ 2. $\frac{4}{11}$ 3. $\frac{4}{17}$ 4. $\frac{12}{7}$ 5. $\frac{89}{31}$
6. Round the numbers given in Exercises 1 through 5 up to two decimals. Round these numbers down to two decimal places.

1.4 Exponents

Exponents provide a convenient notation for representing the product of a number times itself many times. For any signed number a (either positive or negative), we define

$$a^2 = (a)(a)$$
$$a^3 = (a)(a)(a)$$
$$a^4 = (a)(a)(a)(a)$$

and, in general,

$$a^n = \underbrace{(a)(a)(a) \cdots (a)}_{n \text{ times}},$$

where n denotes a positive integer (whole number). The quantity a^n is often read "the nth power of a." Thus

$$5^2 = (5)(5) = 25$$
$$(-4)^3 = (-4)(-4)(-4) = -64$$
$$(-\tfrac{1}{3})^4 = (-\tfrac{1}{3})(-\tfrac{1}{3})(-\tfrac{1}{3})(-\tfrac{1}{3}) = \tfrac{1}{81}$$

and

$$2^{10} = \underbrace{(2)(2)(2)(2) \cdots (2)}_{10 \text{ times}} = 1024.$$

One consequence of this definition is the property

$$\boxed{(a^n)(a^m) = a^{n+m},} \tag{1}$$

1.4 Exponents

where n and m are positive integers. As verification, note that

$$(5^2)(5^3) = [(5)(5)][(5)(5)(5)] = (5)^5 = (5)^{2+3}$$
$$(-2)^4(-2)^3 = [(-2)(-2)(-2)(-2)][(-2)(-2)(-2)] = (-2)^7 = (-2)^{4+3}$$

and

$$(-\tfrac{1}{3})^5(-\tfrac{1}{3})^2 = [(-\tfrac{1}{3})(-\tfrac{1}{3})(-\tfrac{1}{3})(-\tfrac{1}{3})(-\tfrac{1}{3})][(-\tfrac{1}{3})(-\tfrac{1}{3})]$$
$$= (-\tfrac{1}{3})^7 = (-\tfrac{1}{3})^{5+2}.$$

Equation (1) is valid only if the left side of the equation is a number raised to a power times that *same number* raised to a power. The formula is not valid if one a in Eq. (1) is replaced by another number b. In particular, Eq. (1) is *not* applicable to the product $(2)^5(3)^4$.

A second useful property of powers is

$$(a^n)^m = a^{nm}. \qquad (2)$$

Any number a raised to a power n, which is itself raised to a power m, is equal to a raised to the power n times m. Note that

$$(2^2)^3 = [(2)(2)]^3 = [(2)(2)][(2)(2)][(2)(2)] = 2^6 = 2^{(2)(3)}$$

and

$$[(-\tfrac{1}{3})^3]^3 = [(-\tfrac{1}{3})(-\tfrac{1}{3})(-\tfrac{1}{3})]^3$$
$$= [(-\tfrac{1}{3})(-\tfrac{1}{3})(-\tfrac{1}{3})][(-\tfrac{1}{3})(-\tfrac{1}{3})(-\tfrac{1}{3})][(-\tfrac{1}{3})(-\tfrac{1}{3})(-\tfrac{1}{3})]$$
$$= (-\tfrac{1}{3})^9 = (-\tfrac{1}{3})^{(3)(3)}.$$

Using Eqs. (1) and (2) together, we have

$$(2^4)^3(2^2)^4 = 2^{12}2^8 = 2^{12+8} = 2^{20}$$

and

$$[(-4)^3]^5[(-4)^2]^3 = (-4)^{15}(-4)^6 = (-4)^{21}.$$

One must be careful, however, not to confuse these two properties. A common error is to write $(a^n)^m = a^{n+m}$ or $(a^n)(a^m) = a^{nm}$. Both are incorrect and lead to wrong answers.

Equations (1) and (2) can be extended to negative powers if we first give meaning to such quantities. Accordingly, for any nonzero signed number a and any positive integer n, we define

$$\boxed{a^{-n} = \frac{1}{a^n}.} \qquad (3)$$

Therefore

$$5^{-2} = \frac{1}{5^2} = \frac{1}{25}$$

$$(-4)^{-3} = \frac{1}{(-4)^3} = \frac{1}{-64}$$

$$(-\tfrac{1}{3})^{-4} = \frac{1}{(-\tfrac{1}{3})^4} = \frac{1}{\tfrac{1}{81}} = 81$$

and

$$2^{-10} = \frac{1}{2^{10}} = \frac{1}{1024}.$$

It follows from Eqs. (1) and (2) that $2^7 2^{-3} = 2^{7+(-3)} = 2^4$, and $[(-\tfrac{1}{3})^{-4}]^{-5} = (-\tfrac{1}{3})^{(-4)(-5)} = (-\tfrac{1}{3})^{20}$.

It is also useful to define

$$\boxed{a^0 = 1} \qquad (4)$$

for every nonzero* signed number a. Accordingly, $5^0 = 1$, $(-4)^0 = 1$, and $(-\tfrac{1}{3})^0 = 1$.

Using Eqs. (1) and (3), we have $a^n/a^m = a^n(1/a^m) = a^n a^{-m} = a^{n+(-m)} = a^{n-m}$. Therefore

$$\boxed{\frac{a^n}{a^m} = a^{n-m}} \qquad (5)$$

* We do not define $0^0 = 1$, since it leads to contradictions in more advanced mathematics.

for every real number a and all integers n and m. In particular,

$$\frac{5^4}{5^2} = 5^{4-2} = 5^2 = 25$$

$$\frac{2^5}{2^7} = 2^{5-7} = 2^{-2} = \frac{1}{2^2} = \frac{1}{4}$$

and

$$\frac{(-\tfrac{1}{3})^2}{(-\tfrac{1}{3})^5} = (-\tfrac{1}{3})^{2-5} = (-\tfrac{1}{3})^{-3} = \frac{1}{(-\tfrac{1}{3})^3} = \frac{1}{(-\tfrac{1}{27})} = -27.$$

Equations (1) through (5) can be used to simplify tedious multiplication and division operations when each factor can be expressed as the same number raised to a power. For example, if one recognizes that $8 = 2^3$, $512 = 2^9$, $64 = 2^6$, and $1024 = 2^{10}$, then

$$\frac{8(512)}{(64)(1024)} = \frac{2^3 2^9}{2^6 2^{10}} = \frac{2^{3+9}}{2^{6+10}} = \frac{2^{12}}{2^{16}} = 2^{12-16} = 2^{-4} = \frac{1}{2^4} = \frac{1}{16}$$

and, with similar recognition

$$\frac{(81)(\tfrac{1}{6561})(\tfrac{1}{729})}{(\tfrac{1}{243})(\tfrac{1}{27})(\tfrac{1}{9})} = \frac{3^4(1/3^8)(1/3^6)}{(1/3^5)(1/3^3)(1/3^2)} = \frac{3^4 3^{-8} 3^{-6}}{3^{-5} 3^{-3} 3^{-2}} = \frac{3^{4+(-8)} 3^{-6}}{3^{-5+(-3)} 3^{-2}}$$

$$= \frac{3^{-4} 3^{-6}}{3^{-8} 3^{-2}} = \frac{3^{-4+(-6)}}{3^{-8+(-2)}} = \frac{3^{-10}}{3^{-10}} = 3^{-10-(-10)} = 3^0 = 1.$$

Unfortunately, the situation is not so simple if the numbers cannot be expressed as the same number raised to a power. The product $8(511)/(65)(1023)$ is more difficult, since it is not clear how to express each factor as a power of a specific number. It can be done, however, and we show how in Section 1.7.

Equations (1) through (5) are valid for any signed number a and all integers (whole numbers) n and m. We would like to generalize these results to all exponents. That is, we want Eqs. (1) through (5) to be valid even if n and m are not integers. Toward this end, we first give meaning to terms of the form $4^{1/2}$, $8^{1/3}$, and $(120)^{1/10}$.

If Eq. (2) is to remain valid for all n and m, then $(4^{1/2})^2 = 4^{(1/2)(2)} = 4^1 = 4$. Therefore $4^{1/2}$ should be a number whose square equals 4. If $(8^{1/3})^3 = 8^{(1/3)(3)} = 8^1 = 8$, we will want $8^{1/3}$ to be a number whose third power equals 8. If $[(120)^{1/10}]^{10} = (120)^{(1/10)10} = (120)^1 = 120$, then $(120)^{1/10}$ will be a number whose tenth power is 120. It is reasonable therefore to define $a^{1/n}$ as a number whose nth power is a, whenever a is positive and n is a positive integer. Conventionally, $a^{1/n}$ is read "nth root of a" and sometimes is written $\sqrt[n]{a}$*. In particular, $4^{1/2} = \sqrt{4} = 2$, since $2^2 = 4$, and $8^{1/3} = \sqrt[3]{8} = 2$, since $2^3 = 8$. The tenth root of 120 is a bit harder, since it is an infinite decimal. Nonetheless, $(120)^{1/10} = \sqrt[10]{120} = 1.614054\ldots$, since $(1.614054\ldots)^{10} = 120$.

It is not an easy matter to calculate roots. First, most roots are infinite decimals (for example, $\sqrt{2} = 1.414213\ldots$) and, second, formulas are not known except in the simplest cases for performing the calculations. What, for example, is the twenty-first root of π? Numerical procedures are available, however, which give roots to any degree of accuracy desired. Any computer and some hand calculators have no trouble calculating $\pi^{1/21} = 1.056024$ rounded to six decimal places.

A given number a can have more than one root. Both -2 and $+2$ are square roots of 4, since both numbers raised to the second power equal 4. Each number has only one positive nth root, however. To avoid confusion, we let $\sqrt[n]{a}$ denote the positive nth root of a. Accordingly, $\sqrt{4} = 2$, $\sqrt{9} = 3$, $\sqrt[3]{27} = 3$, and $\sqrt[4]{6561} = 9$. We adopt the convention in which a minus sign is used in front of the radical if the negative root is wanted. Then, $-\sqrt{4} = -2$, $-\sqrt{9} = -3$, and $-\sqrt[4]{6561} = -9$. This convention is not used when a number does not have a negative root. In particular, 27 does not have a negative third root, since no negative number raised to the third power can equal $+27$.

We have required a to be positive, since many times nth roots of a negative number make no sense. For instance, there is no real number whose square equals -2; $\sqrt{-2}$ is meaningless. If n is odd, $\sqrt[n]{a}$ can be given meaning for negative a, but we will have no need of such numbers in this text.

Finally, we extend Eq. (3) and define $a^{-1/n} = 1/a^{1/n}$. Then,

$$4^{-1/2} = \frac{1}{4^{1/2}} = \frac{1}{2} \quad \text{and} \quad (6561)^{-1/4} = \frac{1}{(6561)^{1/4}} = \frac{1}{9}.$$

Numbers raised to more complicated fractional powers, like $4^{7/2}$ and $27^{5/3}$ also can be defined mathematically, which we do now for completeness, noting beforehand that such numbers have very limited applications to business problems. In general, we define $a^{p/q} = (a^{1/q})^p$ when a is a positive number and both p and q are positive integers. Therefore $4^{7/2} = (4^{1/2})^7 = 2^7 = 128$, and $(27)^{5/3} = [(27)^{1/3}]^5 = 3^5 = 243$. Note that this definition, $a^{p/q} = (a^{1/q})^p$, preserves the validity of Eq. (2). Also, we could have defined $a^{p/q} = (a^p)^{1/q}$, and this would have been equally correct. For example, $4^{7/2} = (4^7)^{1/2} = (16,384)^{1/2} = 128$, and $(27)^{5/3} = [(27)^5]^{1/3} = (14,348,907)^{1/3} = 243$.

* The second root or square root of a is written $\sqrt{a}$ rather than $\sqrt[2]{a}$ as a matter of convenience.

To handle terms of the form $a^{-p/q}$ we simple extend Eq. (3) and define $a^{-p/q} = 1/a^{p/q}$. Then,

$$4^{-7/2} = \frac{1}{4^{7/2}} = \frac{1}{128} \quad \text{and} \quad (27)^{-5/3} = \frac{1}{(27)^{5/3}} = \frac{1}{243}.$$

The last step is to give meaning to terms of the form a^r, where a and r are any positive numbers, for example, 5^π and $(\sqrt{2})^3$. Unfortunately, this requires calculus. For our purposes, it is sufficient to know that such numbers exist; they can be computed to any degree of accuracy by computers and by some hand calculators, *and* they satisfy Eqs. (1) through (5).

Summarizing our findings, we have the following properties for *all* positive real numbers a, n, and m:

$$a^n a^m = a^{n+m}$$

$$(a^n)^m = a^{nm}$$

$$a^{-n} = \frac{1}{a^n}$$

$$a^0 = 1$$

$$\frac{a^n}{a^m} = a^{n-m}.$$

If, in addition, both n and m are integers, these properties are also true for negative a.

There is one last property of exponents that is useful; it involves the product of two different positive numbers raised to the *same* exponent:

$$\boxed{a^n b^n = (ab)^n.} \tag{6}$$

For example,

$$(5.2)^3(2)^3 = [(5.2)(2)]^3 = (10.4)^3$$
$$(4)^{3.1}(7)^{3.1} = (28)^{3.1}$$

and

$$(1.2)^{-3.4}(1.1)^{-3.4} = [(1.2)(1.1)]^{-3.4} = (1.32)^{-3.4}.$$

Like the other properties, Eq. (6) is also valid for negative a and negative b if n is a positive integer. In particular, $(-3)^5(7)^5 = [(-3)(7)]^5 = (-21)^5$.

1 Elementary Concepts

Equation (6) is used often to simplify nth roots. For example,

$$\sqrt{8} = (8)^{1/2} = [(4)(2)]^{1/2} = 4^{1/2}2^{1/2} = 2\sqrt{2}$$

and

$$\sqrt[3]{270} = (270)^{1/3} = [(27)(10)]^{1/3} = 27^{1/3}10^{1/3} = 3\sqrt[3]{10}.$$

We conclude this section with a warning. Most other properties of exponents ingeniously invented at times of stress, for example, during an examination, are usually not valid. In particular,

$a^n a^m \neq a^{nm}$
$(a^n)^m \neq a^{n+m}$
$a^n b^m \neq (ab)^{n+m}$.

The safest procedure is to understand and memorize the six properties presented here and to *assume* that all other properties are not valid unless you can prove them.

Exercises

In Exercises 1 through 9, simplify each of the given expressions into one exponent.

1. $\dfrac{3^2 3^4}{3^5 3^2}$

2. $\dfrac{7^2 7^{-3} 7^4}{7^8 7^{-2}}$

3. $\dfrac{\pi^4 (\pi^2)^3}{(\pi^{-2})^4 \pi^3}$

4. $[(-\tfrac{1}{2})^2]^4 [(-\tfrac{1}{2})^{-3}]^2 [(-\tfrac{1}{2})^{-4}]^{-5}$

5. $\dfrac{(1.7)^{8.1}(1.7)^{-3.4}}{(1.7)^{-4.1}(1.7)^{3.7}}$

6. $\dfrac{x^3(x^2)^4(x^{-3})^7}{(x^{-3})^{-4} x^5}$

7. $\dfrac{(y^{-3})^{-2} y^4 y^{-1}}{y^2 (y^3)^{-1}}$

8. $\dfrac{(x^4)^2 (1/x)^{-3}}{(1/x)^2}$

9. $\{[(3.1)^{-2}]^{-4}\}^3$

Evaluate the quantities given in Exercises 10 through 21.

10. $9^{3/2}$
11. $16^{-5/4}$
12. $27^{2/3}$
13. $100^{-3/2}$
14. $(3^{1/2})(12^{1/2})$
15. $3^{1/3} 9^{1/3}$
16. $(5)^{-1/2}(20)^{-1/2}$
17. $2^{-3/2} 32^{-3/2}$
18. $\sqrt{\tfrac{9}{4}}$
19. $\sqrt{\tfrac{8}{18}}$
20. $\sqrt{\dfrac{(49)(16)}{25}}$
21. $\sqrt[3]{\dfrac{(27)(8)}{125}}$

1.5 The Quadratic Formula

The properties of exponents can be used to solve certain equations for an unknown number when the unknown quantity is raised to a power, for example, finding z if $z^4 = 16$, or finding x if $x^2 + 2x - 3 = 0$. The simplest sort of problem occurs when the unknown number appears only once in the given equation. Such equations have the form $x^n = c$, where n and c are known positive numbers. The equation $x^7 = 4$ has this form with $n = 7$ and $c = 4$, as does the equation $x^{1.3} = \pi$ with $n = 1.3$ and $c = \pi$. Of course, the unknown number need not be x; any other letter would do. In particular, $y^3 = 27$ is an equation in which the unknown, denoted y, appears only once.

To solve equations of the form $x^n = c$, we raise both sides of the equation to the $1/n$ power. The same result is obtained by taking the nth root of both sides. In either case, one must be careful to perform the same operation on both sides of the equation. If we raise both sides of $x^n = c$ to the $1/n$ power, we obtain $(x^n)^{1/n} = (c)^{1/n}$ which, as a result of Eq. (2) in Section 1.4, becomes $x^1 = c^{1/n}$ or simply $x = c^{1/n}$.

Example 1 Solve $y^3 = 27$ for y.

Solution We raise both sides of this equation to the power $\frac{1}{3}$, obtaining $(y^3)^{1/3} = (27)^{1/3}$ or $y = (27)^{1/3}$. But $(27)^{1/3} = \sqrt[3]{27} = 3$, hence $y = 3$.

Example 2 Solve $x^{1.3} = \pi$.

Solution We raise both sides of this equation to the power $1/1.3$, obtaining $(x^{1.3})^{1/1.3} = \pi^{1/1.3}$ or $x = \pi^{1/1.3}$. As discussed in Section 1.4, it is an extremely difficult task to calculate $\pi^{1/1.3}$. Nonetheless, with the aid of a sophisticated hand calculator, we find $\pi^{1/1.3} = 2.4122\ldots$, so $x = 2.412$ rounded to three decimal places.

If the number n in $x^n = c$ is an *even integer* ($n = 2, n = 4, n = 6$, and so on), there are two real solutions to the given equation. One solution is, as before, $x = c^{1/n}$; the second solution is $x = -c^{1/n}$. Substituting $x = -c^{1/n}$ into the equation $x^n = c$ yields a correct solution, since a negative raised to an even power (which is shorthand for multiplying it by itself an even number of times) is positive. The two solutions are usually given together as $x = \pm c^{1/n}$.

Example 3 Solve $x^4 = 16$.

Solution Since the exponent is an even integer, $n = 4$, there will be two real solutions. Here $x = \pm(16)^{1/4} = \pm\sqrt[4]{16} = \pm 2$.

A more difficult problem arises when the unknown number appears more than once in the given equation, each time raised to a different power, for example, finding x if $2x^5 + 3x^2 - 1 = 0$, or finding y if $y^{10} + 2y^8 - y^3 +$

$7 = 0$. For most equations of this form, the solutions cannot be obtained algebraically, since appropriate methods do not exist. One of the few exceptions is the quadratic equation.

A *quadratic equation* has the form $ax^2 + bx + c = 0$, where a, b, and c are all known numbers and $a \neq 0$. The equation $2x^2 + 5x - 7 = 0$ has this form with $a = 2$, $b = 5$, and $c = -7$. The equation $7x^2 - 2x - 1 = 0$ also has this form with $a = 7$, $b = -2$, and $c = -1$. As usual, the letter used to denote the unknown is not important. The equations $4y^2 - 2y - 3 = 0$ and $4p^2 - 2p - 3 = 0$ are both quadratic equations with $a = 4$, $b = -2$, and $c = -3$. The important feature of a quadratic equation is that the unknown only appears raised to the second and first powers.

We prove at the end of this section that the solutions to equations of the form $ax^2 + bx + c = 0$ are given by the *quadratic formula*

$$x = \frac{-b \pm \sqrt{b^2 - 4ac}}{2a}. \tag{7}$$

To solve any quadratic equation, we simply substitute the values of its coefficients a, b, and c into the quadratic formula and simplify.

Example 4 Solve $x^2 + 2x - 3 = 0$.

Solution This is a quadratic equation with $a = 1$, $b = 2$, and $c = -3$. Substituting these values into the quadratic formula, we obtain

$$x = \frac{-2 \pm \sqrt{(2)^2 - 4(1)(-3)}}{2(1)} = \frac{-2 \pm \sqrt{4 + 12}}{2} = \frac{-2 \pm \sqrt{16}}{2} = \frac{-2 \pm 4}{2}.$$

Using the plus sign, we obtain one solution as $x = (-2 + 4)/2 = 1$. Using the minus sign, we find a second solution as $x = (-2 - 4)/2 = -3$.

Example 5 Solve $4y^2 - 2y = 3$.

Solution We first rewrite this equation as $4y^2 - 2y - 3 = 0$ which is a quadratic equation with $a = 4$, $b = -2$, and $c = -3$. Substituting these values into the quadratic formula, we have

$$y = \frac{-(-2) \pm \sqrt{(-2)^2 - 4(4)(-3)}}{2(4)} = \frac{2 \pm \sqrt{4 + 48}}{8} = \frac{2 \pm \sqrt{52}}{8}.$$

The solutions are $y = (2 + \sqrt{52})/8$ and $y = (2 - \sqrt{52})/8$. With the aid of a calculator, we find $\sqrt{52} = 7.21$ rounded to two decimals. The solutions can be given in decimal form as $y = (2 + 7.21)/8 = 1.15$ and $y = (2 - 7.21)/8 = -0.65$, also rounded to two decimals.

The quadratic formula does not always yield two solutions. If $b^2 - 4ac = 0$, the formula reduces to

$$x = \frac{-b \pm \sqrt{b^2 - 4ac}}{2a} = \frac{-b \pm \sqrt{0}}{2a} = -\frac{b}{2a},$$

and the quadratic equation has only one solution. If $b^2 - 4ac$ is negative, we cannot take its square root, and no real solutions exist. Readers familiar with complex numbers will note that complex solutions exist. Since complex numbers have no use in commercial situations, we do not consider them here.

Example 6 Solve $x^2 - 2x + 1 = 0$.

Solution Here $a = 1$, $b = -2$, and $c = 1$. Substituting these values into the quadratic formula, we obtain

$$x = \frac{-(-2) \pm \sqrt{(-2)^2 - 4(1)(1)}}{2(1)} = \frac{2 \pm \sqrt{4 - 4}}{2} = \frac{2 \pm 0}{2} = 1.$$

The only solution is $x = 1$.

Example 7 Solve $2p^2 + p + 1 = 0$.

Solution Substituting $a = 2$, $b = 1$, and $c = 1$ into the quadratic formula, we calculate

$$p = \frac{-1 \pm \sqrt{(1)^2 - (4)(2)(1)}}{2(2)} = \frac{-1 \pm \sqrt{1 - 8}}{4} = \frac{-1 \pm \sqrt{-7}}{4}.$$

Since $\sqrt{-7}$ is not defined, the given equation has no real solutions.

To prove the quadratic formula, we first rewrite $ax^2 + bx + c = 0$ as $ax^2 + bx = -c$ and divide both sides of the equation by a, obtaining $x^2 + bx/a = -c/a$. If we add $b^2/4a^2$ to both sides of the equation,

$$x^2 + \frac{bx}{a} + \frac{b^2}{4a^2} = -\frac{c}{a} + \frac{b^2}{4a^2},$$

the left side can be rewritten as $[x + (b/2a)]^2$, and the right side can be simplified to $(b^2 - 4ac)/4a^2$. The equation becomes

$$\left(x + \frac{b}{2a}\right)^2 = \frac{b^2 - 4ac}{4a^2}.$$

Taking the square root of both sides, we have

$$x + \frac{b}{2a} = \pm\sqrt{\frac{b^2 - 4ac}{4a^2}}$$

$$x + \frac{b}{2a} = \pm\frac{\sqrt{b^2 - 4ac}}{\sqrt{4a^2}}$$

$$x + \frac{b}{2a} = \frac{\pm\sqrt{b^2 - 4ac}}{2a}.$$

Finally, adding $-b/2a$ to both sides of the last equation, we obtain the solution for x as the quadratic formula.

Exercises

Solve the following equations for the unknown numbers.
1. $x^3 = 8$.
2. $x^3 = 125$.
3. $y^4 = 81$.
4. $p^6 = 64$.
5. $b^{-2} = \frac{1}{4}$.
6. $b^3 = 9$.
7. $p^5 = 1.3$.
8. $y^\pi = 8$.
9. $t^{9.3} = 9.3$.
10. $p^{-1.2} = 3.1$.
11. $x^2 - 5x + 6 = 0$.
12. $2y^2 - 3y - 2 = 0$.
13. $2p^2 + 6p - 4 = 0$.
14. $c^2 - c - 1 = 0$.
15. $x^2 + 6x + 9 = 0$.
16. $y^2 - y - 2 = 0$.
17. $3N^2 + 2N - 1 = 0$.
18. $4N^2 - 2N + 1 = 0$.
19. $5t^2 - t = 1$.
20. $x^2 - 8x = -16$.
21. $x^2 - 2x = 0$.
22. $3b = b^2 - 1$.

1.6 Scientific Notation

For many calculations, it is convenient to first write each number as the product of a number between 0 and 10 and an appropriate power of 10. The result is called *scientific notation*. The primary use of scientific notation to business students is its applications to logarithms as discussed in Section 1.10. Also, all hand calculators automatically round and display numbers in scientific notation when those numbers contain more digits than the machine is capable of accommodating.

Multiplying a number by any power of 10 simply changes the position of the decimal point. In particular,

$$3.412 \times 10^2 = (3.412)(100) = 341.2$$

and

$$9.57 \times 10^{-3} = (9.57)(0.001) = 0.00957.$$

The numbers 341.2 and 3.412 × 10² are both the same, but the second form is in scientific notation. Similarly, the numbers 0.00957 and 9.57 × 10⁻³ are equal, but again the second form is in scientific notation. These two examples provide a method for transforming any number into scientific notation. First rewrite the given number with the decimal point repositioned immediately after the first nonzero number. For example, 5495.12 becomes 5.49512, and 0.00041 becomes 4.1. This result is then multiplied by an appropriate power of 10 determined by the following approach. Count the number of places the decimal was moved to convert the original number to a number between 0 and 10. If the decimal point was moved n places to the left, where n is a positive integer, the appropriate power of 10 will also be n. If the decimal point was moved n places to the right, where n is again a positive integer, the appropriate power of 10 will be $-n$. In particular, we converted 5495.12 to 5.49512 by moving the decimal point three places to the left. Therefore 5495.12 = 5.49512 × 10³. We converted 0.00041 to 4.1 by moving the decimal point four places to the right. Accordingly, 0.00041 = 4.1 × 10⁻⁴. Similarly,

$$952.1 = 9.52 \times 10^2$$
$$1001 = 1.001 \times 10^3$$
$$0.0401 = 4.01 \times 10^{-2}$$
$$0.00004 = 4.0 \times 10^{-5}.$$

The number on the right in each equality is in scientific notation.

Exercises

Write the following numbers in scientific notation.
1. 546
2. 1021
3. 10.21
4. 20,104.412
5. 0.00104
6. 0.01004
7. 10.004
8. 0.953

Convert the following numbers, written in scientific notation, to numbers without exponents.
9. 3.356×10^3
10. 3.356×10^{-3}
11. 3.3×10^3
12. 5.623×10^2
13. 8.63×10^{-2}
14. 4.3×10^5

1.7 Logarithms

We saw in Section 1.4 that multiplication and division can be simplified greatly if each factor is expressed as the same number raised to a power. As an example, we calculated

$$\frac{8(512)}{(64)(1024)} = \frac{2^3 2^9}{2^6 2^{10}} = \frac{2^{12}}{2^{16}} = 2^{12-16} = 2^{-4} = \frac{1}{16}.$$

26 1 Elementary Concepts

Evaluating the term (8)(511)/(65)(1023) was not as simple, since we were unable, in Section 1.4, to rewrite each factor as the same number raised to a power. This can be done, however, with logarithms, which are the subject of this section.

Historically, logarithms were used commercially for evaluating complicated products and quotients. With the advent of hand calculators, this method has become less important. Now the primary use of logarithms to business students is a tool for evaluating data. An application to curve fitting is discussed in Appendix A.4. Other applications occur in statistics and optimization.

Every positive real number N can be expressed as the number 10 raised to a power. That power or exponent is called the *common logarithm of N* and is denoted by log N. For example, 100 can be expressed as 10 raised to the second power; the power 2 is the common logarithm of 100, and we write log 100 = 2. As a second example, consider the number 0.001. This number can be expressed as 10 raised to the power -3, that is, $0.001 = 10^{-3}$; the power -3 is the common logarithm of 0.001, and we write log 0.001 = -3. Still other examples are log 10 = 1, log 10,000 = log 10^4 = 4, and log 1 = log 10^0 = 0.

Common logarithms, or just logarithms for short, are not easy to calculate for most numbers. To find log 2, we first must write 2 as 10 raised to a power. This is no easy matter. Luckily, it has been done. Table C.1 in the Appendix lists logarithms for all numbers between 1.00 and 9.99. This table, like all logarithm tables, needs a bit of explaining. The first two digits of each number are listed in the leftmost column in Table C.1, while the third digit is listed in the topmost row. To find log 2.58 in the table, we first find 2.5 (the first two digits of 2.58) in the left column, and then 8 (the third digit of 2.58) in the top row. The number appearing in both the row corresponding to 2.5 and the column corresponding to 8, 4116, is the logarithm of 2.58—almost. Unfortunately, the main body of a logarithm table does not include decimal points, which will be bothersome at first. The decimal point is always placed in front of the first digit. That is, log 2.58 = 0.4116.

To find log 5.98, we first find 5.9 in the left column of Table C.1, and 8 in the top row. The number listed in both the row corresponding to 5.9 and the column corresponding to 8, 7767, is the logarithm of 5.98 with the decimal point missing. Since the decimal point is always placed in front of the first digit, we have log 5.98 = 0.7767.

To find log 8.27, we locate 8.2 in the left column and 7 in the top row. The number listed in both the row corresponding to 8.2 and the column corresponding to 7 is 9175. Therefore, log 8.27 = 0.9175.

Although the use of Table C.1 quickly becomes routine, one should not lose sight of what logarithms really are. They are the powers to which the number 10 is raised to obtain the original numbers of interest. Writing log 8.27 = 0.9175 means $10^{0.9175} = 8.27$. Similarly, log 2.58 = 0.4116 means $10^{0.4116} = 2.58$.

Table C.1 can be used to find logarithms of all positive numbers (not just those between 1.00 and 9.99) if the numbers are first converted to scientific notation. To find log 3240, first write $3240 = 3.24 \times 10^3$. Then use Table C.1 to find the logarithm of the part of the number not already in exponent form, 3.24. We find $\log 3.24 = 0.5105$. That is, $10^{0.5105} = 3.24$. Combining this result with the properties of exponents, we find $3240 = 3.24 \times 10^3 = 10^{0.5105} 10^3 = 10^{0.5105+3} = 10^{3.5105}$. Since $3240 = 10^{3.5105}$, it follows that $\log 3240 = 3.5105$.

Example 1 Find log 0.00041.

Solution First, $0.00041 = 4.1 \times 10^{-4}$. Using Table C.1, we locate $\log 4.10 = 0.6128$. Therefore $0.00041 = 4.1 \times 10^{-4} = 10^{0.6128} 10^{-4} = 10^{0.6128-4} = 10^{-3.3872}$, and $\log 0.00041 = -3.3872$.

Once the procedure for finding logarithms is understood, it can be simplified by the following rule: To find the logarithm of a positive number in scientific notation, use Table C.1 to find the logarithm of the part of the number not already in exponent form, and then *add* to this result the exponent of 10.

Example 2 Find log 301,000.

Solution First, $301,000 = 3.01 \times 10^5$. The exponent of 10 is 5, and $\log 3.01 = 0.4786$. Therefore $\log 301,000 = 0.4786 + 5 = 5.4786$.

Example 3 Find log 0.128.

Solution First, $0.128 = 1.28 \times 10^{-1}$. The exponent of 10 is -1, and $\log 1.28 = 0.1072$. Therefore $\log 0.128 = 0.1072 + (-1) = -0.8928$.

Table C.1 is by no means complete. First, all logarithm entries have been rounded to four decimal places. The logarithm of 1.28 is really $0.10720996\ldots$ and not 0.1072 as given. For our purposes, however, four-decimal-place accuracy is sufficient. Second, and more serious, Table C.1 does not list the logarithms of all numbers between 1 and 9.99. In particular, the logarithms of 1.285 and 3.0129 are not there. To find logarithms of this sort, we approximate. The number 1.285 is midway between 1.28 and 1.29, so its logarithm should be close to the midpoint between log 1.28 and log 1.29. We find that $\log 1.28 = 0.1072$ and $\log 1.29 = 0.1106$, so we *estimate* $\log 1.285 = 0.1089$. To find log 3.0129, we find $\log 3.01 = 0.4786$ and $\log 3.02 = 0.4800$. Since 3.0129 is approximately one-third the distance between 3.01 and 3.02, we estimate $\log 3.0129 = 0.4790$ which is approximately one-third the distance between log 3.01 and log 3.02.

Example 4 Find log 7454.

Solution First, $7454 = 7.454 \times 10^3$. To find log 7.454, we note that 7.454 is four-tenths the distance between 7.45 and 7.46. Since $\log 7.45 = 0.8722$ and $\log 7.46 = 0.8727$, we estimate $\log 7.454 = 0.8724$. Then, $\log 7454 = 0.8724 + 3 = 3.8724$.

Knowing how to find logarithms of any positive number is only half the problem. If logarithms are to be used for simplifying multiplication and division, we also must know how to reverse the procedure. That is, we must be able to determine a number if its logarithm is known. In one respect, this is easy. If we know log $N = 1.6042$, we also know that $N = 10^{1.6042}$. But we still do not know what $10^{1.6042}$ actually equals.

Table C.1 can be used to find N if log N is known. To simplify the search, note that we need only to find numbers of the form 10^p for p between 0 and 1. Since $10^{1.6042} = 10^{0.6042+1} = 10^{0.6042}10^1$, we can find $10^{1.6042}$ by first finding $10^{0.6042}$ and then multiplying by 10. Similarly, to find $10^{3.9090}$, we need only to find $10^{0.9090}$, since $10^{3.9090} = 10^{0.9090+3} = 10^{0.9090}10^3$; multiplying $10^{0.9090}$ by $10^3 = 1000$ yields $10^{3.9090}$.

To find 10^p for any number p between 0 and 1, find p in the main body of Table C.1. Note the row and column in which p appears and also the two-digit number all the way to the left of that row and the one-digit number all the way above that column. These three digits, the two-digit number *followed* by the one-digit number, are the value of 10^p. This value of 10^p is called the *antilog of p*. That is, $10^p =$ antilog p.

To find $10^{0.6042}$ or, equivalently, the antilog of 0.6042, we first locate 6042 in the table. Recall that the decimal point is always deleted from such entries. The two-digit number all the way to the left of 6042 in Table C.1 is 4.0, while the one-digit number all the way above 6042 is 2. Therefore $10^{0.6042} = 4.02$.

To find $10^{0.9090}$ or, equivalently, antilog 0.9090, first locate 9090 in the table. The two-digit number all the way to the left of 9090 is 8.1, while the one-digit number all the way above 9090 is 1. Therefore antilog $0.9090 = 8.11$.

Example 5 Find antilog 2.5119.

Solution

antilog $2.5119 = 10^{2.5119} = 10^{0.5119+2} = 10^{0.5119}10^2$.

From Table C.1, we find $10^{0.5119} = 3.25$. Therefore antilog $2.5119 = 3.25 \times 10^2 = 325$.

If we need the antilog of a negative number, for example, $10^{-2.2573}$, we follow the same procedure, keeping in mind that all entries in the main body of Table C.1 are between 0 and 1. We could write $10^{-2.2573} = 10^{-0.2573-2} = 10^{-0.2573}10^{-2}$, but this leaves us with $10^{-0.2573}$ which has a negative exponent. Note that the negative exponent -2 is no trouble, since $10^{-2} = 0.01$. To eliminate the negative sign on the troublesome exponent, we add 1 to the exponent and subtract 1 from the exponent. That is, $-2.2573 = -0.2573 - 2 = (-0.2573 + 1) - 1 - 2 = 0.7427 - 3$. Then, using Table C.1, we have antilog $0.7427 = 5.53$ and $10^{-2.2573} = 10^{0.7427}10^{-3} = 5.53 \times 10^{-3} = 0.00553$.

Example 6 Find antilog (-1.1409).

Solution

$$10^{-1.1409} = 10^{-0.1409-1} = 10^{(-0.1409+1)-1-1} = 10^{0.8591-2}$$
$$= 10^{0.8591}10^{-2}.$$

From Table C.1, we find $10^{0.8591} = $ antilog $0.8591 = 7.23$. Therefore $10^{-1.1409} = 10^{0.8591}10^{-2} = (7.23) \times 10^{-2} = 0.0723$ which is the required antilog.

If the value of p is not in the main body of Table C.1, we approximate the value of 10^p by finding the two closest tabular values and estimating. To find $10^{0.4821}$, we search the table for 4821. It is not there. We do find 4814 which corresponds to 3.03 and 4829 which corresponds to 3.04. Since 4821 is about halfway between 4814 and 4829, we estimate $10^{0.4821}$ as 3.035 which is halfway between 3.03 and 3.04.

Example 7 Find antilog (-1.2112).

Solution First we write

$$\text{antilog}(-1.2112) = 10^{-1.2112} = 10^{-0.2112-1} = 10^{(-0.2112+1)-1-1}$$
$$= 10^{0.7888-2} = 10^{0.7888}10^{-2}.$$

Unfortunately 7888 is not in Table C.1. But 7882 and 7889 correspond to 6.14 and 6.15, respectively. Since 7888 is very close to 7889, we approximate $10^{0.7888}$ as 6.149 which is very close to 6.15. Then,

$$10^{-1.2112} = 10^{0.7888}10^{-2} = (6.149) \times 10^{-2} = 0.06149.$$

We are ready to solve the problem posed at the beginning of the section, evaluating $(8)(511)/(65)(1023)$. Using Table C.1, we find $\log 8 = 0.9031$, hence $8 = 10^{0.9031}$; $\log 511 = 2.7084$, hence $511 = 10^{2.7084}$; $\log 65 = 1.8129$, hence $65 = 10^{1.8129}$; and $\log 1023 = 3.0098$, hence $1023 = 10^{3.0098}$. Then,

$$\frac{(8)(511)}{(65)(1023)} = \frac{10^{0.9031}10^{2.7084}}{10^{1.8129}10^{3.0098}} = \frac{10^{0.9031+2.7084}}{10^{1.8129+3.0098}} = \frac{10^{3.6115}}{10^{4.8227}}$$

$$= 10^{3.6115-4.8227} = 10^{-1.2112} = 0.06149.$$

Actually, this computation can be done more easily, as we show in Section 1.8. However, the steps used in this problem form the nucleus of the method to be developed next.

Exercises

In Exercises 1 through 16, find the logarithms of the given numbers.
1. 1.26
2. 12.6
3. 126
4. 0.00126
5. 4.53
6. 4530
7. 0.453
8. 800

9. 74.3   10. 15.1   11. 15.15   12. 0.0112
13. 0.1012   14. 13,933   15. 145,982   16. 0.004114

In Exercises 17 through 32, find the antilogs of the given numbers.
17. 2.6839   18. 0.6839   19. 1.2625   20. 2.2625
21. −0.7375   22. −1.7375   23. 1.7903   24. −1.2097
25. −1.0443   26. 1.8814   27. 0.8416   28. 1.7577
29. −2.1111   30. −0.2991   31. 3.5735   32. −4.5092
33. Use logarithms to evaluate:

(a) $(123)(19)$ (b) $\dfrac{1890}{57}$ (c) $\dfrac{(2110)(3.12)}{0.0112}$

1.8 Computing with Logarithms

Multiplication and division can be performed neatly with logarithms if a few rules are observed.* Actually, these rules are nothing more than the properties of exponentials developed in Section 1.4. They are given here in terms of logarithms, because they are more useful in this form for the type of problem we wish to consider. The proofs of these rules are deferred until the exercises.

Rule 1 $\log AB = \log A + \log B$, where A and B denote real positive numbers.

Accordingly, $\log[(2.1)(3.4)] = \log 2.1 + \log 3.4$, and $\log[(17.29)(45.3)] = \log 17.29 + \log 45.3$. In essence, Rule 1 allows one to rewrite the log of a product as the sum of the individual logarithms of each factor. Since the individual logarithms can be found in Table C.1, the right side of the equation for Rule 1 is usually an efficient means of calculating the left side.

Rule 2 $\log A/B = \log A - \log B$.

Accordingly, $\log(2.1/3.4) = \log 2.1 - \log 3.4$, and $\log(17.29/45.3) = \log 17.29 - \log 45.3$. In essence, this rule changes division into subtraction. Again the right side of the equation for Rule 2 is found easily with the aid of Table C.1, and it provides an efficient means of calculating the left side of the equation for Rule 2.

Rule 3 $\log(A^m) = m \log A$.

Accordingly, $\log(4.51)^{2.1} = (2.1)(\log 4.51)$, and $\log\sqrt{19.1} = \log(19.1)^{1/2} = (\tfrac{1}{2})(\log 19.1)$. Rule 3 reduces powers and roots (which usually involve very difficult operations) to simple multiplication.

---
* Modern calculators have eliminated the need for logarithms in straight multiplication and division. Nonetheless, logarithms, and their properties, retain their importance in other uses, as evidenced in Appendix A.4.

1.8 Computing with Logarithms

The procedure for using logarithms to perform complicated multiplication and division is first to designate the quantity of interest as N. Then take logarithms and use Rules 1 through 3 and Table C.1 to determine a value for log N. Finally, solve for N by taking antilogs.

Example 1 Calculate (2.1)(3.4).

Solution Obviously this does not involve complicated multiplication, but it serves to illustrate the general procedure. We let $N = (2.1)(3.4)$ and take logarithms of both sides. Thus log $N = $ log (2.1)(3.4). Using Rule 1 and Table C.1, we find

$$\log N = \log 2.1 + \log 3.4$$
$$= 0.3222 + 0.5315 = 0.8537.$$

Since log $N = 0.8537$, it follows by taking antilogs that

$$N = \text{antilog } 0.8537 = 7.14,$$

which is the answer to the original problem.

Example 2 Calculate 17.29/45.3.

Solution Designate $N = 17.29/45.3$ and take logarithms of both sides. Then, log $N = $ log (17.29/45.3). Using Rule 2 and Table C.1, we find

$$\log N = \log 17.29 - \log 45.3$$
$$= 1.2378 - 1.6561 = -0.4183.$$

Since log $N = -0.4183$, it follows by taking antilogs that

$$N = \text{antilog } -0.4183 = 10^{-0.4183} = 10^{(-0.4183+1)-1} = 10^{0.5817} 10^{-1}$$
$$= (3.817) \times 10^{-1} = 0.3817,$$

which is the answer to the original problem.

Example 3 Calculate $\sqrt{19.1}$.

Solution Designate $N = \sqrt{19.1} = (19.1)^{1/2}$ and take logarithms of both sides. Then, log $N = $ log $(19.1)^{1/2}$. Using Rule 3 and Table C.1, we calculate

$$\log N = \log (19.1)^{1/2} = (\tfrac{1}{2})(\log 19.1) = (\tfrac{1}{2})(1.2810) = 0.6405.$$

Since log $N = 0.6405$, it follows that

$$N = \text{antilog } 0.6405 = 4.37,$$

which is the answer to the original problem.

Rules 1, 2, and 3 can be used together in the same problem to evaluate complicated quantities.

Example 4 Calculate $\dfrac{(2.13)^3 \sqrt[4]{0.017}}{(1.17)^{2.1}(1.8)^{0.41}}$.

Solution We designate this quantity by N and take logarithms. Then,

$$\log N = \log \frac{(2.13)^3 (0.017)^{1/4}}{(1.17)^{2.1}(1.8)^{0.41}}$$

$$\begin{aligned}
&= \log [(2.13)^3 (0.017)^{1/4}] - \log [(1.17)^{2.1}(1.8)^{0.41}] \quad \text{(Rule 2)}\\
&= [\log (2.13)^3 + \log (0.017)^{1/4}]\\
&\quad - [\log (1.17)^{2.1} + \log (1.8)^{0.41}] \quad \text{(Rule 1)}\\
&= (3)(\log 2.13) + (\tfrac{1}{4})(\log 0.017)\\
&\quad - (2.1)(\log 1.17) - (0.41)(\log 1.8) \quad \text{(Rule 3)}\\
&= 3(0.3284) + (\tfrac{1}{4})(-1.7696) - 2.1(0.0682) - 0.41(0.2553) \quad \text{(Table C.1)}\\
&= 0.9852 - 0.4424 - 0.1432 - 0.1047\\
&= 0.2949.
\end{aligned}$$

Since $\log N = 0.2949$, it follows that $N = 1.972$.

Exercises

Use logarithms to evaluate the quantities given in Exercises 1 through 14.

1. $(4.7)(0.21)$
2. $(4.7)(8.1)$
3. $(4.3)(16.9)(0.013)$
4. $\dfrac{9.3}{2.7}$
5. $\dfrac{0.013}{0.121}$
6. $\sqrt[3]{4}$
7. $\dfrac{6}{9}$
8. $\dfrac{(3.1)(2.91)}{(6.4)}$
9. $(3.12)^{1.3}$
10. $(16.9)^{1/3}(18.1)^{0.7}$
11. $\dfrac{(45.1)(19.3)}{(6.2)(31.9)}$
12. $\dfrac{(19.25)(18.37)}{(6.217)^2}$
13. $\dfrac{(141.2)^{1.1}(95.3)^{0.01}}{(15.2)^{1/2}(101)^{1.7}}$
14. $\dfrac{(1852)^2(211)^3}{(95)^5(21)^2}$

15. Prove Rule 1. *Hint:* Designate $a = \log A$ and $b = \log B$. First show that $A = 10^a$ and $B = 10^b$, hence $AB = 10^a 10^b = 10^{a+b}$. Deduce that $\log AB = a + b$.
16. Prove Rule 2. *Hint:* Designate $a = \log A$ and $b = \log B$. First show that

$A = 10^a$ and $B = 10^b$, hence $A/B = 10^a/10^b = 10^{a-b}$. Then deduce that log $A/B = a - b$.

17. Prove Rule 3. *Hint:* Designate $a = \log A$. First show that $A = 10^a$, hence $A^m = (10^a)^m = 10^{am}$. Then deduce that $\log A^m = am$.

1.9 Cartesian Coordinate System

Consider the map illustrated in Figure 1.1 with streets running in either a north-south direction or an east-west direction. By using the center of Broad and Market Streets as a reference (perhaps a motorist has stopped there for directions), it is easy to locate any other point on the map. The intersection of Elm Lane and Maple Street is two blocks west and one block south of the reference point. The light at Freeman Street and Valley Road is three blocks east and two blocks north of the reference point. In each case, the new point on the map is uniquely determined from the reference point by two numbers and their directions.

The cartesian coordinate system is a generalized version of the previous map. To construct this new system, two intersecting perpendicular lines (forming an angle of 90 degrees with each other) are first drawn, as illustrated in Figure 1.2. The horizontal line is often called the *x*-coordinate axis (or just the *x*-axis for short), while the vertical line is often called the *y*-coordinate axis (or *y*-axis for short). The intersection of these two axes is the *origin*, and it represents the reference point of the system.

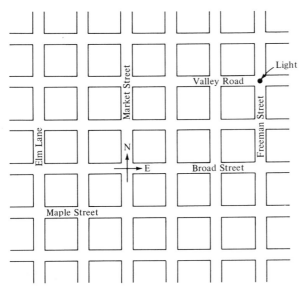

FIGURE 1.1

34 1 Elementary Concepts

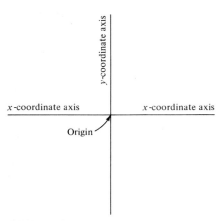

FIGURE 1.2

Each axis is marked in fixed units of length, as illustrated in Figure 1.3. Units to the right of the origin on the x-axis and units above the origin on the y-axis are assigned positive values. Units to the left of the origin on the x-axis and units below the origin on the y-axis are assigned negative values. The complete system is the *cartesian* (or rectangular) *coordinate system*.

Note that arrows have been appended to the positive portions of the x- and

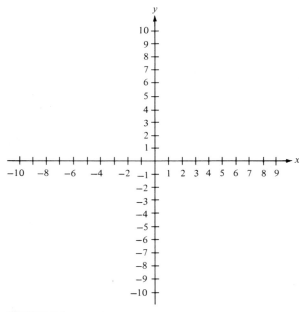

FIGURE 1.3

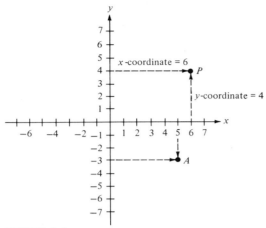

FIGURE 1.4

y-axes in Figure 1.3. These arrows simply indicate, visually, the direction of increasing values of x and y. We can think of the x-axis as representing the east-west direction on a map, and the y-axis as representing the north-south direction. The arrows then indicate the directions east and north. Moving in a positive x-direction corresponds to moving east. Moving in a negative y-direction corresponds to moving south. Motion in the directions of north and west is defined by moving in the positive y- and negative x-directions, respectively.

The usefulness of the cartesian coordinate system is that any point on the plane can be located from the origin by two numbers. Directions need not be specified, since they are inherent in the signs of the numbers. As an example, consider the point P shown in Figure 1.4. To reach this point from the origin, one must move 6 units along the x-axis in the positive direction and then 4 units in the positive y-direction along a line segment parallel to the y-axis beginning at $x = 6$. The point P is located by the two numbers 6 and 4, if we agree that the first number, 6, denotes moving 6 units along the x-axis and the second number, 4, denotes moving 4 units in the y-direction.

Every point on the plane can be given by two numbers. By convention, the first number always indicates movement along the x-axis, while the second number always indicates movement parallel to the y-axis. Positive numbers denote movement in the positive direction (in the direction of the arrows); negative numbers indicate movement in the negative direction. The two numbers defining the point of interest are separated by a comma and enclosed in parentheses. Point P in Figure 1.4 is given by (6, 4), and point A is given by (5, −3). To reach A from the origin (our reference point), we move 5 units along the x-axis in the positive x-direction (the first number is +5), and then we move 3 units in the negative y-direction (the second number is −3).

The two numbers defining a given point are called the *coordinates* of the

36 1 Elementary Concepts

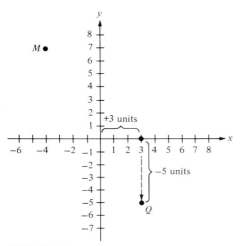

FIGURE 1.5

point. Not surprisingly, the first number is often called the *x-coordinate* and the second number the *y-coordinate*. For point A in Figure 1.4, the x-coordinate is 5, and the y-coordinate is -3.

Example 1 Locate and plot the point having coordinates $(3, -5)$.

Solution The point of interest is reached by first moving 3 units along the x-axis in the positive direction from the origin and then moving 5 units in the negative y-direction. The point is plotted as Q in Figure 1.5.

Example 2 Determine the coordinates of the point M shown in Figure 1.5.

Solution To reach M from the origin, we must move 4 units in the negative x-direction (which implies an x-coordinate of -4) and then 7 units in the positive y-direction (which implies a y-coordinate of $+7$). The coordinates are $(-4, 7)$. Recall that the x-coordinate is always given before the y-coordinate.

A cartesian coordinate system divides the plane into four sections. Each section is called a *quadrant*, labeled I, II, III, or IV as illustrated in Figure 1.6. A point in the first quadrant has both a positive x- and a positive y-coordinate. A point in the second quadrant has a negative x-coordinate and a positive y-coordinate, while a point in the third quadrant has a negative x-coordinate and a negative y-coordinate. A point in the fourth quadrant has a positive x-coordinate but a negative y-coordinate.

In practice two modifications are often made in the cartesian coordinate system as we have defined it. First, the letters used to label the axes need not be x and y but can be any other two letters that are convenient. In a problem dealing with the price and demand of a certain product, it may be more revealing to label the axes P for price and D for demand. Regardless of the letters used, the first component of a point always refers to movement along the horizontal

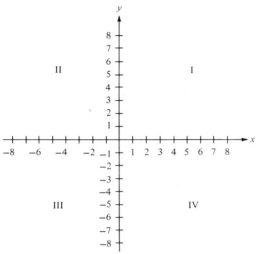

FIGURE 1.6

axis, while the second component refers to movement in a direction parallel to the vertical axis.

The second modification is that the same scale need not be used on both coordinate axes. The scale on the horizontal axis can, and frequently does, differ from the scale used on the vertical axis. Figure 1.7 illustrates a coordinate

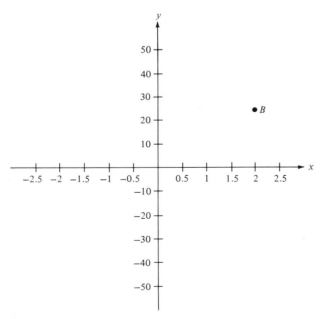

FIGURE 1.7

system with two scales. The coordinates of point B are (2, 25). The only restriction in this modification is that each scale, once chosen, must be marked off consistently in equal units.

Exercises

1. Consider the points labeled A through K in Figure 1.8.
    (a) Determine the coordinates of each point.
    (b) Which points are located in quadrant I?
2. Redo Exercise 1 for the points A through H in Figure 1.9.
3. Consider the points shown in Figure 1.10.
    (a) Draw a coordinate system that places A in quadrant I, B in quadrant II, and C in quadrant IV.
    (b) Draw a coordinate system that places A, B, and C in quadrant I and D in quadrant IV.
    (c) Draw a coordinate system that places all points in quadrant III.
4. Plot each of the following points on the same coordinate system:
    (a) $(3, -1)$ (b) $(2, 5)$ (c) $(-5, 2)$ (d) $(-6, -6)$.
5. Plot each of the following points on the same coordinate system:
    (a) $(3, -10)$ (b) $(2, 50)$ (c) $(-5, 20)$ (d) $(-6, -60)$.
6. Plot each of the following points on the same coordinate system:
    (a) $(250, 45)$ (b) $(500, -10)$ (c) $(400, 20)$ (d) $(-100, -5)$.
7. Determine the second coordinate of every point on the x-axis. What is the first coordinate of every point on the y-axis?

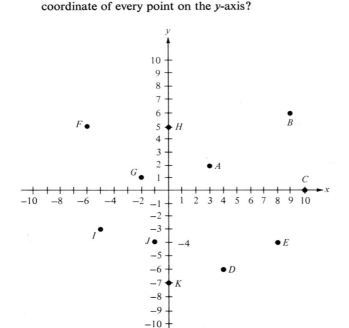

FIGURE 1.8

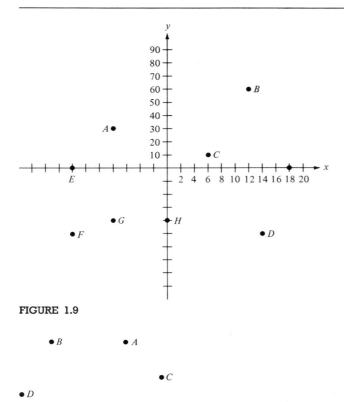

FIGURE 1.9

FIGURE 1.10

8. Construct a cartesian coordinate system with each axis scaled the same. Draw a straight-line segment between the origin and the point having coordinates (10, 10). Determine the angle this line makes with the positive x-axis.
9. Construct a cartesian coordinate system and draw a line parallel to the x-axis. What do all points on this line have in common?

1.10 Graphs

In Sections 1.2 and 1.5 we reviewed methods for solving one equation in one unknown. When the unknown was raised to the first power only, we used the properties of signed numbers to find its one value. When the unknown was raised to a power, we used the properties of exponents and, when appropriate, the quadratic formula to find the solutions. Occasionally, we found more than one solution, but still a finite number. In this section, we consider one equation with two unknowns, and the situation becomes more complicated.

The problem associated with two unknowns in one equation is to find the value or values of both unknowns that simultaneously satisfy the given equation, for example, to find numerical values for x and y that satisfy $y = 2x^2 + 1$, or to find values for x and y that satisfy $y = 3x - 2$. The difficulty is that such equations generally have infinitely many solutions.

Some of the solutions can be found by arbitrarily selecting values for one of the unknowns and substituting these values into the given equation. The result is one equation in one unknown, which often can be solved for the remaining unknown. As an example, we consider the equation $y = 3x - 2$. Arbitrarily setting $x = 1$ and substituting it into the equation, we find $y = 3(1) - 2$ or $y = 1$. Therefore $x = 1$ and $y = 1$ is one solution; these values together satisfy the given equation. To obtain another solution, we arbitrarily select another value for x. Setting $x = 2$, we substitute it into the given equation, obtaining $y = 3(2) - 2$ or $y = 4$. Therefore, $x = 2$ and $y = 4$ is a second solution. These values together also satisfy the given equation. Continuing in this manner, we next try $x = 0$ and find $y = -2$; then we try $x = -1$ and find $y = -5$; and we try $x = 3$, obtaining a corresponding value of $y = 7$. These results are collected together in Table 1.1. Each y-value is listed immediately to the right of its corresponding x-value.

TABLE 1.1

x	y
1	1
2	4
0	-2
-1	-5
3	7

The main objection to this procedure is that it provides only *some* of the solutions, not all of them. Yet, it is not likely that we can ever do better algebraically, since one equation in two unknowns generally has infinitely many solutions. If we are willing to use geometrical methods, however, the prospects are brighter.

Every solution of one equation in two unknowns is a pair of numbers, perhaps $x = 2$ and $y = 4$ or $x = 3$ and $y = 7$. Using the coordinate system developed in Section 1.9, we can represent a solution, that is, a pair of numbers, as a point on the plane. Two different solutions are plotted as two different points. In particular, the solutions listed in Table 1.1 are plotted in Figure 1.11.

Geometrically, the set of *all* solutions of one equation in two unknowns is a curve on the plane. The exact form of this curve often can be determined by looking at a plot of some of the solution points and making an educated guess. In particular, the points shown in Figure 1.11 appear to lie on a straight line; it seems likely therefore that the geometric solution of $y = 3x - 2$ is the dashed line drawn in Figure 1.12. We have drawn the line dashed to indicate that, at this stage, we cannot be certain that every point on the straight line has

1.10 Graphs

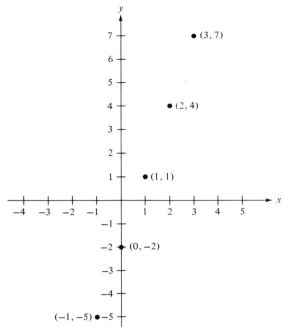

FIGURE 1.11

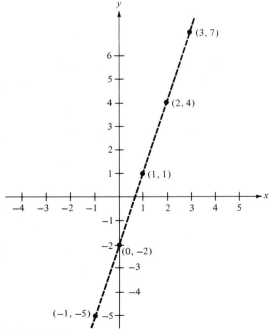

FIGURE 1.12

42   1 Elementary Concepts

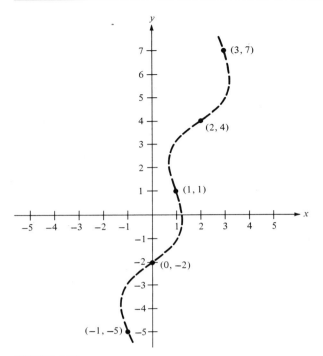

FIGURE 1.13

coordinates that are solutions to the given equation. For example, the curve illustrated in Figure 1.13 also contains the points listed in Table 1.1.

One way to be certain that Figure 1.12 is correct is to plot more points than the five listed in Table 1.1. (By using methods presented in Chapter 2, it can easily be verified that Figure 1.12 is in fact the correct graph of the equation $y = 3x - 2$ without plotting additional points.) For now, however, we must plot as many points as necessary to gain a reasonable idea of the shape of the curve before drawing it.

Let us summarize our steps. We were given one equation with two unknowns, $y = 3x - 2$, and we were interested in the set of all solutions. As a first step, we algebraically located some solutions by arbitrarily picking values for one of the unknowns, substituting these values into the given equation, and solving the resulting equation for the corresponding values of the other unknown. The second step was to plot on a cartesian coordinate system every solution we found. The last step was to determine the curve that contained the plotted points. At this stage, we simply fill in, using a dashed line, the curve that appears to fit the plotted points best. This procedure is called *graphing*, and the final curve is the *graph of the equation* under consideration. The line plotted in Figure 1.12 is the graph of the equation $y = 3x - 2$.

Example 1   Find the graph of the equation $y = 2x^2 + 1$.

Solution We begin by finding some solutions. Arbitrarily setting $x = 0$ and substituting it into the given equation, we find $y = 2(0)^2 + 1$ or $y = 1$. Therefore $x = 0$ and $y = 1$ is one solution. To find a second solution, we set $x = 1$ and substitute it into $y = 2x^2 + 1$. Then, $y = 2(1)^2 + 1$ or $y = 3$, so $x = 1$ and $y = 3$ is another solution. Continuing in this manner, we generate Table 1.2. Next we plot these points on a coordinate system, which has been done in Figure 1.14 Finally, we determine the curve that contains these points. This curve appears to be the one drawn in Figure 1.15, which is the graph of $y = 2x^2 + 1$.

TABLE 1.2

x	y
0	1
1	3
-1	3
2	9
-2	9
3	19
-3	19

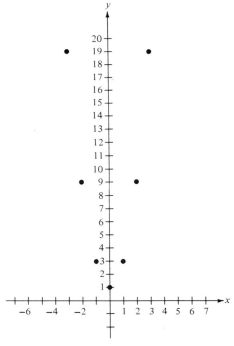

FIGURE 1.14

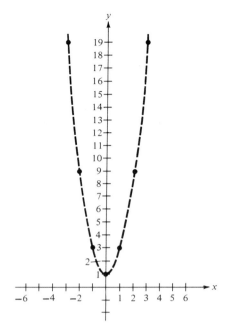

FIGURE 1.15

A few observations are now in order. First, the graph of an equation is only a geometric representation of the solutions of one equation in two unknowns. The actual solutions are the coordinates of each and every point on the graph. Figure 1.12 is the geometric representation of the solutions to $y = 3x - 2$. The solutions themselves are the coordinates of the points on this curve. In particular, the point $(\frac{1}{2}, -\frac{1}{2})$ is on the curve, so $x = \frac{1}{2}$ and $y = -\frac{1}{2}$ is a solution to $y = 3x - 2$, which can be verified by direct substitution. Additionally, we see that $(-4, -14)$ is a point on the curve, so $x = -4$ and $y = -14$ is still another solution.

Second, we determined the curve in both previous examples by looking at only a few points. This involves the assumption that the curve behaves nicely between the plotted points. Formal justification of this assumption, however, requires the material considered in Chapters 2 and 7. For now, we continue to assume that a curve can be drawn from a sufficient number of its points. Just how many points is sufficient depends on the curve and the foresight of the plotter. One can never plot too many points. A good rule of thumb is, "Too many is always better than too few."

Finally, note that in both examples we first guessed values for x and then solved for y. The reverse procedure is equally correct. Solving $y = 3x - 2$, we could set $y = 2$, substitute it into the equation, and obtain $2 = 3x - 2$. It follows that $x = \frac{4}{3}$, and $x = \frac{4}{3}$ and $y = 2$ is indeed a solution. Similarly, we could have guessed values for y and then solved for x in $y = 2x^2 + 1$. In particular, setting $y = 2$, we find $2 = 2x^2 + 1$, $2x^2 = 1$, $x^2 = \frac{1}{2}$, and $x = \pm\sqrt{\frac{1}{2}}$. Two solutions are $x = \sqrt{\frac{1}{2}}$, $y = 2$ and $x = -\sqrt{\frac{1}{2}}$, $y = 2$. The choice of which unknown is assigned values is made by the solver. It is dictated by personal preference and the ease with which the second unknown can be found. Given the equation $y^5 - 2y^2 = x + 1$, it is easy to pick values of y and solve for x, but quite difficult to pick values of x and solve for y.

Example 2 Graph the equation $y = x^3 - x^2 + 1$.

Solution Arbitrarily selecting values for x and solving for y, we generate Table 1.3. Plotting these points, we obtain Figure 1.16.

TABLE 1.3

x	y
0	1
1	1
-1	-1
2	5
-2	-11
$\frac{1}{2}$	$\frac{7}{8}$

Example 3 Graph the equation $x^2 + y^2 = 25$.

Solution Arbitrarily selecting values for y and solving for x, we generate Table 1.4. Plotting these points, we obtain Figure 1.17.

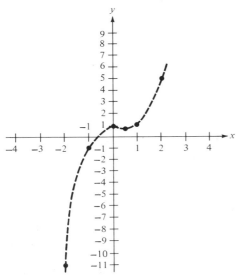

FIGURE 1.16

TABLE 1.4

y	x
0	±5
5	0
−5	0
3	±4
−3	±4
4	±3
−4	±3

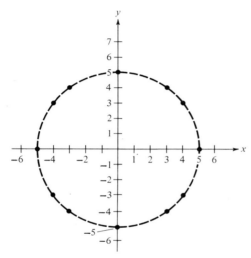

FIGURE 1.17

Exercises

Plot the graphs of the equations in Exercises 1 through 10.
1. $2x - 3y = 5$.
2. $y = x - 2$.
3. $6x - 2y = 3$.
4. $x^2 - y = 0$.
5. $y = 2x^2$.
6. $3y - 4x = 7$.
7. $y = x^3 - 2x^2 + x$.
8. $y = 3x - 4$.
9. $x^2 - y^2 + 4 = 0$.
10. $y = \sqrt{x}$.
11. Graph $y = x$ and $y = -x$ on the same axes. How do these curves differ?
12. Graph the equation $y = 2x + 5$. Select x-values of 0, 1, 3, 5, 7 and -1, -3, -5, -7. Determine from the graph the values of y when x is 2 and when x is -8.
13. Determine if the points having coordinates $(-2, 6)$, $(0, 2)$, and $(1, 9)$ lie on the graph of the equation $y = 3x^2 + 4x + 2$.
14. Graph the two equations $y = 6x + 3$ and $y = 5x - 2$ on the same coordinate system. Determine the point of intersection of these curves from the resulting graph.

15. Redo Exercise 14 for $y = 5x^2 - 2$ and $y = x + 3$.
16. Show graphically that the two equations $y = x^2 + 1$ and $y = -x^2$ have no points in common.

1.11 Sigma Notation

In various parts of this text, we will need the sum of large numbers of terms. Sometimes we will need the actual total, and then we will have to sum the numbers physically. Other times, however, we will only need to indicate the appropriate sum. An example is the statement, "Yearly expenditures are the sum of weekly expenditures." Here are we not explicitly calculating the expenditure over the year but simply indicating that it is the sum of 52 numbers. In cases like this, there is a useful mathematical notation for indicating the appropriate sum.

Consider the case of a teacher who has a list of seven grades and wants their sum. If we denote the first grade as G_1, the second grade as G_2, and so on through the seventh grade which we denote as G_7, the final sum can be given symbolically as

$$\text{Sum} = G_1 + G_2 + G_3 + G_4 + G_5 + G_6 + G_7. \tag{8}$$

Of course, if we had 52 items to add as opposed to only 7, writing an expression similar to Eq. (8) would be tedious indeed.

A more convenient way to represent the right side of Eq. (8) symbolically is $\sum_{i=1}^{7} G_i$. The capital Greek letter sigma ($\sum$) denotes a sum. The quantity $i = 1$ below the sigma indicates that the sum starts with G_1, that is, G_i with $i = 1$. The addition is finished when the subscript on G reaches the value listed above the sigma, which in this case is 7. Intermediate values in the sum are obtained by replacing the subscript i on individual G terms with consecutive integers between the two numbers listed below and above the sigma.

The notation $\sum_{i=3}^{8} S_i$ indicates a sum. The sum starts with S_3, since the quantity below the sigma is $i = 3$, and it includes all values of S through S_8, since the number above the sigma is 8. That is,

$$\sum_{i=3}^{8} S_i = S_3 + S_4 + S_5 + S_6 + S_7 + S_8.$$

Similarly,

$$\sum_{i=1}^{5} T_i = T_1 + T_2 + T_3 + T_4 + T_5$$

$$\sum_{i=2}^{4} x_i = x_2 + x_3 + x_4$$

and

$$\sum_{i=9}^{9} y_i = y_9.$$

In general,

$$\sum_{i=m}^{n} q_i = q_m + q_{m+1} + q_{m+2} + \cdots + q_n \tag{9}$$

is the *sigma notation* for the sum of quantities q_i ranging from q_m through q_n successively. Obviously, the left side of Eq. (9) is more compact than the expanded form given on the right side.

Example 1 Give the expanded form of $\sum_{i=4}^{8} (x_i - i)$.

Solution We want the sum of terms of the form $x_i - i$ beginning with $i = 4$ and continuing successively through $i = 8$. Therefore

$$\sum_{i=4}^{8} (x_i - i) = (x_4 - 4) + (x_5 - 5) + (x_6 - 6) + (x_7 - 7) + (x_8 - 8).$$

Example 2 Give the expanded form of $\sum_{i=1}^{6} [i/(i + 1)]$.

Solution

$$\sum_{i=1}^{6} \left(\frac{i}{i+1}\right) = \left(\frac{1}{1+1}\right) + \left(\frac{2}{2+1}\right) + \left(\frac{3}{3+1}\right) + \left(\frac{4}{4+1}\right) + \left(\frac{5}{5+1}\right) + \left(\frac{6}{6+1}\right).$$

Example 3 Determine the sigma notation for the sum

$$\left(\frac{3^2}{2} + 1\right) + \left(\frac{4^2}{2} + 1\right) + \left(\frac{5^2}{2} + 1\right) + \left(\frac{6^2}{2} + 1\right) + \cdots + \left(\frac{100^2}{2} + 1\right).$$

Solution Each term in the sum is of the form $(i^2/2) + 1$, where i is an integer from 3 to 100 inclusive. The sum can be given as

$$\sum_{i=3}^{100} \left(\frac{i^2}{2} + 1\right).$$

Example 4 Weekly expenditures for a given year are denoted as W_1 through W_{52} successively. Develop a formula for the yearly expenditure.

Solution Denote the yearly expenditure as Y. Since it is the sum of the individual weekly expenditures, $Y = \sum_{i=1}^{52} W_i$.

All the examples so far have used the subscript i. Any other letter would do equally well. Thus

$$\sum_{j=1}^{4} x_j = x_1 + x_2 + x_3 + x_4$$

and

$$\sum_{k=1}^{10} k^2 = 1^2 + 2^2 + 3^2 + 4^2 + 5^2 + 6^2 + 7^2 + 8^2 + 9^2 + 10^2.$$

Sometimes we are given data and we would like to indicate that some portion of these data is to be summed. Test scores for a particular student are 60, 70, 75, 80, 82, 83, 87, 90. If we do not wish to consider the first and last scores for some reason, but only the sum of the intermediate grades, we can indicate this sum by $\sum_{i=2}^{7} G_i$. Of course this notation is good only if we understand that G_1 signifies the first score, G_2 the second score, and G_7 the next to the last score. In general, we always assume that data are ordered as they appear.

Finally, when we write a sigma without any numbers below or above it, we mean that the sum is to include all possible terms. It should be clear from the context which terms are possible. For example, certain data may include pairs of numbers, and we may wish to multiply the members of each pair and then sum over all the pairs. If the number of data points is not known in advance, we can still indicate the desired sum by $\sum x_i y_i$. For the data listed in Table 1.5, this sum is

$$\sum x_i y_i = x_1 y_1 + x_2 y_2 + x_3 y_3 + x_4 y_4 \quad \text{(since there are four pairs of data points)}$$
$$= (1)(2) + (2)(4) + (3)(6) + (4)(8) = 60.$$

TABLE 1.5

x	y
1	2
2	4
3	6
4	8

Exercises

1. Write the expanded form of the following expressions:

    (a) $\sum_{i=1}^{3} (x_i)^2$

    (b) $\sum_{i=3}^{11} 2x_i$

(c) $\sum_{k=2}^{7} (x_k + y_k)$ (d) $\sum_{j=99}^{105} (3M_j + 4)$.

2. Write the expanded form of the following expressions:

(a) $\sum_{k=1}^{10} k$ (b) $\sum_{m=2}^{6} \left(\frac{m+2}{m+3}\right)$

(c) $\sum_{j=0}^{7} (j+1)$ (d) $\sum_{i=1}^{15} 1$

(e) $\sum_{i=0}^{6} (-1)^i$ (f) $\sum_{p=7}^{14} (p-10)^2$.

3. Write the following expressions in sigma notation.
   (a) $3(2)^2 + 3(3)^2 + 3(4)^2 + 3(5)^2 + \cdots + 3(29)^2$
   (b) $2(3)^2 + 3(3)^2 + 4(3)^2 + 5(3)^2 + \cdots + 29(3)^2$
   (c) $2(3)^2 + 2(3)^3 + 2(3)^4 + 2(3)^5 + \cdots + 2(3)^{29}$
   (d) $3(2)^2 - 3(3)^2 + 3(4)^2 - 3(5)^2 + \cdots + 3(28)^2 - 3(29)^2$.

4. Calculate the following sums for the data given in Table 1.6.

(a) $\sum_{i=1}^{5} x_i$ (b) $\sum_{j=1}^{5} y_j$

(c) $\sum_{i=1}^{3} x_i$ (d) $\sum_{k=1}^{5} (x_k + y_k)$

(e) $\sum_{k=1}^{5} x_k + \sum_{k=1}^{5} y_k$ (f) $\sum_{m=2}^{4} x_m y_m$.

(g) What can you conclude about (d) and (e)?

5. Calculate the following sums for the data given in Table 1.7.
   (a) $\sum x_i$ (b) $\sum y_i$
   (c) $\sum (x_i)^2$ (d) $\sum (x_i - 2)$
   (e) $\sum x_i y_i$ (f) $(\sum x_i)(\sum y_i)$.
   (g) What can you conclude about (e) and (f)?

6. Write each of the following expressions in expanded form and verify that they are equal.

(a) $\sum_{i=1}^{6} \left(\frac{1}{i}\right)$ (b) $\sum_{i=2}^{7} \left(\frac{1}{i-1}\right)$ (c) $\sum_{i=0}^{5} \left(\frac{1}{i+1}\right)$.

TABLE 1.6

i	x	y
1	0	6
2	1	7
3	2	8
4	-1	3
5	-4	-2

TABLE 1.7

i	x	y
1	0	3
2	8	2
3	-2	6
4	5	9
5	-3	10
6	7	1

7. Prove the following identities by converting each side to expanded form

   (a) $c\left(\sum_{i=1}^{n} x_i\right) = \sum_{i=1}^{n} (cx_i)$

   (b) $\sum_{i=1}^{n} (x_i + y_i) = \sum_{i=1}^{n} x_i + \sum_{i=1}^{n} y_i$

   (c) $\sum_{i=1}^{m} x_i + \sum_{i=m+1}^{n} x_i = \sum_{i=1}^{n} x_i.$

8. Determine the validity of the statement

   $$\sum_{i=1}^{n} (x_i y_i) = \left(\sum_{i=1}^{n} x_i\right)\left(\sum_{i=1}^{n} y_i\right).$$

9. Derive a formula using sigma notation for the average of a set of grades $G_1, G_2, G_3, \ldots, G_n$.

Chapter 2

Equations and Curves

Equations are a convenient and concise way of representing relationships between quantities such as sales and advertising, profit and time, cost and number of units manufactured, and so on. They are used to "model" or represent real-world situations.

The idea and usefulness of a model are not new; we use models every day without recognizing them as such. A road map, for example, is a model. It does not give weather, traffic conditions, or road hazards, but it is still useful in planning or making decisions about a trip. Molecular diagrams in chemistry books, organization charts in management books, and road signs indicating a sharp curve or a steep incline are other examples of useful models representing actual conditions in a simplified, compact way. So too, mathematical equations are used as models in the world of business.

Many business situations can be adequately represented or modeled by equations. Although these equations, like a road map, may not reveal all the relationships between the quantities under investigation, they often contain enough information for one to make meaningful observations and practical decisions.

2.1 Linear Equations and Straight Lines

We begin our study with the simplest and yet one of the most important equations in both business and mathematics: the linear equation whose graph is a straight line. Formally, a linear equation is defined as follows.

Definition 2.1 A *linear equation* in two variables x and y is an equation of the form

$$cx + dy = e, \tag{1}$$

where c, d, and e are known numbers with c and d not both zero.

Definitions are very precise mathematical statements. Unfortunately this precision often makes a definition seem very complicated, which it is not. Usually, a few moments of thought is all that is needed to convert the given statement to an understandable concept. As an example, let us return to our first definition.

Definition 2.1 simply states that any equation having the form of (that is, looks like) the equation $cx + dy = e$, where the letters c, d, and e are replaced by numbers (for example, $3x + 7y = 10$), is called a *linear* equation. The definition does not give any clues as to what a linear equation means physically—that will come later. It does say that, if an equation looks like Eq. (1), we will *call* it a linear equation.

Examples of linear equations are

$$6x + 2y = 15 \tag{2}$$

$$-x + 7y = 0 \tag{3}$$

$$\tfrac{1}{2}x - \tfrac{3}{4}y = 1.7 \tag{4}$$

$$y = 0. \tag{5}$$

Equation (2) has the required form with $c = 6$, $d = 2$, and $e = 15$, while in Eq. (3) $c = -1$, $d = 7$, and $e = 0$. If we first rewrite Eq. (4) as $\tfrac{1}{2}x + (-\tfrac{3}{4})y = 1.7$, then $c = \tfrac{1}{2}$, $d = -\tfrac{3}{4}$, and $e = 1.7$. Equation (5) also has the form of Eq. (1) with $c = e = 0$ and $d = 1$.

Example 1 Determine whether or not $x^2 + y^2 = 4$ is a linear equation.

Solution No, it is not. Here both quantities x and y are squared, whereas Eq. (1) requires x and y to appear *by themselves* multiplied *only* by known numbers.

Example 2 Determine whether or not $1/x + 2y = 0$ is a linear equation.

Solution No, it is not. Here x appears as $1/x$, not as x multiplied by a known number as required in Eq. (1).

Equation (1) is indeed very precise. An equation is called a linear equation if and only if it has the form of a constant times one quantity plus a constant times another quantity equal to a constant. No x^2 terms, no $\sqrt{y}$ terms, and no xy terms are allowed.

It should also be stressed that the letters x and y in Eq. (1) are *not* fixed; any two letters can be used. Thus $6C + 2N = 15$ is a linear equation in the quantities C and N [see Eq. (2)], and $\frac{1}{2}p - \frac{3}{4}q = 1.7$ is a linear equation in the quantities p and q [see Eq. (4)].

Example 3 Determine whether or not $C = 115 + 40N$ is a linear equation.

Solution If we first rewrite this equation as $C + (-40)N = 115$, we see that it is a linear equation in the quantities C and N, since it represents the constant 1 times C plus the constant -40 times N equal to the constant 115.

Example 4 A new car dealer determines that the daily cost of operating a showroom can be separated into a fixed cost of $115 covering insurance, rent, lighting, and so on, and a variable salary cost. Each salesperson is paid $40 per day, but the total number of salespeople who work changes from day to day. Show that the equation relating cost to the number of salespeople at work on any given day is a linear equation.

Solution The quantities we are trying to relate in this problem are the daily cost and the number of salespeople who work on any given day. Let C denote daily cost and N represent the number of salespeople working on any given day. Then $C = 115 + 40N$ which, from Example 3, is a linear equation.

Note in Example 4 that, if we knew how many salespeople were working on a given day, we could easily determine the total cost of operating the store. We would simply substitute the number of salespeople into the equation where the letter N appears, obtaining one equation in the unknown quantity C. Using the techniques of solving one equation in one unknown (see Section 1.2), we could then solve for the value of C. Similarly, if we knew the cost of operating the store on a particular day (the value of C), we could then use the equation to find the number of employees who worked that day. We illustrate this with another example.

Example 5 The number of cans of cat food sold weekly in a particular store is related to the price charged. Let D denote the number of cans sold and P the price per can (in cents) and assume the equation relating these quantities is

$$D = 182 - 7P. \tag{6}$$

Determine (a) whether or not the equation relating D and P is a linear equation, (b) the number of cans sold when the price is 20¢ per can, and (c) the price on the day that 77 cans are sold.

Solution (a) Rewriting Eq. (6) as $D + 7P = 182$, we see that it is a linear equation. Using Eq. (6) (or the rewritten form) we can determine the number of cans sold given the price, or the price charged given the number of cans sold. (b) For a price of 20¢, we substitute $P = 20$ into Eq. (6) and compute $D = 182 - 7(20) = 182 - 140 = 42$ cans sold. (c) If the store sold 77 cans, substituting $D = 77$ into Eq. (6) and solving for P, we have

$77 = 182 - 7P$
$7P = 105$
$P = 15$¢ per can.

Equations having the form of Eq. (1) are singled out as a special class of equations because they have several useful properties. One property is that all linear equations have graphs that are straight lines. Let us plot the graphs of a few examples to verify this property.

Example 6 Graph Eq. (2), $6x + 2y = 15$.

Solution We first plot some points satisfying the equation by picking arbitrary values of either x or y and finding the corresponding values of the other quantity in the equation. Arbitrarily, we select x-values and solve for the corresponding y-values. When $x = 0$, Eq. (2) becomes $6(0) + 2y = 15$, hence $y = \frac{15}{2} = 7.5$. When $x = 1$, Eq. (2) becomes $6(1) + 2y = 15$ and, solving for y, we obtain $2y = 15 - 6$, $2y = 9$, and $y = \frac{9}{2} = 4.5$. Continuing in this manner, we generate Table 2.1 from which we obtain Figure 2.1.

Example 7 Graph Eq. (3), $-x + 7y = 0$.

Solution Again we first plot points on the curve by picking arbitrary values of either x or y and then finding their corresponding y or x values. Arbitrarily selecting x-values, we compute as follows. When $x = 0$, Eq. (3) becomes $-(0) + 7y = 0$, hence $y = 0$. When $x = 1$, Eq. (3) becomes $-(1) + 7y = 0$ from which we obtain $y = \frac{1}{7}$. Continuing, we generate Table 2.2 and Figure 2.2.

TABLE 2.1

x	y
0	7.5
1	4.5
2	1.5
3	-1.5
-1	10.5

2.1 Linear Equations and Straight Lines **57**

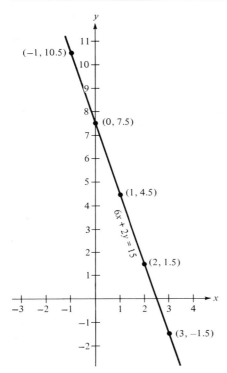

FIGURE 2.1

TABLE 2.2

x	y
0	0
1	$\frac{1}{7}$
2	$\frac{2}{7}$
3	$\frac{3}{7}$
7	1

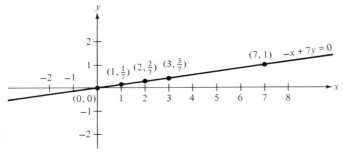

FIGURE 2.2

In Examples 6 and 7 we arbitrarily selected x-values and then solved for the corresponding y-values. We would have obtained the same graphs had we first picked y-values and then solved for the corresponding x-values.

Example 8 Redo Example 6 by picking y-values and solving for x.

Solution With $y = 0$, Eq. (2) becomes $6x + 2(0) = 15$ from which we obtain $6x = 15$, and $x = 2.5$. When $y = 3$, Eq. (2) becomes $6x + 2(3) = 15$, hence $6x + 6 = 15$, $6x = 9$, and $x = 1.5$. Continuing in this manner, we generate Table 2.3, all of whose points lie on the graph in Figure 2.1.

TABLE 2.3

x	y
2.5	0
1.5	3
0.5	6
3.5	−3
$\frac{13}{6}$	1

Wherever we have an equation in two quantities, say x and y, we can select different values for one of the quantities and then use the equation to determine the corresponding values of the other quantity. By choosing different values of one quantity we are obviously *varying* or changing the values of that quantity. Accordingly, the quantities themselves, say x and y, are referred to as *variables* rather than quantities. Since the term "variable" is standard mathematical terminology, we will use it in this text. Therefore Eq. (1) is a linear equation in the variables x and y. The equation in Example 3 contains the variables C and N.

Now, knowing that the graph of a linear equation is a straight line and that a straight line is uniquely determined by two distinct points, we can simplify our graphing procedures. Rather than finding many points on the graph, as we did in Examples 6 and 7, we can find only *two* points and then draw a straight line through them.

Example 9 Graph Eq. (4), $\frac{1}{2}x - \frac{3}{4}y = 1.7$.

Solution Since Eq. (4) is linear, we need only two points that satisfy it in order to graph the line. Rewriting the equation in decimal form, $0.5x - 0.75y = 1.7$, we arbitrarily choose two values of x, say $x = 0$ and $x = 1$, and find their corresponding y-values. When $x = 0$, the equation becomes

$$(0.5)(0) - (0.75)y = 1.7$$
$$-(0.75)y = 1.7$$
$$y = -\frac{1.7}{0.75} = -2.27.$$

When $x = 1$, the equation becomes

$$(0.5)(1) - (0.75)y = 1.7$$
$$-(0.75)y = 1.7 - 0.5$$
$$-(0.75)y = 1.2$$
$$y = -\frac{1.2}{0.75} = -1.6.$$

Plotting these two points, $x = 0, y = -2.27$ and $x = 1, y = -1.6$, we obtain Figure 2.3. Then, drawing a straight line through the points, we obtain Figure 2.4 as the graph of Eq. (4).

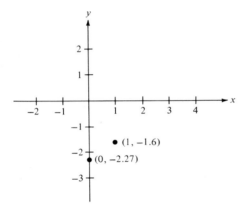

FIGURE 2.3

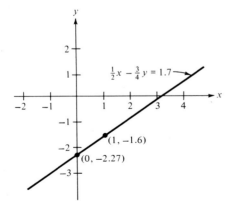

FIGURE 2.4

60 2 Equations and Curves

Example 10 Graph $15p + 10q = 150$.

Solution We first note that this equation is linear in the variables p and q. Since its graph is a straight line, we can sketch it easily by first finding two points on the line and then drawing a straight line through these points. When $p = 0$, the equation becomes $15(0) + 10q = 150$, $10q = 150$, and $q = 15$. When $p = 2$, the equation becomes $(15)(2) + 10q = 150$ from which we obtain $30 + 10q = 150$, $10q = 150 - 30$, $10q = 120$, and $q = 12$. Thus two points on this line are $p = 0$, $q = 15$ and $p = 2$, $q = 12$. Graphing these points (see Figure 2.5), we can draw the straight line given in Figure 2.6.

Warning: The method of plotting two points and then drawing a straight line through them is valid only *if* the equation is known to be linear. When applied to equations that are not linear equations, for example, $x^2 + y^2 = 4$, this procedure results in erroneous graphs.

Finally, we consider the special straight lines given by Eq. (1) when either c or d equals zero. Such lines can be simplified into the following equations:

$$x = h \tag{7}$$

$$y = k, \tag{8}$$

where $h = e/c$ and $k = e/d$ are known constants. Equation (5) is an example of Eq. (8) with $k = 0$.

Example 11 Graph the equation $y = 5$.

Solution A casual approach may be to assume that x is always zero and $y = 5$, but this would be wrong and almost completely opposite the actual situation.

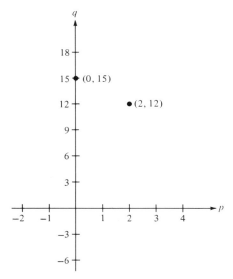

FIGURE 2.5

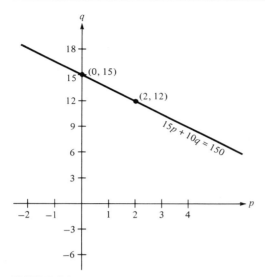

FIGURE 2.6

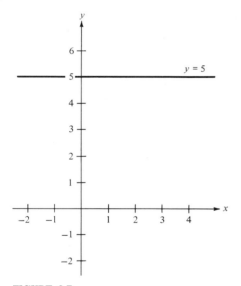

FIGURE 2.7

The equation $y = 5$ provides absolutely no constraints on x, so in fact x must be arbitrary. *It can be anything.* The only way we could conclude that x is zero would be to have the equation $x = 0$, which is not the case here.

From a closer look at the given equation, we see that $y = 5$ can be written as

$$0x + 1y = 5. \tag{9}$$

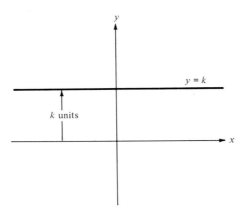

FIGURE 2.8

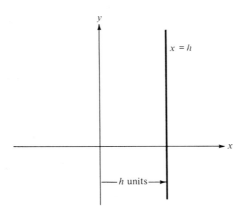

FIGURE 2.9

Thus, as long as $y = 5$, any value of x will satisfy Eq. (9). In particular, $x = 1$, $y = 5$ and $x = -3$, $y = 5$ are two points that satisfy either Eq. (9) or the given equation. Plotting these points and then drawing a straight line through them (see Figure 2.7) we find the graph of $y = 5$ is a straight line parallel to the x-axis having all y-coordinates equal to 5.

By generalizing Example 11, it follows that the graph of Eq. (8), given in Figure 2.8, is a straight line parallel to the x-axis having all y-coordinates equal to k, while the graph of Eq. (7), given in Figure 2.9, is a straight line parallel to the y-axis having all x-coordinates equal to h. In particular, the line $x = 0$ is the y-axis, and the line $y = 0$ is the x-axis.

Exercises

1. Determine which of the following equations are linear:
   (a) $2x = y$ (b) $2x = 1/y$
   (c) $xy = 4$ (d) $x = 4$
   (e) $2x - 3y = 0$ (f) $y = 4x$
   (g) $y = 4x^2$ (h) $x - 2 = 3y$
   (i) $1/x + 1/y = 2$ (j) $x = y$.

2. A large television manufacturer has determined that the number of television sets sold, denoted by N, is related directly to the amount of money spent on advertising. In particular, every million dollars in advertising expenditures results in an additional 50,000 television sets being sold although 10,000 sets would be sold with no advertising. Let E denote the amount of money (in millions of dollars) committed to advertising. Show that the equation relating N to E is linear.

3. After carefully studying used car price lists, Mr. Henry has determined that his particular model car, purchased yesterday for $5000, will depreciate $1000 per year. Let V denote the value of Mr. Henry's car at any given time and let t denote time measured in years. Show that the equation relating V to t is linear.

4. It has been determined by the Chubby Cat Food Corporation that the amount of cat food, denoted by A, sold on any given day in Newark, New Jersey, is given by the equation $A = 200 - 5p$, where p represents the price of each can (in nickels). Determine the number of cans of cat food the company can expect to sell if it prices each can at 50¢.

5. Graph the following equations:
   (a) $2x + 3y = 6$ (b) $-2x + 3y = 6$
   (c) $2x - 3y = 6$ (d) $2x + 3y = -6$
   (e) $3x + 2y = 6$ (f) $x = 7$
   (g) $10x - 5y = 50$ (h) $x = y$.

6. Graph the following equations in the given variables:
   (a) $2P + 3Q = 6$ (b) $-N + 2M = 10$
   (c) $N = 1 + 2M$ (d) $5r - 2s = 0$.

7. The sales (in millions of dollars) of a particular company are given by $S = 2E + 3.5$, where E represents advertising expenditures (in millions of dollars). Determine the amount of money that must be committed to advertising in order to realize gross sales of $10 million.

8. The current value V of a particular model automobile originally purchased for $5000 is given by $V = -1250t + 5000$, where t denotes time (in years). Determine (a) the date when the car will have lost all its resale value, and (b) the date when the car will be worth exactly half its original purchase price.

9. Using the data in Exercise 4, determine the price at which cans of cat food cease to be saleable.

10. The number of incoming telephone calls N received daily by a local answering service is given by $N = 5t + 100$, where t denotes time (in days). The current staff can handle 500 calls per day. Determine how long it will take for the number of incoming calls to exceed the capacity of the current staff.

2.2 Properties of Straight Lines

Definition 2.2 Let (x_1, y_1) and (x_2, y_2) be any two distinct points on the same straight line (or, alternatively, satisfying the same linear equation). The *slope* of the line is

$$\text{Slope} = \frac{y_2 - y_1}{x_2 - x_1}. \tag{10}$$

Let us see what this definition says and what it does not say. The definition does not indicate what a slope represents; a graphical interpretation is required for this. Definition 2.2 simply tells us how to calculate something called a slope given the coordinates of any two points on a straight line. Given the two points, we first must subtract y-values, then subtract x-values in the same order as we subtracted their corresponding y-values, and finally divide the subtracted x-values into the subtracted y-values. The number we obtain is called the slope.

Example 1 Find the slope of the line $6x + 2y = 15$.

Solution In Example 6 in Section 2.1 we found that two points on this line are $(0, 7.5)$ and $(1, 4.5)$. Therefore, letting $x_1 = 0$, $y_1 = 7.5$ and $x_2 = 1$, $y_2 = 4.5$, and substituting into Eq. (10), we have

$$\text{Slope} = \frac{4.5 - 7.5}{1 - 0} = \frac{-3}{1} = -3.$$

One should check that the value of the slope is the same if we select $(0, 7.5)$ as the (x_2, y_2) point and $(1, 4.5)$ as the (x_1, y_1) point. Definition 2.2 simply requires that the x-value of each point be subtracted in the same order as the y-value prior to the division operation.

Example 2 Find the slope of the linear equation $y - 2x = 1$.

Solution Definition 2.2 requires that we select two points satisfying the given equation $y - 2x = 1$. Arbitrarily selecting x-values of 1 and 4 (any choice would do) we calculate the corresponding y-values as 3 and 9. Thus two points satisfying the equation are $(1, 3)$ and $(4, 9)$. Letting $(1, 3)$ be the (x_1, y_1) point and $(4, 9)$ the (x_2, y_2) point, and substituting into Eq. (10), we have

$$\text{Slope} = \frac{9 - 3}{4 - 1} = \frac{6}{3} = 2.$$

In order to have a graphical interpretation of the slope of a straight line consider Figure 2.10. P_1 and P_2 are two points on this straight line selected at random. As indicated, the difference in the y values, $y_2 - y_1$, represents the vertical distance between the two points P_1 and P_2, while the difference in the

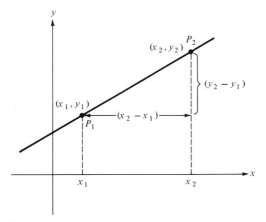

FIGURE 2.10

x-values represents the horizontal distance between the two points. A specific example should clarify this point. In Example 2, we showed that the line satisfying the equation $y - 2x = 1$ had a slope of 2. This curve, along with the two points P_1 and P_2 used to calculate the slope, is illustrated in Figure 2.11.

From the graph it is evident that, had we started at point P_1 and moved

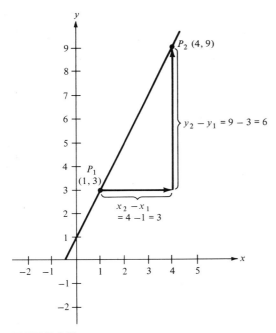

FIGURE 2.11

to point P_2 along the line, we would have traveled a distance of 3 units to the right and up a distance of 6 units as indicated by the arrows. The 6 and 3 are the respective differences in y- and x-values from point P_1 to point P_2. The ratio 6/3, or simply 2, is the slope of the line.

Recall that we selected points P_1 and P_2 in Figure 2.11 at random. Had we selected two other points on the curve and calculated the slope, we still would have found the slope to be 2. The reason for this is that any two points on the curve illustrated in Figure 2.11 are related in a particular way. By starting at one point on the curve, every other point on the line can be reached by increasing or decreasing y- and x-values in the ratio 2/1. If we start at point P_1 and increase x by 1 unit, we must increase y by 2 units to land back on the curve. Should we increase x by 5 units, we must increase y by 10 units to remain on the curve.

The significance of the slope is that it gives us the particular ratio relating points on a given line. As such, the slope respresents the *rate of change in y associated with a change in x*, and this has interesting and important commercial applications.

Example 3 It is known that the monthly sales of a particular model automobile S (in units) are related to the advertising expenditures E (in millions of dollars) by the equation $S = 20{,}000 + 5000E$. Determine the rate of change in sales with respect to advertising expenditures.

Solution Rewriting the given equation as $S - 5000E = 20{,}000$, we observe it is linear so its graph is a straight line. The rate of change is just the slope. To find the slope, we first need two distinct points on the line. Arbitrarily choosing $E_1 = 0$ and $E_2 = 1$, we find the corresponding values of S as $S_1 = 20{,}000$ and $S_2 = 25{,}000$. The rate of change in sales with respect to advertising expenditures is

$$\frac{S_2 - S_1}{E_2 - E_1} = \frac{25{,}000 - 20{,}000}{1 - 0} = 5000.$$

Whenever E is increased by 1 unit (in this case $1 million), the monthly sales will be increased by 5000 units (in this case cars).

Graphically, a line has a positive slope if the angle between the line and the positively directed horizontal axis is between 0 and 90 degrees, whereas a line has a negative slope if the angle between the line and the positively directed horizontal axis is between 90 and 180 degrees. That is, a line with a positive slope slants upward to the right, whereas a line with a negative slope slants downward to the right. See Figures 2.12 and 2.13. A line with a large positive slope is steeper than a line with a smaller positive slope, and a line with a large negative slope is steeper downward than a line with a less negative slope.

A line parallel to the x-axis has zero slope, since any two points on the line have the same y-coordinate (see Figure 2.8). A line parallel to the y-axis does

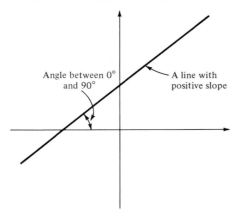

FIGURE 2.12

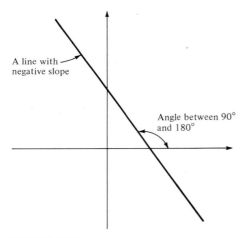

FIGURE 2.13

not have a slope. Any two points on such a line have the same x-coordinate and substituting this value into Eq. (10) would result in a denominator of zero which is undefined arithmetically.

Given a straight line $cx + dy = e$ with $d \neq 0$, we can rewrite the equation as $dy = -cx + e$ or $y = (-c/d)x + (e/d)$. Setting $m = -c/d$ and $b = e/d$, we obtain the equation

$$y = mx + b. \tag{11}$$

Equation (11) is an alternative representation for all straight lines not parallel to the y-axis. For lines parallel to the y-axis [see Eq. (7)], $d = 0$ in Eq. (1) and the above manipulations are invalid. For most lines, however, Eq. (11) is valid, and it has several advantages over Definition 2.1. One such advantage is that the slope is easily obtained as m, the coefficient of x.

Example 4 Show that the slope of a straight line given by Eq. (11) is m.

Solution If (x_1, y_1) and (x_2, y_2) are any two points on the line given by Eq. (11), they must satisfy Eq. (11). Necessarily, $y_1 = mx_1 + b$ and $y_2 = mx_2 + b$. Then,

$$\text{Slope} = \frac{y_2 - y_1}{x_2 - x_1} = \frac{(mx_2 + b) - (mx_1 + b)}{x_2 - x_1} = \frac{mx_2 - mx_1}{x_2 - x_1} = \frac{m(x_2 - x_1)}{x_2 - x_1} = m.$$

The quantity b in Eq. (11) also has geometric significance; it represents the value at which the line crosses the y-axis and is commonly referred to as the *y-intercept*.

A second advantage of Eq. (11) over Eq. (1) is the ease with which the equation of a line can be obtained from its points.

Example 5 Find the equation of the line containing the two points $(1, -3)$ and $(3, 5)$.

Solution Since these points are on the line of interest, we can compute the slope of this line by Definition 2.2. Thus

$$m = \frac{y_2 - y_1}{x_2 - x_1} = \frac{5 - (-3)}{3 - 1} = \frac{8}{2} = 4$$

and, from Eq. (11), the line will have the form $y = 4x + b$. To find b, we substitute either point into this last equation. Using the point $(1, -3)$, we obtain $-3 = 4(1) + b$, $-3 - 4 = b$, and $b = -7$. Note that, if we had used the point $(3, 5)$, we would have found $5 = 4(3) + b$, $5 - 12 = b$, and again $b = -7$. The line is $y = 4x - 7$.

Example 6 At a recent sales meeting of the Lincoln Hamburger Company, Figure 2.14 was displayed to emphasize the growth in profit. Find the equation relating profit P to time t (in years).

Solution Obviously the graph indicates a straight-line relationship between P and t, so we seek an equation of the form

$$P = mt + b. \tag{12}$$

Here $t = 0$ corresponds to January 1, 1971, $t = 1$ corresponds to January 1, 1972, and $t = 5$ corresponds to January 1, 1976. To find the slope, we need two

2.2 Properties of Straight Lines

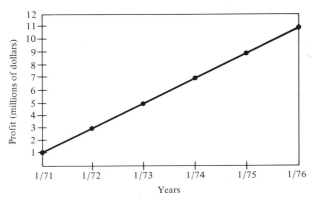

FIGURE 2.14

points on the line. These can easily be obtained from the graph as $P = 1$ when $t = 0$ and $P = 11$ when $t = 5$. Thus

$$m = \frac{P_2 - P_1}{t_2 - t_1} = \frac{11 - 1}{5 - 0} = \frac{10}{5} = 2.$$

The y-intercept is also easily obtained from the graph as $b = 1$. Note that the y-axis here corresponding to $t = 0$ is January 1, 1971. Alternatively, we can obtain b by first substituting $m = 2$ into Eq. (12), obtaining $P = 2t + b$, and then substituting a point on the line into this last equation to obtain b. Following this procedure and using the point $(P = 1, t = 0)$ we again find $1 = 2(0) + b$ and $b = 1$. Either way the equation of interest is $P = 2t + 1$.

Example 7 A recent survey conducted for the Chubby Cat Food Corporation resulted in Table 2.4. Find the equation relating the number of cans sold daily (N) to the price of each can (P).

Solution Graphing these five data points, we see that the relationship appears to be a straight line, so we seek an equation of the form*

$$N = mP + b, \tag{13}$$

TABLE 2.4

Number of cans sold	18,000	17,500	17,000	16,000	14,000
Price (¢)	20	25	30	40	60

---

* We have made the assumption here that the data conform to a straight line, and that, if we were to obtain data points between those given, they would correspond to the same line.

where N denotes the number of cans sold and P denotes the price per can (in cents). Any two of the five given data points will suffice to calculate m. Taking $N_1 = 18{,}000$, $P_1 = 20$, $N_2 = 17{,}500$, and $P_2 = 25$, we compute

$$m = \frac{N_2 - N_1}{P_2 - P_1} = \frac{17{,}500 - 18{,}000}{25 - 20} = \frac{-500}{5} = -100.$$

Equation (13) becomes $N = -100P + b$. To obtain b, we substitute any one of the data points, say $P_1 = 20$, $N_1 = 18{,}000$, into this last equation and find

$$18{,}000 = -100(20) + b$$
$$18{,}000 + 2000 = b$$
$$b = 20{,}000.$$

Thus $N = -100P + 20{,}000$.

The last equation has interesting connotations when $P = 0$. Then, $N = -100(0) + 20{,}000 = 20{,}000$, so that, even if the Chubby Cat Food Corporation gave its product away free, only 20,000 cans would be consumed daily. Given that not everyone owns a cat and that each cat can eat only one can per day, this result is not unexpected.

Example 8 Table 2.5 is the result of years of data collecting by the American Citrus Corporation. Find the equation relating the number of orange trees that bear fruit G to the number of trees planted N.

TABLE 2.5

Number of orange trees planted	120	140	160	180
Number that bear fruit	114	133	152	171

Solution Graphing these four data points, we see that the relationship appears to be a straight line, so we seek an equation of the form

$$G = mN + b. \tag{14}$$

Two points on the line are $N_1 = 120$, $G_1 = 114$ and $N_2 = 140$, $G_2 = 133$. The slope of Eq. (14) is

$$m = \frac{G_2 - G_1}{N_2 - N_1} = \frac{133 - 114}{140 - 120} = \frac{19}{20} = 0.95.$$

With $m = 0.95$, Eq. (14) becomes $G = 0.95N + b$. To obtain b, we substitute any corresponding values of G and N, say $N_1 = 120$ and $G_1 = 114$, into this last equation and find

$114 = (0.95)(120) + b$
$114 = 114 + b$
$b = 0.$

Thus $G = 0.95N$.

Exercises

1. Find the slopes of the straight lines given in Exercise 1 in Section 2.1.
2. Find the straight line containing the given points:
   (a) (1, 2) and (2, 5)   (b) (7, −3) and (−1, −8)
   (c) (−1, 2) and (4, 2)   (d) (1, 0) and (0, 1)
   (e) (2, −1) and (2, 4).
3. Figure 2.15 illustrates the cumulative monthly attendance at a local amusement park for the past year. Find the equation relating attendance A to time t (in months).

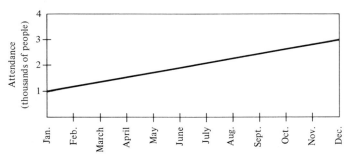

FIGURE 2.15

4. Figure 2.16 illustrates the cumulative weekly sales receipts of a supermarket over the past year. (a) Find the equation relating gross income I to time t for the first 20-week period. (b) Find the equation relating I to t for the last 32 weeks of the year.

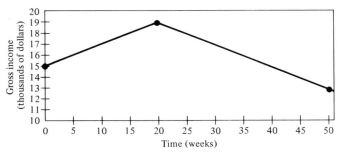

FIGURE 2.16

TABLE 2.6	
Number of free samples distributed	Number of subsequent purchases
1000	2050
1500	2075
2000	2100
3000	2150
5000	2250

TABLE 2.7	
Number of bulbs produced	Number of defective bulbs
25,000	75
50,000	150
60,000	180
75,000	225
90,000	270

5. After test-marketing a new bleach, the White-All Bleach Company collected the data given in Table 2.6. Plot the points in Table 2.6 to verify that the relationship between the number of subsequent purchases P and the number of samples distributed S is a straight line, and then find the equation of that line.
6. Quality-control tests on the manufacture of light bulbs resulted in Table 2.7. Plot the points in this table to verify that the relationship between the number of defective bulbs D and the number of bulbs produced N is a straight line, and then find the equation of that line.

2.3 Break-Even Analysis

Linear equations are useful in a business setting for determining the relationships between short-run costs and short-run sales revenues. Traditionally, the term *short run* refers to a time period during which both cost and price per item remain constant. Over any extended time frame inflation, supply and demand, and other economic factors act to change the cost and price. Over the short run, generally a year or less, these economic conditions have little influence.

Every manufacturing process involves costs which are often separated into fixed costs and variable costs. Fixed costs include rent, insurance, property taxes, and other expenses which are present regardless of the number of items produced. Over the short run they are fixed, and they exist even if no items are produced. Variable costs, however, are those expenses directly attributable to the manufacture of the items themselves, such as labor and raw materials. Variable costs depend directly on the number of items produced—the more items, the higher the variable costs. If we restrict ourselves to short-run conditions, the cost per item will be fixed and the variable cost will be the cost of manufacturing each item times the number of items produced.

If we designate the total cost by C, the cost of manufacturing an item by a, the number of items being manufactured by x, and the fixed cost by b, we can write

$$C = ax + b. \tag{15}$$

That is, the total cost is the sum of the variable cost and the fixed cost. The numbers a and b are assumed known and fixed, hence Eq. (15) is a linear equation in C and x.

Example 1 A company manufacturing electronic calculators has recently signed contracts with its suppliers and the labor unions representing its employees. For the duration of these contracts, the cost of manufacturing each calculator will be $30. The company estimates that the fixed costs for this period will be $10,000. Determine the total cost for this process.

Solution Using Eq. (15) with $a = 30$ and $b = 10,000$, we have $C = 30x + 10,000$. If 500 calculators are produced, the cost will be $C = 30(500) + 10,000$ or $25,000. If no calculators are produced, the total cost will be $C = 30(0) + 10,000$ or $10,000, which is the fixed cost.

If we continue to restrict ourselves to the short run, the price of each item produced and sold will also be a constant. The sales revenue is simply the number of items sold times the price of each item. Designating the sales revenue by R and the price per item by p, we have

$$R = px, \tag{16}$$

where x again represents the number of items sold. Since p is assumed fixed and known, Eq. (16) is a linear equation in R and x.

Example 2 The manufacturing company in Example 1 can sell all the calculators it produces and has fixed the price of each calculator at $40. Determine the sales revenue that can be expected.

Solution Using Eq. (16) with $p = 40$, we have $R = 40x$. In particular, if 500 calculators are sold, the sales revenue will be $R = 40(500)$ or $20,000.

From Examples 1 and 2, we note that a production run of 500 calculators will result in a total cost of $25,000 and a sales revenue of only $20,000. The company will experience a loss of $5000. Such embarrassing situations can be avoided with *break-even analysis*. As the name suggests, such analysis involves finding the level of sales below which it will be unprofitable to produce items and above which sales revenue exceeds costs so that a profit is made. This level is the *break-even point*.

The break-even point occurs when total cost exactly equals sales revenue. If we restrict ourselves to the short run and assume that all items produced can be sold, the break-even point can be obtained by setting the right side of Eq. (15) equal to the right side of Eq. (16). That is, $C = R$, hence

$$ax + b = px. \tag{17}$$

Equation (17) is one equation in the one unknown, x, which can be solved algebraically by the methods in Section 1.2.

For the electronic calculator described in Examples 1 and 2, we found $C = 30x + 10,000$ and $R = 40x$. The break-even point occurs when $C = R$ or when $30x + 10,000 = 40x$. This is Eq. (17) for the calculator manufacturer. Solving for x, we find that the break-even point is $x = 1000$ units. Any production below 1000 units will result in a loss, while any production above 1000 units will produce a profit.

Example 3 A lamp manufacturer determines that the manufacturing costs associated with each lamp are $5 and that the fixed costs are $7000. Determine the break-even point if each lamp sells for $7. Assume that each lamp made can be sold.

Solution The total cost for this process [Eq. (15)] is $C = 5x + 7000$. Sales revenue is given by Eq. (16) as $R = 7x$. The break-even point is the value of x for which $C = R$ or $5x + 7000 = 7x$. Solving for x, we obtain $x = 3500$ lamps as the break-even point.

Example 4 A dress manufacturer determines that the production costs associated directly with each dress are $9 and that the fixed costs are $25,000. Determine the break-even point if each dress sells for $14. Assume that all dresses manufactured can be sold.

Solution The total cost for this process is given by Eq. (15) as $C = 9x + 25,000$. The sales revenue is given by Eq. (16) as $R = 14x$. The break-even point is the solution of the equation $9x + 25,000 = 14x$, or $x = 5000$ dresses.

Break-even problems can be solved graphically as well as algebraically. The procedure is to plot Eqs. (15) and (16) on the same graph, as in Figure 2.17.

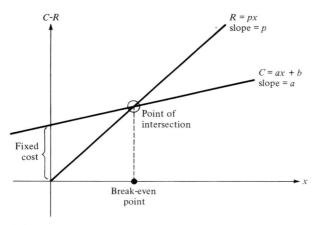

FIGURE 2.17

Since both Eqs. (15) and (16) are linear equations, their graphs are straight lines. Note that the horizontal axis is x, the number of units produced and sold, whereas the vertical axis is either C or R, depending on which equation is being considered. It follows that the unit sales price p is the slope of the line defined by Eq. (16), and the cost of manufacturing each unit a is the slope of the line defined by Eq. (15). The y-intercept corresponding to Eq. (15) is simply the fixed cost.

The break-even point is the value of x for which $C = R$, which in Figure 2.17 is the value of x corresponding to the point of intersection of the two lines.

Example 5 Graphically determine the break-even point for the manufacturing process described in Example 3.

Solution In Example 3, we determined the equations $C = 5x + 7000$ and $R = 7x$. Using the graphing procedures given in Section 2.2, we plot each line on the same graph in Figure 2.18. The x-component of the point of intersection of these two lines is read directly from the graph as $x = 3500$ which is the same break-even point found algebraically in Example 3.

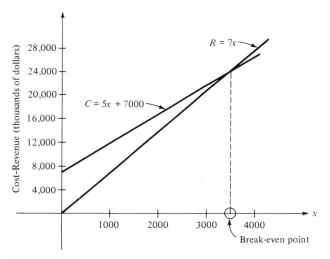

FIGURE 2.18

Exercises

1. A publisher of a current economics textbook determines that the manufacturing costs directly attributable to each book are $10 and that the fixed costs are $20,000. The publisher sells each book for $12 per copy.
   (a) Determine the equation relating the total cost to the number of books published.
   (b) Determine the equation relating the sales revenue to the number of books published.

(c) What will the profit be if 3000 books are published?
(d) Algebraically determine the break-even point for this process.
2. Find the break-even point in Exercise 1 graphically. Which method do you prefer?
3. A manufacturer of staplers determines that the manufacturing costs directly attributable to each stapler are $2 and that the fixed costs are $15,000. Each stapler sells for $3. Determine the break-even point for this process both algebraically and graphically.
4. Using the results from Exercise 3, determine (a) the total cost of the process at the break-even point, (b) the total sales revenue of the process at the break-even point, and (c) the profit at the break-even point.
5. A manufacturer of light bulbs determines that each bulb costs 15¢ in direct expenses and that the process as a whole incurs fixed costs of $1500.
   (a) Determine the break-even point if each bulb sells for 25¢.
   (b) Determine the break-even point if each bulb sells for 50¢.
   (c) Does it make sense that the answer in part (b) is smaller than that in part (a)?
6. A manufacturer of Lucite pipe holders has determined that the firm has a break-even point of 2500 units. Determine the price of each holder if each item costs $3 to manufacture and the process involves fixed costs of $5000.
7. A manufacturer of specialty bookends has determined that the break-even point for the manufacturing process is 6000 units. Determine the cost of producing each bookend set, if the fixed costs are $24,000 and each set sells for $12.
8. A manufacturer of automatic fire alarm systems determines that its total cost is is given by $C = 1000x^2 + 5000x + 10{,}000$. (Note that this equation is no longer linear.) Each system sells for $12,000. Determine the break-even point for this process (a) graphically, and (b) algebraically.

2.4 Quadratic Equations and Curves

A curve slightly more complex than a straight line is one involving a quadratic (variable squared) term.

Definition 2.3 A *quadratic equation* is an equation of the form

$$y = ax^2 + bx + d, \tag{18}$$

where a, b, and d are known numbers with a not equal to zero.

Examples of quadratic curves are

$$y = 2x^2 - \tfrac{1}{2} \tag{19}$$

$$y + x = x^2 \tag{20}$$

$$N^2 = 2C + 4. \tag{21}$$

Equation (19) has the required form with $a = 2$, $b = 0$, and $d = -\frac{1}{2}$. If we rewrite Eq. (20) as $y = x^2 - x$, then $a = 1$, $b = -1$, and $d = 0$. Similarly, rewriting Eq. (21), we obtain $C = \frac{1}{2}N^2 - 2$ which has the form $C = aN^2 + bN + d$ with $a = \frac{1}{2}$, $b = 0$, and $d = -2$.

As in the case of linear equations, the letters y and x used in Eq. (18) are irrelevant; any other two letters [see Eq. (21)] are equally appropriate. The essential point is the form of the relationship between the variables. That is, a quadratic equation is one in which one variable can be written as the sum of a constant times the second variable squared plus a constant times the second variable plus a constant.

Example 1 Determine whether or not the equation $x^2 + y^2 = 4$ is a quadratic equation.

Solution No. Here both variables appear squared, which is not the form of Eq. (18). Further, if we first solve for y, obtaining $y = \sqrt{4 - x^2}$, we introduce a square root which also does not meet the required form.

Whereas the graphs of linear equations are straight lines, the graphs of quadratic equations are parabolas, as illustrated in the following example.

Example 2 From past experience, a small dress manufacturer knows that profit P (in thousands of dollars) is related to number of pieces N (in thousands of units) by the equation $P = -0.05N^2 + 5N - 10$, under the assumption that all dresses produced will be sold. Graph the equation and explain the physical significance of the curve.

Solution To graph the equation, we first plot points. Arbitrarily choosing values of N, substituting these values into the given equation, and finding the corresponding values of P, we obtain Table 2.8 from which Figure 2.19 follows.

If no dresses are produced, $N = 0$, the manufacturer will lose \$10,000 which represents the fixed costs such as rent, insurance, and depreciation. As dresses are manufactured and sold, profits increase. Two thousand dresses, corresponding to $N = 2$, will generate almost enough capital to cover expenses. Fifty thousand dresses will result in a maximum profit of

TABLE 2.8

N	P
0	−10
2	−0.2
5	13.75
25	83.75
50	115
60	110
100	−10

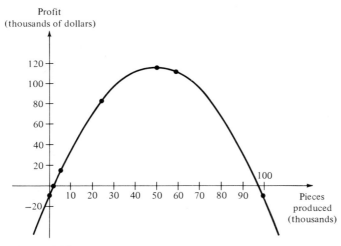

FIGURE 2.19

$115,000. The situation changes, however, with a production schedule in excess of 50,000 units. Additional dresses will require more machines, operators, internal paperwork, and record keeping which, given the physical space limitations of the plant, will reduce efficiency.

In Example 2, we arbitrarily chose values of N and then solved for the corresponding P values. As in the case of linear equations, we could have chosen values of P and then solved for N. Unlike the situation encountered with linear equations, however, this second approach is more difficult. In particular, if we had substituted $P = 0$ into the equation in Example 2, we would have had to solve the resultant equation $0 = -0.05N^2 + 5N - 10$ for the variable N. The values of N can be obtained from the quadratic formula introduced in Chapter 1, but the calculations are involved. It is easier to first pick N and then compute the corresponding values of P. In general, whenever we wish to graph a quadratic equation, it is easier to select values of the variable that appears squared [x in Eq. (18) and N in Example 2] and then use the given equation to find the value of the second variable, rather than the other way around. Sometimes, however, we have no choice.

Consider a clothing store chain that has determined its daily demand D for a particular style of shirt is related to the price per shirt P (in dollars) by the equation $D = -\frac{1}{5}P^2 + 125$. How many shirts can the store expect to sell if it charges $10 per shirt?

Since we are interested in the demand corresponding to a price $P = 10$ (in dollars), we substitute this value of P into the given equation to obtain

$$D = -\tfrac{1}{5}(10)^2 + 125 = -\tfrac{100}{5} + 125 = -20 + 125 = 105.$$

The chain can expect to sell 105 shirts.

Continuing, we now pose the following question: If the chain has 120 shirts in stock and wishes to discontinue that particular model, what price should it charge in order to deplete the inventory at the end of the day? This problem is essentially the reverse of the original question. There we were given the price and asked to calculate the demand, whereas now we are given the desired demand and asked to calculate the price that will generate this demand. Substituting $D = 120$ into the demand-price equation, we obtain

$$120 = -\tfrac{1}{5}P^2 + 125$$
$$\tfrac{1}{5}P^2 = 5$$
$$P^2 = 25$$
$$P = \pm \$5.$$

Since $P = -\$5$ is unrealistic, the appropriate price is \$5.

In general, for an equation of the form of Eq. (18) the problem of solving for the quadratic variable (the variable that is squared) given the value of the linear term (the variable that is not squared) reduces to finding a solution of the equation

$$ax^2 + bx + c = 0, \tag{22}$$

where a, b, and c are known constants with $a \neq 0$. The answer is given by the quadratic formula introduced in Chapter 1. The solutions to Eq. (22) from the quadratic formula, are

$$x_1 = \frac{-b + \sqrt{b^2 - 4ac}}{2a} \quad \text{and} \quad x_2 = \frac{-b - \sqrt{b^2 - 4ac}}{2a}.$$

Example 3 Find the values of x that satisfy

$$3x^2 - 12x + 1 = 0.$$

Solution Using the quadratic formula with $a = 3$, $b = -12$, and $c = 1$, we obtain

$$x_1 = \frac{-(-12) + \sqrt{(-12)^2 - 4(3)(1)}}{2(3)} = \frac{12 + \sqrt{144 - 12}}{6}$$
$$= \frac{12 + \sqrt{132}}{6} = \frac{12 + 11.49}{6} = \frac{23.49}{6} = 3.92$$

and

$$x_2 = \frac{-(-12) - \sqrt{132}}{6} = \frac{12 - 11.49}{6} = 0.09.$$

Example 4 Based on observations of prices, the demand D for oranges at a local fruit stand satisfies the equation $D = -\frac{1}{2}P^2 - 5P + 600$, where P is the price per orange (in cents). On a given Saturday morning, the store has 200 oranges in stock. Determine the price the store should charge for oranges if it wishes to deplete its inventory by the end of the day.

Solution We seek the price that will create a demand of $D = 200$. Note that, if the store gives the oranges away free ($P = 0$), it will create a demand of 600 oranges, which will certainly deplete the current stock but will not result in an attractive income. Alternatively, a price of $P = 25¢$ will only create a demand of

$$D = -\tfrac{1}{2}(25)^2 - 5(25) + 600 = 162.5,$$

which will not meet the objective.

Substituting $D = 200$ into the demand-price equation, we find that P must satisfy

$$200 = -\tfrac{1}{2}P^2 - 5P + 600$$
$$\tfrac{1}{2}P^2 + 5P - 400 = 0.$$

Using the quadratic formula with $a = \tfrac{1}{2}$, $b = 5$, and $c = -400$, we obtain

$$P_1 = \frac{-5 + \sqrt{(5)^2 - 4(\tfrac{1}{2})(-400)}}{2(\tfrac{1}{2})} = \frac{-5 + \sqrt{25 + 800}}{1}$$
$$= -5 + \sqrt{825} = -5 + 28.7 = 23.7$$

and

$$P_2 = \frac{-5 - \sqrt{825}}{1} = -5 - 28.7 = -33.7.$$

The oranges should be priced at 23¢, which will create a demand of

$$D = -\tfrac{1}{2}(23)^2 - 5(23) + 600 = 220.$$

A price of 24¢ will only create a demand of

$$D = -\tfrac{1}{2}(24)^2 - 5(24) + 600 = 192,$$

which will not deplete the stock.

Example 5 A warehouse has 12,000 cans of a discontinued tennis ball which it wishes to distribute in a week. From past experience, it is known that the demand D per week (in cans) is related to the price P (in dollars) by the equation

$$D = -3000P^2 - 13{,}500P + 27{,}000.$$

Determine the price that will result in zero inventory.

Solution We seek the price P that will yield $D = 12{,}000$. Substituting this value into the given equation, we find that P must satisfy

$$12{,}000 = -3000P^2 - 13{,}500P + 27{,}000$$
$$3000P^2 + 13{,}500P - 15{,}000 = 0$$
$$3P^2 + 13.5P - 15 = 0.$$

Using the quadratic formula with $a = 3$, $b = 13.5$, and $c = -15$, we obtain

$$P_1 = \frac{-13.5 + \sqrt{(13.5)^2 - 4(3)(-15)}}{2(3)} = \frac{-13.5 + \sqrt{182.25 + 180}}{6}$$

$$= \frac{-13.5 + \sqrt{362.25}}{6} = \frac{-13.5 + 19.03}{6} = \frac{5.53}{6} = 0.92$$

and

$$P_2 = \frac{-13.5 - \sqrt{362.25}}{6} = \frac{-13.5 - 19.03}{6} = -5.42.$$

The tennis balls should be wholesaled at 92¢ per can.

Exercises

1. Determine which of the following equations are quadratic curves:
   (a) $x^2 - x = y$
   (b) $y^2 = 3$
   (c) $x^2 - 2x + 2 = 0$
   (d) $y - x^2 = 0$
   (e) $y + x = 3$
   (f) $N^2 = 2D + 5$
   (g) $R = 2S^2$
   (h) $\sqrt{y} = x$.

2. Graph the following quadratic curves:
   (a) $y = x^2$
   (b) $y = -x^2$
   (c) $y = -x^2 - x + 4$
   (d) $y = x^2 + x - 4$
   (e) $D = 2P^2 - 8$.

3. Use the quadratic formula to find the values of x that satisfy the following equations:
   (a) $x^2 - x - 6 = 0$
   (b) $3x^2 - 2x - 5 = 0$
   (c) $4x^2 - 7 = 0$
   (d) $-\frac{1}{3}x^2 + x + 1 = 0$
   (e) $x^2 - x = 4$.

4. Ogden Motors has 15 identical model automobiles which it wants to sell within a month. From past experience, it is known that the demand D per month is related

to the price P (in dollars) by the equation $D = -0.04P^2 + 10,000$. Determine the maximum price that will result in no inventory at the end of a month.

5. Ogden Motors uses the formula $V = (-0.05t^2 - 0.05t + 0.8)N$ to compute the book value of used cars, where V denotes the present value (in dollars), t denotes the age of the car (in years), and N denotes the new car price (in dollars).
   (a) Find the value of a $5000 automobile after 3 years.
   (b) Find the value of a $5000 automobile immediately after it leaves the showroom.
   (c) Determine the time when a $4000 automobile will be worth $1500.
   (d) Determine the time when a $4000 automobile will be worth $1000.
   (e) How long will it take for a car's value to reach one-half its original price?

6. A manufacturer has determined that the yearly profit P (in dollars) is directly related to the number of units sold N (in hundreds) by the formula $P = -N^2 + 1100N - 100,000$.
   (a) Graph this equation by plotting the points corresponding to $N = 0$, 100, 200, 500, 600, 900, and 1000.
   (b) Determine the profit if no units are sold.
   (c) Determine the profit if more than 100,000 units are sold.
   (d) How many units must be sold if a profit of $100,000 is desired?
   (e) From the graph in part (a), estimate the number of units sold that will yield the greatest profit.

2.5 Polynomial Equations and Curves

Having considered equations involving only linear terms (linear equations) and those containing squared terms (quadratic equations), we now generalize to higher-degree equations.

Definition 2.4 An *nth degree polynomial equation* is an equation of the form

$$y = a_n x^n + a_{n-1} x^{n-1} + \cdots + a_1 x + a_0, \tag{23}$$

where n is a known positive integer and $a_n, a_{n-1}, \ldots, a_1, a_0$ are known numbers with $a_n \neq 0$.

Examples of polynomial curves are

$$y = 3x^4 - 2x^3 + 5x^2 - 7x + 1 \tag{24}$$

$$y = x^3 - 2x \tag{25}$$

$$y = -x^5 + 1. \tag{26}$$

Equation (24) is a fourth-degree polynomial equation having the form $y = a_4 x^4 + a_3 x^3 + a_2 x^2 + a_1 x + a_0$ with $n = 4$, $a_4 = 3$, $a_3 = -2$, $a_2 = 5$, $a_1 = -7$, and $a_0 = 1$. Equation (25) is a third-degree polynomial equation;

it has the form of Eq. (23) with $n = 3$, $a_3 = 1$, $a_2 = 0$, $a_1 = -2$, and $a_0 = 0$. Equation (26) has the form of (23) with $n = 5$, $a_5 = -1$, $a_4 = a_3 = a_2 = a_1 = 0$, and $a_0 = 1$ and is a fifth-degree polynomial equation.

In particular, a second-degree polynomial has the form $y = a_2x^2 + a_1x + a_0$, while a first-degree polynomial equation has the form $y = a_1x + a_0$. From Eqs. (11) and (18), we recognize these equations as quadratic equations and linear equations, respectively.

Example 1 A large men's toiletry company plans to market a new brand of shaving cream. From past experience with other shaving creams, the company expects that the gross cumulative profit P (in millions of dollars) from this new brand will be related to time t (in years) by the equation $P = -0.014t^3 + 0.26t^2 - 0.128t - 1.5$. Graph this curve and determine the anticipated income at the end of the first year.

Solution Using the given equation with $t = 1$, we compute the anticipated income at the end of the first year as $P = -0.014(1)^3 + 0.26(1)^2 - 0.128(1) - 1.5 = -1.38$ or a loss of $1.38 million. In order to graph the equation, we plot additional points by arbitrarily choosing different values of t and calculating corresponding values of P. In this way, we obtain Table 2.9 from which Figure 2.20 follows. We select values of t and solve for values of P, rather than the reverse, since the resulting equation is easier to solve. The doubtful reader should try selecting a value of P and solving the equation for t.

In Figure 2.20 the curve between $t = -2$ and $t = 17$, depicts a life cycle typical of many brands on the market. A product becomes available approximately 2 years after a decision by management to produce and market it. The decision to introduce a new product triggers 2 years of research and development and initial production expenditures resulting in a loss to the company of $1.5 million by $t = 0$, the time the product is first retailed. An introductory period then follows for approximately 5 years, during which the product begins to establish a market. Nonetheless, even for the first 3 years of the product's life, the company still incurs a loss. During this period, sales are not sufficient to offset previous losses plus current advertising and production expenses. Following the introductory period, the product enjoys a rapid growth until the twelfth year of its existence, when gross profit peaks at $10.35 million. At this point, sales begin to decline, and the product, if kept on the market, will incur losses adversely affecting the cumulative profit. By the eighteenth year the company will have lost all the money it had previously made.

Observe that Figure 2.20 is useful only between $t = -2$ and $t = 12$. Obviously, the product cannot generate income before it is conceived, as indicated in the graph for values of t less than -2, and it is unlikely that

TABLE 2.9

t	-2	-1	0	$\frac{1}{4}$	$\frac{1}{2}$	1	2	3	4	8	12	13	15	17
P	-0.09	-1.10	-1.5	-1.52	-1.50	-1.38	-0.83	0.08	1.25	6.95	10.21	10.02	7.83	2.68

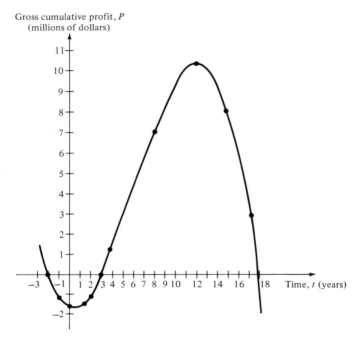

FIGURE 2.20

management will continue production at a loss after the twelfth year. We discuss the more general case of restricting intervals in Chapter 6.

Polynomial curves of degree 3 or higher have limited commercial application. Nonetheless, polynomial curves can be useful on occasion, and we return to them when we discuss differentiation.

Exercises

1. Determine which of the following equations represent polynomial curves and, for those that do, give the degree:
   (a) $y = x^5 - 2x^2$
   (b) $y^2 = x^3 + 1$
   (c) $y - x^2 = x^4$
   (d) $y^5 = 1$
   (e) $y = x^2 - 2x + 5$
   (f) $y = \sqrt{x} + 1$.

2.6 Exponential Equations and Curves

An extremely useful curve in business problems, especially those dealing with compound interest, is one in which a variable appears as an exponent. In Chapter 3, we consider the application of compound interest and exponential curves to the time value of money. Here we only define the curves.

Definition 2.5 An *exponential equation* is an equation of the form

$$y = a(b^x), \qquad (27)$$

where a and b are known numbers with b positive.

Examples of exponential curves are $y = 5(3^x)$ with $a = 5$ and $b = 3$, $y = -17(\pi^x)$ with $a = -17$ and $b = \pi = 3.14159\ldots$, and $y = 15.2(5.7^x)$ with $a = 15.2$ and $b = 5.7$. If $b = 1$, Eq. (27) becomes $y = a(1^x) = a(1) = a$ which is a straight line in the form of Eq. (8).

For most commercial applications, a is also positive and Eq. (27) has one of the two general shapes depicted in Figures 2.21 and 2.22. Note that in both cases the graph never reaches the x-axis but approaches that axis from one direction. This property is one of the primary characteristics of exponential curves.

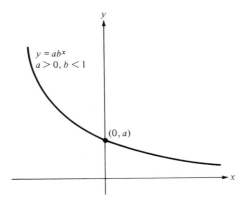

FIGURE 2.21

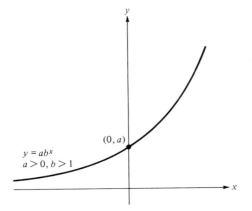

FIGURE 2.22

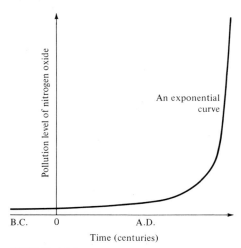

FIGURE 2.23

Many natural phenomena occur which can be accurately modeled or represented by exponential equations. Examples of such situations are pollution levels, the use of natural resources, and radioactive decay of certain materials. In practice, phenomena such as these are very misleading. Their graphs stay relatively constant or flat for many years, very much like the graph of a linear equation. As the value of the exponent builds, however, the value of the y-variable suddenly "takes off" beyond any expectation based on a linear model of the situation. Such a situation is presented in Figure 2.23, illustrating the pollution level of nitrogen oxide versus time (in centuries).

Exercises

Determine which of the following equations represent exponential curves:
1. $y = 2(7^x)$.
2. $y = (2^x)7$.
3. $y = 2(x^7)$.
4. $y = 2(7^{x^2})$.
5. $y = (7^x)^2$.
6. $y = -2(\frac{1}{2})^x$.
7. $y = 2(-\frac{1}{2})^x$.
8. $y = \sqrt{5}(\sqrt{3})^x$.
9. $y = 9(1.1)^x$.

Chapter

The Value of Money

One area of business where equations are used constantly is finance. Here questions center on the time value of money, either the future value of money invested in the present or the present value of money receivable in the future. One may have to determine whether a $5000 investment now that promises a $1500 return for the next 5 years is better than a $4500 investment now with a guaranteed return of $2000 every other year for the next 8 years. Or, more personally, one may want to know how much money should be deposited monthly in a savings account earning 5% interest if the goal is to accumulate $20,000 after 18 years to pay for a college education.

As we shall see, mathematical equations model these situations perfectly. That is, we can use an appropriate equation to obtain the exact answer. The mathematical underpinnings of all questions involving the value of money are based on the equation governing compound interest.

3.1 Compound Interest

One of the most widely used and important concepts in commercial financing, investment analysis, and accounting is compound interest. The defining property of *compound interest* is that, once interest is paid on an initial investment (in a savings bank, for example), the interest then is added to the original investment so that it too earns interest in succeeding time periods. In effect, interest is paid on the interest.

As an example, if $1000 is invested in a bank paying 6% interest compounded annually, in 1 year the principal earns 6% of $1000 or (0.06)(1000) = $60. The new principal is $1060 (the original investment of $1000 plus the $60 interest payment), and the second year's interest payment will be based on this new amount. Interest for the second year will be 6% of $1060 or (0.06)(1060) = $63.60, and the balance at the end of the second year will total $1123.60. Interest payments for the third year will be computed for this new balance. The results of all interest computations through the fifth year have been collected into Table 3.1.

TABLE 3.1

1.	Original investment	$1000.00
2.	Interest for first year (6% of line 1)	60.00
3.	Principal during the second year (line 1 plus line 2)	$1060.00
4.	Interest for second year (6% of line 3)	63.60
5.	Principal during the third year (line 3 plus line 4)	$1123.60
6.	Interest for third year (6% of line 5)	67.42
7.	Principal during the fourth year (line 5 plus line 6)	$1191.02
8.	Interest for fourth year (6% of line 7)	71.46
9.	Principal during the fifth year (line 7 plus line 8)	$1262.48
10.	Interest for fifth year (6% of line 9)	75.75
11.	Principal at the end of the fifth year (line 9 plus line 10)	$1338.23

Obviously we could continue Table 3.1 and find the principal at the end of any year. But this can be time-consuming, especially if we are interested in the principal after 25 or 30 years. Luckily there exists a formula that allows us to calculate such principals with very little work.

Let us return to Table 3.1 and approach the problem from a different direction. For notational simplicity, we let $P(1)$ denote the principal after the first year, $P(2)$ the principal after the second year, $P(3)$ the principal after the third year, and so on. The original investment is denoted by $P(0)$. It follows from line 3 in the table that

$$P(1) = P(0) + (0.06)P(0).$$

That is, the principal after 1 year equals the initial investment $P(0)$ plus 1 year's interest. Factoring $P(0)$ from the right yields

$$P(1) = (1 + 0.06)P(0). \tag{1}$$

In Table 3.1, $P(1) = \$1060$. The interest for the second year is $(0.06)P(1)$, and the principal after 2 years is (see line 5)

$$P(2) = P(1) + (0.06)P(1)$$
$$P(2) = (1 + 0.06)P(1). \qquad (2)$$

Substituting Eq. (1) into Eq. (2), we have

$$P(2) = (1 + 0.06)(1 + 0.06)P(0)$$
$$P(2) = (1 + 0.06)^2 P(0). \qquad (3)$$

Equations (1) and (3) begin to form a pattern. We could guess that $P(3) = (1 + 0.06)^3 P(0)$, that $P(4) = (1 + 0.06)^4 P(0)$, and that $P(25) = (1 + 0.06)^{25} P(0)$. We could even try to generalize this pattern to other interest rates, for example, 7 or 8%. This too can be done. In fact, if we let i denote the interest and $P(n)$ the principal after the nth year,

$$\boxed{P(n) = (1 + i)^n P(0).} \qquad (4)^*$$

Such situations are schematically represented in Figure 3.1.

Equation (4) simplifies compound interest calculations. To calculate the principal at the end of the nth year, we simply add the interest rate to 1, raise

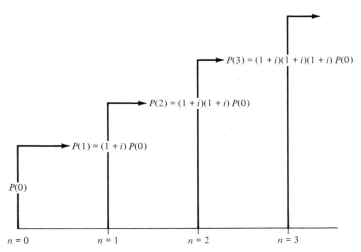

FIGURE 3.1

* Equation (4) is an exponential curve having the form of Eq. (25) in Chapter 2 with $a = P(0)$, $b = (1 + i)$, $y = P(n)$, and x replaced by n.

this sum to the nth power, and multiply this result by the original investment. Furthermore, extensive tables similar to the one given in Appendix C.3 are available for finding the value of $(1 + i)^n$ with different values of n and i. In particular, if the quantity $(1 + 0.06)^5$ is needed, go to Appendix C.3 and locate the column corresponding to $i = 0.06 = 6\%$ and the row corresponding to $n = 5$. The entry in that row and column, 1.338226, is $(1 + 0.06)^5$ rounded to six decimal places.

Example 1 One thousand dollars is invested in a time savings account that pays 6% interest compounded annually. Determine the balance after 5 years and then after 25 years.

Solution Here $P(0) = 1000$ and $i = 0.06$, and we seek the principal after 5 years, or $P(5)$. Setting $n = 5$ in Eq. (4) we calculate $P(5) = (1 + 0.06)^5 \times (1000) = (1.338226)(1000) = \$1338.23$. Compare this figure with the last dollar figure tabulated in Table 3.1.

To find the principal after 25 years, we set $n = 25$ in Eq. (4). Again $P(0) = 1000$ and $i = 0.06$, hence $P(25) = (1 + 0.06)^{25}(1000)$. To find $(1 + 0.06)^{25}$ we go to Appendix C.3 and locate the column corresponding to $i = 0.06$ and the row corresponding to $n = 25$. The entry in that row and column, 4.291871, is $(1 + 0.06)^{25}$. Therefore

$$P(25) = (1 + 0.06)^{25}(1000) = (4.291871)(1000) = \$4291.87.$$

Certainly, it was easier to obtain $P(25)$ this way than to continue Table 3.1 through another 20 years.

Interest rates are generally quoted on an annual basis, although in practice the interest often is compounded semiannually, monthly, weekly, or daily. The time between successive interest computations is called the *conversion period*. For conversion periods less than 1 year, the *interest rate per conversion period* is the annual rate divided by the number of conversion periods per year. If we let r designate the annual interest rate, the interest rate per conversion period is $r/2$ for semiannual payments, $r/4$ for quarterly payments, $r/12$ for monthly payments, and $r/365$ for daily payments. In such cases, the annual rate is called the *nominal rate*. If no conversion period is stated, it is assumed to be annual and the nominal and actual rates are identical.

Equation (4) is valid even when the conversion periods are less than 1 year. Simply take i as the interest rate per conversion period and $P(n)$ as the balance after n conversion periods. In particular, if the interest is 8% compounded quarterly, $i = 0.08/4 = 0.02$ which is the rate per quarter. Also, $P(10)$ denotes the principal after 10 conversion periods which, in this case, is 10 quarters or $2\frac{1}{2}$ years.

Example 2 Ten thousand dollars is invested in a time savings account that pays 8% interest compounded quarterly. Determine the balance after 5 years.

Solution Since interest is paid quarterly, we take one-quarter of a year as our basic time period. Then, the balance after 5 years is given by the balance after 20 quarters or $P(20)$. The rate applied each quarter is the annual rate divided by 4 or $0.08/4 = 0.02$. Using Eq. (4) with $n = 20$, $i = 0.02$, and $P(0) = 10,000$ we obtain $P(20) = (1 + 0.02)^{20} 10,000$. From Appendix C.3, we find

$$(1 + 0.2)^{20} = 1.485947.$$

Therefore $P(20) = (1.485947)(10,000) = \$14,859.47$.

Example 3 Ten thousand dollars is invested in a savings account that pays 8% interest compounded semiannually. Determine the balance after 5 years.

Solution Since interest is paid semiannually, we take one-half of a year as our basic time period. Accordingly, the balance after 5 years is given by the balance after 10 half-years or $P(10)$. The rate applied each half-year is the annual rate divided by 2 or $0.08/2 = 0.04$. Using Eq. (4) with $n = 10$, $i = 0.04$, and $P(0) = 10,000$, we obtain $P(10) = (1 + 0.04)^{10}(10,000)$. From Appendix C.3, we find $(1 + 0.04)^{10} = 1.480244$. Therefore $P(10) = (1.480244)(10,000) = \$14,802.44$.

Example 4 Ten thousand dollars is invested in a savings account that pays 8% interest compounded monthly. Determine the balance after 5 years.

Solution Here $i = 0.08/12 = 0.00667$ and we seek $P(60)$, the balance after 60 months or 5 years. Using Eq. (4), we obtain $P(60) = (1 + 0.00667)^{60}(10,000)$. Unfortunately, the value of $(1 + i)^{60}$ is not tabulated for $i = 0.00667$ in Appendix C.3. Although more extensive tables are available, we can use a modern calculator to calculate directly $(1 + 0.00667)^{60} = 1.490142$; hence $P(60) = (1.490142)(10,000) = \$14,901.42$.

Exercises

1. Determine the balance after 4 years resulting from $2000 being deposited in a savings account that yields 5% annual interest compounded annually.
2. Determine the balance after 4 years resulting from $2000 being deposited in a savings account that yields 6% annual interest compounded monthly.
3. Ms. Brown borrows $2500 from a friend who charges 8% interest compounded annually. Determine her debt after 3 years.
4. Redo Exercise 3 with the interest compounded quarterly.
5. Ms. Brown invests $2500 in a venture that pays 10% interest compounded quarterly. Determine her balance after 3 years.
6. Redo Exercise 5 with the interest compounded semiannually.
7. Mr. Johnson deposits $500 in a savings account that pays 6% interest compounded annually. How much will he have after 25 years?
8. Redo Exercise 7 with the interest compounded semiannually.
9. Determine the balance after 3 years resulting from $900 being deposited in a savings account that pays 6% interest compounded monthly.

10. Determine the balance after 1 year resulting from $400 being deposited in a savings account that yields 5% interest compounded daily. Set up but do not solve.
11. Redo Exercise 10 for the balance after 4 years.

Many banks use an *approximate year* rather than a calendar year for certain interest computations. Every month is assumed to have exactly 30 days, resulting in an approximate year of 360 days. Use approximate time for Exercise 12.

12. Mr. Henry deposits $1000 in a savings account that yields 6% interest compounded daily. Determine his balance after 3 years. Set up but do not solve.
13. Redo Exercise 12 using exact time.

3.2 Time Value of Money

Let us rewrite Eq. (4) once again and look at it a little more closely:

$$P(n) = (1 + i)^n P(0). \qquad [4]$$

Given an initial deposit $P(0)$ and the interest rate i, we can use Eq. (4) to find the value of the investment anytime in the future. Effectively, $P(0)$ represents the present value of our money, while $P(n)$ denotes the future value of this money. To emphasize these time relationships, it is usual to let FV denote the future value of a sum of money and let PV denote the present value of a sum of money. Then Eq. (4) can be rewritten as

$$FV = (1 + i)^n PV. \qquad (5)$$

Equations (4) and (5) are identical. The notation used in Eq. (5), however, reinforces the time dependencies. Given the present value of a lump sum investment and the interest rate i (per conversion period) we can use Eq. (5) to calculate the future value of this investment n conversion periods later.

Equation (5) is particularly important when there are several different ways to invest the same amount of money and we must determine which one of the choices will be the most profitable. For example, suppose we have $10,000 in available cash and are invited to invest this money in a land venture with an expected return of $15,000 in $5\frac{1}{2}$ years. If we decide against the land venture, we can deposit our money in time savings certificates with a guaranteed interest of 8% per year compounded quarterly. What should we do?

To answer this question intelligently, we must compare the amounts each investment will return at the *same point in time*, in this case after $5\frac{1}{2}$ years. The future value of the land venture is fixed at $15,000. What is the future value of the money if it is deposited in the bank?

Since the interest is compounded quarterly, $i = 0.08/4 = 0.02$ per quarter.

The present value of our investment is $10,000, and we seek the future value after $n = 22$ quarters. Using Eq. (5) and Appendix C.3 we calculate the future value of the bank investment as

$$FV = (1 + 0.02)^{22}(10,000) = (1.545980)(10,000) = \$15,459.80.$$

Clearly the bank is a more profitable investment by $459.80.

Example 1 Mr. James' barber would like Mr. James to lend him $2000 so that the barber shop can be modernized. The barber promises to pay Mr. James $2600 at the end of 2 years. How does this investment compare with investing the same money in the bank for 2 years at 10% compounded semiannually?

Solution At the end of 2 years, the future value of the barber shop investment is $2600. To determine the future value of $2000 deposited in the bank, we use Eq. (5) with $i = 0.10/2 = 0.05$, $n = 4$, and $PV = 2000$. Then,

$$FV = (1 + 0.05)^4(2000) = (1.215506)(2000) = \$2431.01.$$

Therefore the barber shop investment offers the greater future value at the end of 2 years.

Example 2 Ms. Everet has $1000 to deposit. One bank offers 7% interest compounded annually, while a second bank offers 6% interest compounded monthly. Which bank should she choose if she wants the greatest return after 4 years?

Solution To determine the future value of $1000 at the first bank we use Eq. (5) with $i = 0.07$, $n = 4$, and $PV = \$1000$. Then

$$FV = (1 + 0.07)^4(1000) = (1.310796)(1000) = \$1310.80.$$

To determine the future value of $1000 at the second bank, we use Eq. (5) with $i = 0.06/12 = 0.005$, $n = 48$, and $PV = 1000$. Then,

$$FV = (1 + 0.005)^{48}(1000) = (1.270489)(1000) = \$1270.48.$$

Ms. Everet would do better at the first bank.

Equation (5) gives the future value of money in terms of the present value. If we divide both sides of Eq. (5) by $(1 + i)^n$, we obtain

$$PV = \frac{FV}{(1 + i)^n}$$

or

$$PV = (1 + i)^{-n}FV. \qquad (6)$$

Equation (6) gives the present value of money in terms of the future value. As we shall see, Eq. (6) is often more useful than Eq. (5) in many financial decision-making situations. The values of $(1 + i)^{-n}$ are tabulated for different values of i and n in Appendix C.4, similar to the tabulations in Appendix C.3. For example, if we need $(1 + 0.05)^{-5}$, we go to Appendix C.4 and locate the column corresponding to $i = 0.05$ and the row corresponding to $n = 5$. The entry in that row and column, 0.783526, is $(1 + 0.05)^{-5}$.

Example 3 Determine the present value of $15,000 due in 5 years at 5% compounded annually.

Solution Here we have no money available in the present but we will receive $15,000 in the future. Therefore $FV = 15,000$. Using Eq. (6) with $n = 5$, we calculate $PV = (1 + 0.05)^{-5}(15,000)$. Using Appendix C.4, we locate $(1 + 0.05)^{-5} = 0.783526$, hence

$$PV = (0.783526)(15,000) = \$11,752.89.$$

The present value is the amount required *now* that, dollar for dollar, is equivalent to the stated future value. It follows from Example 3 that, at 5% compounded annually, $11,752.89 now is equivalent to $15,000 in 5 years. In other words, if $11,752,89 is deposited now at 5% compounded annually, it will grow to $15,000 in 5 years.

Example 4 Mr. Kakowski's bank is offering secured saving certificates for $4\frac{1}{2}$ years at 8% interest compounded quarterly. Certificates can be purchased in $100 amounts with a minimum deposit of $1000. How much should Mr. Kakowski invest if he wants a final return of $20,000 (to pay for his child's college education)?

Solution We are given $FV = 20,000$, $n = 18$, and $i = 0.08/4 = 0.02$, and we seek PV. Substituting these values into Eq. (6), we obtain

$$PV = (1 + 0.02)^{-18}(20,000) = (0.700159)(20,000) = \$14,003.18,$$

where the value of $(1 + 0.02)^{-18} = 0.0700159$ is taken directly from Appendix C.4. Since certificates can be purchased in $100 lots only, Mr. Kakowski can either invest $14,000 now and receive $19,995.44 at maturity or invest $14,100 and receive $20,138.27.

Often one has to choose between several different investment opportunities each having a known future value. Here present values can be more useful than future values. This is especially true if the given future values mature at different times. For example, is an investment that returns $50,000 at the end of 8 years better than one that returns $43,500 at the end of 5 years? In all problems such as these, we must bring the potential profits of each investment to the same point in time before an intelligent comparison can be made. The most convenient time is the present, since we can use Eq. (6) to convert all future values, no matter when they occur, to the present.

Example 5 Mr. Kingsly plans to sell his bakery and retire. Two of his employees wish to buy the bakery, but they do not have any immediate cash. They expect to make money from operating the bakery, so each makes an offer. Employee A wants the business for $50,000 payable at the end of 8 years. Employee B wants the business for $43,500 due at the end of 5 years. Which offer is better at an interest rate of 5% per year?

Solution Since each offer matures at a different date, we translate each to its present worth and then compare.

Employee A:

$$PV = (1 + 0.05)^{-8}(50{,}000) = (0.676839)(50{,}000)$$
$$= \$33{,}841.95$$

Employee B:

$$PV = (1 + 0.05)^{-5}(43{,}500) = (0.783526)(43{,}500)$$
$$= \$34{,}083.38$$

Since the present value of employee B's offer is higher, it is the better offer. Here the present value represents the equivalent cash settlement *now*. That is, $50,000 due in 8 years is equivalent to $33,841.95 now at 5%. Similarly, $43,500 due in 5 years is equivalent to $34,083.38 now.

Example 6 Before Mr. Kingsly can decide between the two offers in Example 4, he receives a third offer. Another buyer will pay $10,000 immediately plus another $29,200 in 4 years. How does this offer compare with the other two?

Solution The present worth of the $10,000 down payment is $10,000, since Mr. Kingsly would receive the money immediately. To obtain the present value of the future payment we use Eq. (6) with the same interest rate given in Example 4.

$$PV = (1 + 0.05)^{-4}(29{,}200) = (0.822702)(29{,}200) = \$24{,}022.90.$$

The present value of the entire offer is 10,000 + 24,022.90 = $34,022.90 which is better than the offer from employee A but not as good as the offer from employee B.

Example 7 The Die-Cast Corporation has three offers for its die-casting equipment. The first buyer is willing to purchase the equipment for $50,000, payable at the end of 8 years. The second buyer is willing to spend $39,000, consisting of an immediate payment of $14,000 now and $25,000 due in 6 years, while the third buyer will purchase the equipment for $30,000 payable immediately. Determine the best offer for the equipment assuming all potential purchasers can meet their obligations and the Die-Cast Corporation can deposit all money received in a bank account paying 8% interest.

Solution Since each offer matures at a different date, we first compute their respective present values.

First buyer:

$$PV = (1 + 0.08)^{-8}(50,000) = (0.540269)(50,000)$$
$$= \$27,013.45.$$

Second buyer:

$$PV = (1 + 0.08)^{-6}(25,000) + 14,000$$
$$= (0.630170)(25,000) + 14,000$$
$$= 15,754.25 + 14,000$$
$$= \$29,754.25.$$

Third buyer:

$$PV = \$30,000.$$

The third offer is best, since its present value is the highest.

Exercises

1. Find the future value of $15,000 after 4 years at 6% interest compounded annually.
2. Find the future value of $12,000 after 7 years at 5% interest compounded annually.
3. Find the future value of $12,000 after 7 years at 8% interest compounded quarterly.
4. Find the future value of $8000 after 10 years at 4% interest compounded semiannually.
5. Find the present value of $15,000 due in 4 years at 6% interest compounded annually.
6. Find the present value of $12,000 due in 7 years at 5% interest compounded annually.

7. Find the present value of $12,000 due in 7 years at 8% interest compounded quarterly.
8. Find the present value of $8000 due in 10 years at 4% interest compounded semiannually.
9. With the birth of their child, the Boswells decide to deposit a sum of money in government bonds paying 4% annual interest compounded annually. Their objective is to accumulate enough money in 18 years to cover their child's educational expenses which they estimate will be $22,000. Determine the amount that they should invest now in order to meet their objective.
10. How much money should be deposited in 8% savings certificates compounded semiannually if the desired objective is $10,000 after $4\frac{1}{2}$ years?
11. With the birth of his daughter, Mr. Tuck decides to place a sum of money in a time savings account which yields 6% compounded semiannually. If the objective is to accumulate $5000 for his daughter's wedding by her twentieth birthday, determine the amount of the deposit. How much money will be available if Mr. Tuck's child does not marry until her twenty-fifth birthday?
12. A man has $1000 to deposit. Should he put it in a bank offering 2% interest compounded quarterly or one offering 4% interest compounded annually?
13. Ms. Field's financial advisor has recommended that she invest $20,000 in a new housing development with an anticipated return of $28,000 in 6 years. The advisor claims that this is a better investment than keeping her money in the bank at 6% interest compounded annually. Is the advisor correct?
14. Dr. Baxter has $10,000 for investment purposes. She can put it into a friend's business with an expected return of $14,000 in 3 years, or she can put it in the bank at 8% interest compounded quarterly. Which opportunity is the most profitable?
15. Mr. Jones has two buyers for his business. Buyer A will pay $10,000 immediately and another $25,000 in 7 years. Buyer B will pay $7000 immediately and another $26,000 in 5 years. Which is the best offer if the interest rates are 7% compounded annually?
16. A store owner has three buyers for his business. Buyer A will pay $20,000 now and another $5000 at the end of 4 years. Buyer B will pay $15,000 now and another $10,000 at the end of 3 years. Buyer C will pay $10,000 now and another $18,000 at the end of 6 years. Which is the best offer if the interest rate is 8% compounded annually?
17. An individual has three possible opportunities for investing the same amount of money. The first will return $8000 in 4 years, the second will return $7000 in 2 years, and the third will return $10,000 in 7 years. Which opportunity is the most attractive if the interest rate is 4% compounded annually?
18. Redo Exercise 17 with an interest rate of 8% compounded annually.
19. Dr. Baxter has two land ventures available to her for $7000 each. Venture A will return $15,000 in 6 years, whereas venture B will return $16,500 in 8 years. Which opportunity is the most attractive at current interest rates of 6% compounded annually?
20. The owner of a local dress shop offers a buyer two options: pay $20,000 immediately or pay $37,000 at the end of 10 years. Interest rates are 7% compounded annually.
    (a) What is the future value of each offer after 10 years?

(b) What is the present value of each offer?
(c) Which is the best offer for the buyer?
21. Solve Eq. (5) for i and show that

$$i = \left(\frac{FV}{PV}\right)^{1/n} - 1.$$

22. Using the results from Exercise 21, determine the annual interest rate required to convert $10,000 to $15,000 in 5 years.
23. Using the results from Exercise 21, determine the annual interest rate required to convert $1000 to $1350 in 3 years.
24. Using the results from Exercise 21, determine the annual interest rate required to double an investment after 5 years.
25. Use logarithms to solve Eq. (5) for n and show that

$$n = \frac{\log (FV/PV)}{\log (1 + i)}.$$

26. Determine the time required to convert $1000 to $1500 at 6% interest compounded annually. Use Exercise 25.
27. Redo Exercise 26 with the interest compounded semiannually.
28. Redo Exercise 26 with an interest rate of 8% compounded annually.
29. Determine the time it will take for an initial investment of $6000 to triple if it is invested in a savings plan that pays 8% annual interest compounded quarterly. Use the results from Exercise 25.

3.3 Present Value of a Cash Flow

In both Sections 3.1 and 3.2 we concerned ourselves with *lump sum* investments. Either we calculated the future value of one deposit invested now, or we calculated the present value of one payment due in the future. We have not considered investments requiring a *set* of payments due at *different* times, a situation known as a *cash flow*.

As an example of a cash flow problem, suppose we have an investment which will return $500 in 1 year, another $300 in 3 years, and a final $400 in 4 years, with interest rates of 5% compounded annually. What is the present value of such an opportunity? That is, what is the cash equivalent now of the entire transaction?

A simple approach is to compute the present value of each of the individual payments using Eq. (6),

$$PV = (1 + i)^{-n} FV, \qquad [6]$$

and then sum these individual present values to obtain the present value of the entire investment.

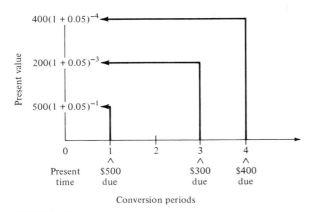

FIGURE 3.2

The first payment of $500 is due in 1 year. We compute its present value from Eq. (6) and Appendix C.4 as $PV_1 = (1 + 0.05)^{-1}(500) = \$476.19$. The second payment of $300 is due in 3 years. Again using Eq. (6) and Appendix C.4, we find its present value as $PV_2 = (1 + 0.05)^{-3}(300) = \$259.15$. Similarly, the present value of the last payment is $PV_3 = (1 + 0.05)^{-4}(400) = \$329.08$. Summing, we obtain the present value of the entire investment as $PV_1 + PV_2 + PV_3 = 476.19 + 259.15 + 329.08 = \$1064.42$.

In most present-value problems, a *time diagram* illustrating the contributions to the total present value from the individual payments is helpful. The time diagram for this cash flow is given in Figure 3.2.

This approach of summing individual present values is the general procedure for calculating present values of all cash flows. Simply use Eq. (6) to find the present values of each payment and then sum the results. One modification, however, is usual. Rather than finding the present value, it is more common to find the *net present value* which is the present value *minus* the cost of the investment. The net present value measures the additional money over and above the cost of the investment, which would have to be deposited in a bank to equal the returns guaranteed by the investment.

Example 1 Ms. Tilson invests $1000 in a friend's business. Her friend guarantees a return of $500 in a year, $300 in 3 years, and $400 in 4 years. Determine the net present value of this investment if the current interest rate is 5% compounded annually.

Solution We previously calculated the present value of this investment as $1064.42. Since the investment costs $1000, the net present value is

Net $PV = 1064.42 - 1000.00 = \$64.42$.

100 3 The Value of Money

Therefore this investment is profitable, since Ms. Tilson would need an additional $64.42 (that is, she would have to deposit $1064.42 instead of $1000) if she wanted the same return from a bank.

Example 2 Determine the net present values of the two investments described on page 87 if the interest rate is 8% per annum compounded quarterly. The first is a $5000 investment which returns $1500 every year for the next 5 years while the second is a $4500 investment which returns $2000 every other year for the next 8 years.

Solution First, since the interest is compounded quarterly, we take one-quarter of a year as our basic time unit and $i = 0.08/4 = 0.02$. Time diagrams for both investments are given in Figures 3.3 and 3.4, respectively. The individual

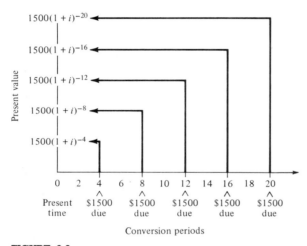

FIGURE 3.3

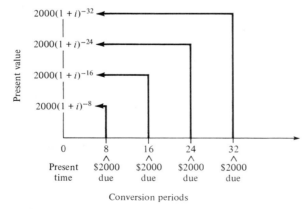

FIGURE 3.4

present values for the first investment are

$(1 + 0.02)^{-4}(1500) = \$1385.77$
$(1 + 0.02)^{-8}(1500) = 1280.24$
$(1 + 0.02)^{-12}(1500) = 1182.74$
$(1 + 0.02)^{-16}(1500) = 1092.67$
$(1 + 0.02)^{-20}(1500) = \underline{1009.46}$
$\phantom{(1 + 0.02)^{-20}(1500) =\ }\text{Total} = \$5950.88.$

Since the cost of participating in this investment is $5000, the net present value is

Net $PV = 5950.88 - 5000$
$\phantom{\text{Net } PV\ } = \$950.88.$

The individual present values for the second investment are

$(1 + 0.02)^{-8}(2000) = \$1706.98$
$(1 + 0.02)^{-16}(2000) = 1456.89$
$(1 + 0.02)^{-24}(2000) = 1243.44$
$(1 + 0.02)^{-32}(2000) = \underline{1061.27}$
$\phantom{(1 + 0.02)^{-32}(2000) =\ }\text{Total} = \$5468.58.$

Since the cost of participating in the second investment is $4500, the net present value is

Net $PV = 5468.58 - 4500.00 = \$968.58.$

To achieve the same return from a bank that we can obtain from the first investment, we must add $950.88 to the $5000 cost. Similarly, we must add $968.58 to the cost of the second investment if we expect the bank to equal its return. Since the bank requires more money in the second investment than it does in the first to meet their respective returns, the second investment, the one with the highest net present value, is the more attractive.

Suppose we want the present value of an investment that returns $25 every month for the next 4 years at 6% interest compounded monthly. We could find it by calculating the present value of each payment and then adding the results, but this would require a good deal of work, since over a 4-year period there are 48 payments. It would be useful to have a formula instead. Fortunately such a formula exists if all the payments are equal and if the payment dates coincide with the conversion dates for paying interest.

Figure 3.5 illustrates an investment that returns R dollars every conversion period for the next n periods. If we let C_0 denote the original cash outlay re-

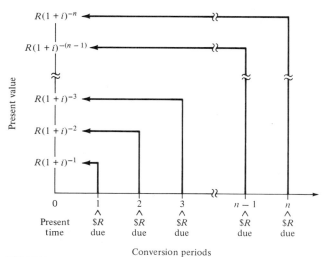

FIGURE 3.5

quired to participate in the investment,

$$\text{Net } PV = R(1+i)^{-1} + R(1+i)^{-2} + R(1+i)^{-3} + \cdots \\ + R(1+i)^{-(n-1)} + R(1+i)^{-n} - C_0.$$

Note that each term containing R represents the present value of *one* of the future payments. In particular, $R(1+i)^{-1}$ is the present value of the first payment, $R(1+i)^{-2}$ is the present value of the second payment, and $R(1+i)^{-n}$ is the present value of the last payment. Factoring R from these terms, we have

$$\text{Net } PV = R[(1+i)^{-1} + (1+i)^{-2} + (1+i)^{-3} + \cdots \\ + (1+i)^{-(n-1)} + (1+i)^{-n}] - C_0. \qquad (7)$$

The sum in brackets is commonly denoted as $a_{\overline{n}| i}$. That is,

$$a_{\overline{n}| i} = (1+i)^{-1} + (1+i)^{-2} + (1+i)^{-3} + \cdots \\ + (1+i)^{-(n-1)} + (1+i)^{-n},$$

and Eq. (7) can be simplified to

$$\boxed{\text{Net } PV = R a_{\overline{n}| i} - C_0.} \qquad (8)*$$

*If present value is required instead of net present value, Eq. (8) is still applicable with C_0 eliminated. That is, present value $PV = R a_{\overline{n}| i}$.

Appendix C.5 contains entries for $a_{\overline{n}|i}$ for various values of i and n. For example, to find $a_{\overline{6}|0.05}$, which corresponds to an interest rate of 5% over six conversion periods, go to Appendix C.5 and locate the column that corresponds to $i = 0.05$. Then find the row corresponding to $n = 6$. The element in this row and column, 5.075692, is $a_{\overline{6}|0.05}$.

The value of $a_{\overline{6}|0.05}$ is the sum of the bracketed terms in Eq. (7) with $n = 6$ and $i = 0.05$. Explicitly,

$$a_{\overline{6}|0.05} = (1 + 0.05)^{-1} + (1 + 0.05)^{-2} + (1 + 0.05)^{-3} + (1 + 0.05)^{-4} + (1 + 0.05)^{-5} + (1 + 0.05)^{-6}$$
$$= 5.075692.$$

Note that each of the individual terms $(1 + 0.05)^{-1}$ through $(1 + 0.05)^{-6}$ could have been found in Appendix C.4 and then summed. Obviously, it is more convenient to calculate the sum directly from Appendix C.5.

Example 3 Determine the net present value of an investment costing $800 which will return $25 at the end of every month for the next 4 years if current interest rates are 6% compounded monthly.

Solution Since interest is compounded monthly, the basic period of time is 1 month. Therefore $i = 0.06/12 = 0.005$. Here $R = 25$, $n = 48$, and $C_0 = 800$. Using Eq. (8), we have

$$\text{Net } PV = (25)a_{\overline{48}|0.005} - 800. \qquad (9)$$

In Appendix C.5 we first locate the column corresponding to $i = 0.005 = \frac{1}{2}\%$ and then the row corresponding to $n = 48$. We find $a_{\overline{48}|0.005} = 42.580318$. Substituting this value into Eq. (9) we compute

$$\text{Net } PV = 25(42.580318) - 800$$
$$= 1064.51 - 800$$
$$= \$264.51.$$

Example 4 Determine the net present value of an investment costing $900 which will return $45 every quarter for the next $5\frac{1}{2}$ years if current interest rates are 5% per annum compounded quarterly.

Solution Since interest rates are compounded quarterly, we take one-quarter of a year as our basic time period. Then $i = 0.05/4 = 0.0125$. Here $R = 45$, $n = 22$, and $C_0 = 900$. Using Eq. (8) and Appendix C.5, we calculate
$$\text{Net } PV = (45)a_{\overline{22}|0.0125} - 900$$
$$= (45)(19.130563) - 900$$
$$= 860.88 - 900$$
$$= -\$39.12,$$

which is negative. Therefore we can generate the same return from a bank with an initial investment of $39.12 *less* than the $900 amount required for the investment. Obviously, it is more profitable not to partake in the investment and to deposit the available cash in a savings account.

Equation (8) is indeed easy to use when it applies. One must be careful, however, to apply it correctly. First, Eq. (8) is valid *only* if the payment dates and the conversion dates coincide. This is not the case in Example 2 where the first investment has four conversion dates between successive payment dates. Second, the payments must all be equal, which is not the case in Example 1. Third, one must remember that i denotes the rate per conversion period, which generally is not the annual rate, and that n denotes the number of conversion periods in the investment, which usually differs from the number of years of the investment. Nonetheless, when Eq. (8) is applicable, it saves a good deal of work. It even can be combined with Eq. (6) to determine net present values of investments involving both time and single lump sum payments.

Example 5 Dr. Ericson invests $700 in a corporate bond that returns $20 in dividends every half-year for the next 20 years plus an additional $1000 at the end of the twentieth year. What is the net present value of this investment if interest rates are 8% per annum compounded semiannually?

Solution We divide the problem into two parts: one involving the $20 payments, and the other involving the $1000 final payment. We find the net present value by first calculating the present value (not net) of each part of the investment, then summing to obtain the total present value, and finally subtracting from this result the cost of the investment.

The present value (not net) of all dividends is obtained from Eq. (8) with the C_0 term omitted. Here $R = 20$, $n = 40$, and $i = 0.08/2 = 0.04$. Therefore

$$PV = (20)a_{40\rceil 0.04} = 20(19.792774) = \$395.86.$$

The present value of the $1000 lump sum payment due in 20 years or 40 half-years can be obtained directly from Eq. (6).

$$PV = (1 + 0.04)^{-40}(1000) = (0.208289)(1000) = \$208.29.$$

The *net* present value of the entire transaction is

Net $PV = 395.86 + 208.29 - 700 = -\$95.85,$

which is negative. As such, Dr. Ericson would have done better by depositing his $700 in a bank and not investing in the bond.

Appendix C.5 is by no means extensive or complete, and occasionally one needs values of $a_{\overline{n}\rceil i}$ that are not tabulated there, for example, $a_{\overline{120}\rceil 0.005}$. We show in Appendix B.1, that $a_{\overline{n}|i}$ is equal to the algebraic quantity

$$a_{\overline{n}|i} = \frac{1 - (1+i)^{-n}}{i}, \qquad (10)$$

for all values of i and n. With the aid of a sophisticated modern calculator, we can use Eq. (10) to determine $a_{\overline{n}|i}$ for any value not in Appendix C.5, as well as for all values that are tabulated.

Example 6 Determine $a_{\overline{120}|0.005}$.

Solution Using Eq. (10) and a modern calculator, we compute

$$a_{\overline{120}|0.005} = \frac{1 - (1+0.005)^{-120}}{0.005} = \frac{1 - 0.549632733}{0.005} = 90.073453.$$

Example 7 Use Eq. (10) and Appendix C.4 to determine $a_{\overline{15}|0.07}$.

Solution

$$a_{\overline{15}|0.07} = \frac{1 - (1+0.07)^{-15}}{0.07} = \frac{1 - 0.362446}{0.07} = 9.107914,$$

which agrees with the entry in Appendix C.5.

Exercises

If required entries are not found in Appendix C.5, substitute appropriate values in Eq. (10) and solve if a suitable calculator is available.

1. Mr. Samuels is invited to invest $3500 in a venture that will return $750 in 1 year, another $1100 in 2 years, and a final $2000 in 4 years. Determine whether or not this is a profitable investment if current interest rates are 4% per annum compounded annually.
2. The guaranteed returns of three different investment opportunities are listed in Table 3.2. Which one is the most desirable at an annual interest rate of 8% compounded quarterly if each investment requires an initial cash outlay of $2400?

TABLE 3.2

	Guaranteed return ($)				
	1 year later	2 years later	3 years later	4 years later	5 years later
Investment A	1000	1000	1000	1000	1000
Investment B	0	2500	0	0	2500
Investment C	600	800	2200	800	600

3. Determine the present value of an opportunity that will return $500 in half a year, $1000 in a year and a quarter, and $2000 in 3 years if the current interest rate is 6% per annum compounded monthly.
4. Determine the present value of a $50 return every quarter for the next 10 years at 4% annual interest compounded quarterly.
5. Mrs. Wilson has $2000 to invest. Either she can deposit this money in a time savings plan that will pay 5% annual interest or she can lend the money to a friend who will repay her $750 each year for the next 3 years. Which opportunity is more profitable assuming that interest rates remain at their current level?
6. Determine whether or not the investment opportunity described in Exercise 5 is profitable if the interest rate is 8%.
7. Ms. Johnson is offered two investment opportunities. The first will return $500 at the end of the year for the next 20 years, while the second will return $1000 at the end of the year for the next 7 years. Which is more profitable at an annual interest rate of 6%?
8. A man can initially invest $20,000 and receive $2300 at the end of each quarter for the next 3 years, or he can invest $18,000 now and receive $1500 at the end of each quarter for the next 4 years. Which opportunity is more attractive at an annual interest rate of 8% compounded quarterly?
9. Determine the present value of an investment that will return $500 at the end of each year for the next 15 years at an annual interest rate of 5%.
10. A man can invest $50,000 now and receive $2000 at the end of each month for the next 2 years plus an additional $12,000 at the end of the second year, or he can invest $70,000 now and receive $3500 at the end of each month for the next 2 years. Which opportunity is the most attractive at 6% interest per annum compounded monthly?
11. Determine the present value of an opportunity that will return $50 at the end of every month for the next 10 years, plus an additional $1000 at the end of each year for the first 5 years, if the interest rate is 6% per annum compounded monthly.
12. Redo Exercise 11 if the $1000 is paid at the beginning of the year for the first 5 years.

3.4 Future Value of an Annuity

In Section 3.3, we found the present value of a cash flow. By similar procedures we can compute the future value of a cash flow: Simply calculate the future value of each individual payment using Eq. (5),

$$FV = (1 + i)^n PV, \quad [5]$$

and then sum the results. For most investment analyses, however, future values have limited appeal.

Every potential investment can be evaluated two ways. Either one can determine the cash value of the investment in the future (future value) or one

3.4 Future Value of an Annuity

can determine the cash value of the investment now (present value). Invariably, the motivation is greater for, "What is it worth now?" in contrast to, "What will it be worth tomorrow?" The one notable exception is annuities.

Definition 3.1 An *annuity* is a set of equal payments made at equal intervals of time.

Installment loans, mortgages, life insurance premiums, social security benefits, and Christmas clubs are all examples of annuities. In each, one party, be it an individual or a corporation, pays to another party a set of equal payments called *periodic installments* or *rents*, denoted hereafter as R, at equal periods of time designated *rent periods* or *payment intervals*.

Annuities are classified as *ordinary* or *due*. With an ordinary annuity, payments are made at the end of each rent period, whereas with an annuity due, payments are made at the beginning of the period. An example of the former is an automobile loan for which the first monthly payment is not required until 30 days after the loan has been consumated. A life insurance policy requiring immediate payment of the first premium is an example of an annuity due. An annuity is *simple* if the conversion period at which interest is paid coincides with the rent period of the annuity. Simple annuities are easy to formulate mathematically as already evidenced by Eq. (8).

Example 1 Paul Astor decides to save for a new stereo by depositing $30 at the end of each month in his savings account which pays 6% interest compounded monthly. How much will he have at the end of 6 months.

Solution A time diagram for this situation is given in Figure 3.6. The payments are due at the end of each period, hence this is an example of an ordinary annuity. Observe that, since the first payment is not made until the end of the first month, it will draw interest for only 5 months. Similarly, the last payment, made at the end of the 6-month interval of interest, will draw no interest but will of course contribute to the final sum. Using Appendix C.3, with $i = 0.06/12 = 0.005$, we compute

$(1 + 0.005)^5(30) = (1.025251)(30) = \$30.76$
$(1 + 0.005)^4(30) = (1.020151)(30) = \$30.60$
$(1 + 0.005)^3(30) = (1.015075)(30) = \$30.45$
$(1 + 0.005)^2(30) = (1.010025)(30) = \$30.30$
$(1 + 0.005)^1(30) = (1.005)(30) = \$30.15$
$(1 + 0.005)^0(30) = (1.000)(30) = \$30.00.$

The *net future value* which is defined as the sum of the individual future values is

Net $FV = 30.76 + 30.60 + 30.45 + 30.30 + 30.15 + 30 = \$182.26$.

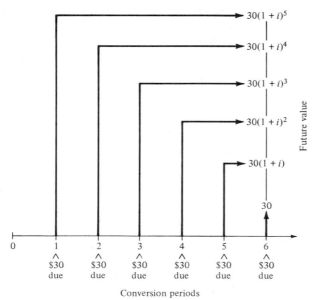

FIGURE 3.6

Example 2 The Jacksons decide to save $100 every half-year, beginning immediately, for a carpet they anticipate buying in 4 years. How much will they have at the end of the fourth year if they make eight deposits in an account yielding 8% interest compounded semiannually.

Solution A time diagram for this situation is given in Figure 3.7. Since the payments are due at the beginning of each time period, this is an example of an annuity due. In particular, the first payment draws interest for all eight periods, and the last payment made at the beginning of the eighth period also earns interest. Using Appendix C.3, with $i = 0.08/2 = 0.04$, we calculate

$(1 + 0.04)^8(100) = (1.368569)(100) = \$136.86$
$(1 + 0.04)^7(100) = (1.315932)(100) = \$131.59$
$(1 + 0.04)^6(100) = (1.265319)(100) = \$126.53$
$(1 + 0.04)^5(100) = (1.216653)(100) = \$121.67$
$(1 + 0.04)^4(100) = (1.169859)(100) = \$116.99$
$(1 + 0.04)^3(100) = (1.124864)(100) = \$112.49$
$(1 + 0.04)^2(100) = (1.0816)(100) = \$108.16$
$(1 + 0.04)^1(100) = (1.04)(100) = \$104.00.$

Summing these individual contributions, we obtain

Net $FV = \$958.29.$

3.4 Future Value of an Annuity

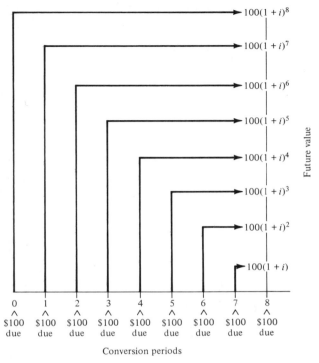

FIGURE 3.7

Since simple annuities involve equal payments with the payment dates coinciding with the conversion dates, we may hope for a formula similar to Eq. (8), which would simplify our calculations. Fortunately, such a formula exists.

Figure 3.8 illustrates the ordinary annuity for which R dollars is due at the end of every conversion period for the next n periods. Clearly, the net future value of the investment at the end of the nth period is

Net $FV = R + R(1 + i) + \cdots + R(1 + i)^{n-3} + R(1 + i)^{n-2} + R(1 + i)^{n-1}$.

Factoring R, we have

$$\text{Net } FV = R[1 + (1 + i) + \cdots + (1 + i)^{n-3} + (1 + i)^{n-2} + (1 + i)^{n-1}]. \tag{11}$$

The sum in the brackets, denoted by $s_{\overline{n}|i}$, is tabulated in Appendix C.6 for various values of i and n. Appendix C.6 is arranged identically to Appendix C.5. For example, to find $s_{\overline{6}|0.03}$, first locate the column in Appendix C.6 corre-

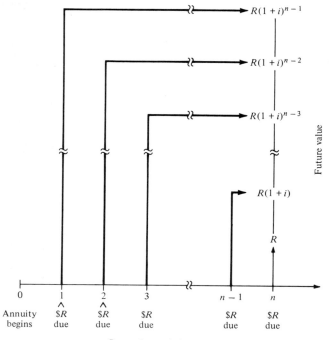

FIGURE 3.8

sponding to $i = 0.03 = 3\%$, and then find the row corresponding to $n = 6$. The entry in that row and column, 6.468410, is $s_{\overline{6}|\,0.03}$. That is,

$$s_{\overline{6}|\,0.03} = 1 + (1 + 0.03)^1 + (1 + 0.03)^2 + (1 + 0.03)^3 \\ + (1 + 0.03)^4 + (1 + 0.03)^5$$

$$= 6.468410.$$

With $s_{\overline{n}|\,i}$, Eq. (11) can be simplified to

$$\boxed{\text{Net } FV = R s_{\overline{n}|\,i}.} \tag{12}$$

Example 3 Redo Example 1 using Eq. (12).

Solution Again $R = 30$, $i = 0.06/12 = 0.005$, and $n = 6$. From Appendix C.6, we find $s_{\overline{6}|\,0.005} = 6.075502$. Then

Net $FV = 30(6.075502) = \$182.27$.

Note how much easier it is to do the problem this way in contrast to the solution to Example 1. The penny difference is due to roundoff error.

Example 4 Determine the net future value after 20 years of an ordinary annuity of $80 at the end of each half-year at 5% per annum compounded semiannually.

Solution Here $R = 80$, $n = 40$, and $i = 0.05/2 = 0.025$. From Appendix C.6, we find $s_{\overline{40}|\,0.025} = 67.402554$. Using Eq. (12), we obtain

Net $FV = (80)(67.402554) = \$5392.20$.

If $R = 1$, Eq. (12) becomes Net $FV = s_{\overline{n}|\,i}$. Therefore $s_{\overline{n}|\,i}$ is often called the *future value of an annuity* at $1 per period. For similar reasons, the quantity $a_{\overline{n}|\,i}$ introduced in Section 3.3 is called the *present value of an annuity* at $1 per period.

As is the case with all formulas, one must be careful to use Eq. (12) only when it is valid. Equation (12) can be used only for simple ordinary annuities. It is not valid when payment dates and conversion dates do not coincide exactly, and it is not valid when an annuity is due.

For an annuity due the first payment is made at the beginning, and no payment is due on the last day. The general case is illustrated in Figure 3.9. Now,

Net $FV = R(1 + i) + R(1 + i)^2 + \cdots$
$\qquad + R(1 + i)^{n-2} + R(1 + i)^{n-1} + R(1 + i)^n.$

Factoring $R(1 + i)$, we have

Net $FV = R(1 + i)[1 + (1 + i) + \cdots$
$\qquad + (1 + i)^{n-3} + (1 + i)^{n-2} + (1 + i)^{n-1}].$

Since the sum in brackets is $s_{\overline{n}|\,i}$, we can simplify the expression for the net future value of an annuity due to

$$\text{Net } FV = R(1 + i)s_{\overline{n}|\,i}. \qquad (13)$$

Equation (13) reflects the fact that for an annuity due each payment draws interest for one more period than is the case in an ordinary annuity.

Example 5 Redo Example 2 using Eq. (13).

Solution Again $R = 100$, $n = 8$, and $i = 0.08/2 = 0.04$. From Appendix C.6,

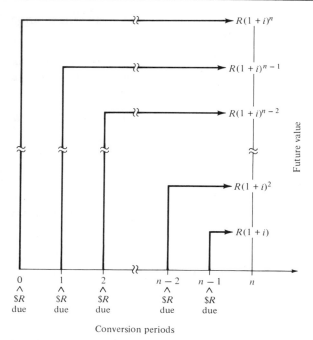

FIGURE 3.9

we find $s_{\overline{8}|0.04} = 9.214226$. Therefore

Net $FV = (100)(1 + 0.04)(9.214226) = \$958.28$.

Compare this method of solution to that used in Example 2. Which is easier? The 1¢ difference is due to roundoff errors in the solution to Example 2.

Example 6 Determine the net future value after 11 years of an annuity due of \$45 per quarter at 4% per annum compounded quarterly.

Solution Here $R = 45$, $n = 44$, and $i = 0.04/4 = 0.01$. From Appendix C.6, we find $s_{\overline{44}|0.01} = 54.931757$. Using Eq. (13) we obtain

Net $FV = 45(1 + 0.01)(54.931757) = \$2496.65$.

Appendix C.6 is neither extensive or complete. In Appendix B.1, we show that $s_{\overline{n}|i}$ can be given by the formula

$$s_{\overline{n}|i} = \frac{(1 + i)^n - 1}{i},$$
(14)

for all values of n and i. With the aid of a modern calculator, we can use Eq. (14) to calculate $s_{\overline{n}|i}$ for any value not in Appendix C.6, as well as those values that are tabulated.

Example 7 Determine $s_{\overline{240}|0.003}$.

Solution Using Eq. (14) and a modern calculator, we compute

$$s_{\overline{240}|0.003} = \frac{(1 + 0.003)^{240} - 1}{0.003} = \frac{2.052220 - 1}{0.003} = 350.7400.$$

Example 8 Determine $s_{\overline{40}|0.08}$, using Appendix C.3 and Eq. (14).

Solution

$$s_{\overline{40}|0.08} = \frac{(1 + 0.08)^{40} - 1}{0.08} = \frac{21.724521 - 1}{0.08} = 259.056513,$$

which agrees within roundoff error with the entry in Appendix C.6. The entries in Appendix C.6 were calculated with a computer using double-precision arithmetic; they are accurate to a full six decimal places.

Exercises

If required entries are not found in Appendix C.6, substitute appropriate values into Eq. (14) and use a suitable calculator if available.

1. Determine the total value of an annuity of $1000 per year for 10 years at 4% interest compounded annually for (a) an ordinary annuity, and (b) an annuity due.
2. Determine the total value of an annuity of $40 per quarter for 3 years at 4% interest compounded quarterly for (a) an ordinary annuity, and (b) an annuity due.
3. Determine the total value of an annuity of $50 per quarter for 6 years at 5% interest compounded quarterly for (a) an ordinary annuity, and (b) an annuity due.
4. Determine the total value after 2 years of an ordinary annuity of 10¢ per day at 5% interest compounded daily.
5. Mr. Neulander deposits $10 at the end of each week in a Christmas club account at 5% interest compounded weekly. How much will be in the account at the end of 52 weeks?
6. To provide for his child's education, Mr. O'Toole deposits $100 every June 30 and December 31 for 17 years. Determine the value of the annuity just after his last payment, which occurs on December 31, if his bank pays 4% interest compounded semiannually.
7. Stephen Klein decides to save for the down payment on a new car by depositing $20 every month in his savings account which bears 6% interest compounded monthly. If he makes his first deposit on January 1, 1976, and his last deposit on December 1, 1978, how much will he have accumulated for the down payment by January 1, 1979?
8. A man buys municipal bonds which yield $300 in interest every half year beginning June 30, 1971. He deposits this in a savings account paying 5% interest compounded

semiannually. How much will he have on December 31, 1977, if the last interest payment is on that date?

9. In 1950, Ms. Anlicker decided to invest $100 at the end of each year into a no-load mutual fund. Determine the value of her holdings just after her 1975 contribution if the fund's stock has increased its market value at 8% per year.

3.5 Effective Interest

Obviously, interest compounded quarterly generates more money than the same interest compounded annually, since the interest itself is drawing interest for some of the time. In particular, if we invest $100 at 8% compounded quarterly, our balance at the end of 1 year or four quarters will be $FV = (1 + 0.02)^4(100) = \$108.24$, of which $8.24 is the result of compound interest. To generate the same amount with interest compounded annually we would have to receive 8.24%.

Definition 3.2 *Effective interest* is the rate that must be compounded annually to generate the same interest as the nominal rate compounded over its stated conversion period.

To determine effective interest, we first calculate the future value of $1 after 1 year at the stated interest rate. We then subtract the original $1 deposit from this result, leaving the actual interest earned. This last figure also is the effective interest rate in decimal form. In particular, let E denote the effective interest and let N represent the total number of conversion periods per year for the stated rate. If the stated rate is compounded quarterly, $N = 4$, whereas if the stated rate is compounded daily, $N = 365$. It follows from Eq. (5) that the value of $1 at the end of a year or N conversion periods is $FV = (1 + i)^N(1)$. Therefore

$$E = (1 + i)^N - 1. \tag{15}$$

Example 1 Determine the effective rate for a 12% interest-bearing account compounded (a) quarterly, (b) monthly, (c) weekly, and (d) daily.

Solution We use Appendix C.3 when the appropriate values of $(1 + i)^N$ are listed; if not, we compute directly with a modern calculator.

(a) The actual quarterly interest rate is $i = 0.12/4 = 0.03$, hence $E = (1 + 0.03)^4 - 1 = 1.125509 - 1 = 0.125509$ or 12.55%.
(b) The actual monthly interest rate is $i = 0.12/12 = 0.01$, hence $E = (1 + 0.01)^{12} - 1 = 1.126825 - 1 = 0.126825$ or 12.68%.

(c) The actual weekly interest rate is $i = 0.12/52 = 0.002308$, hence $E = (1 + 0.002308)^{52} - 1 = 1.127359 - 1 = 0.127359$ or 12.74%.
(d) The actual daily interest rate is $i = 0.12/365 = 0.000328767$, hence $E = (1 + 0.000328767)^{365} - 1 = 1.127475 - 1 = 0.127475$ or 12.75%.

Observe from the previous example that the effective rate increases as N, the number of conversion periods in a year, increases. Question: As N becomes very large, does the corresponding effective rate also become large? Answer: No. In particular, let $i = r/N$ in Eq. (15), where again r denotes the nominal annual rate of interest and N designates the number of conversion periods per year. We seek

$$E = \lim_{N \to \infty} \left(1 + \frac{r}{N}\right)^N - 1.$$

It can be shown (although not considered here) that

$$\lim_{N \to \infty} \left(1 + \frac{r}{N}\right)^N = e^r,$$

where e is the irrational number 2.718281828459045 rounded to 15 decimals, commonly known as *Euler's number*. Therefore the largest effective interest that can be generated from $r\%$ nominal interest compounded continuously is

$$E_c = e^r - 1, \tag{16}$$

which is referred to as the *continuous conversion rate*. For the 12% nominal rate in Example 1, we compute with the aid of a calculator

$$E_c = e^{0.12} - 1 = (2.71828)^{0.12} - 1 = 1.127497 - 1 = 0.127497 \text{ or } 12.7497\%,$$

a rate not too different from that achieved by daily compounding.

Euler's number occurs often in commercial situations, most often in the equation

$$y = e^x.$$

We return to this equation in Chapter 6. For now, let us note that Eq. (17) is an exponential equation in the form of Eq. (27) in Chapter 2 with $a = 1$ and $b = e = 2.71828\ldots$.

Exercises

In Exercises 1 through 8 use a modern calculator to find effective interest for the stated rates.

1. 6% compounded quarterly.
2. 6% compounded daily.
3. 6% compounded continuously.
4. 4% compounded semiannually.
5. 4% compounded continuously.
6. 10% compounded annually.
7. 10% compounded quarterly.
8. 10% compounded continuously.

Chapter

Matrix Operations

To this point, we have considered only single equations. Linear, quadratic, and exponential equations were discussed in Chapter 2. In Chapter 3, we developed the individual equations that arose naturally from a study of the time value of money. We have not considered situations involving two or more equations simultaneously, and so we do this now.

4.1 Matrices

Numbers are used constantly in everyday life. Yet, there are many occasions when one number is not sufficient. As an example, a shoe chain consisting of four stores may use one number, $44,000, to report its annual net profits. If, however, the chain wanted the net profits of each store, then four numbers would be necessary [$21,000, $14,000, −$7000, $16,000]. Here $21,000 is the annual profit for the first store, $14,000 is the profit for the second store, the third entry signifies a loss of $7000 for the third store, and $16,000 represents the fourth store's profit.

Similarly, if an individual wanted to list a stock portfolio consisting of 100 shares of General Motors, 50 shares of IBM, and 200 shares of Exxon, more than one number would be needed. An appropriate list would be [100, 50, 200].

Whenever a list of numbers is required, it is common to give the list in a row, separating each number by a comma and enclosing the entire list within brackets. We did this in both the examples above: [$21,000, $14,000, −$7000, $16,000] and [100, 50, 200]. Such configurations are called *row vectors*. Examples of other row vectors are [9, 22], [21, 0, 0, 5, 0], and [2, $\sqrt{5}$, e].

As one might expect, lists can be given equally well in columns. Whenever this is done, the resulting configuration is called a *column vector*. Examples of column vectors are

$$\begin{bmatrix} \$21{,}000 \\ \$14{,}000 \\ -\$7{,}000 \\ \$16{,}000 \end{bmatrix}, \quad \begin{bmatrix} 100 \\ 50 \\ 200 \end{bmatrix}, \quad \begin{bmatrix} 9 \\ 22 \end{bmatrix}, \quad \text{and} \quad \begin{bmatrix} 21 \\ 0 \\ 0 \\ 5 \\ 0 \end{bmatrix}.$$

Column vectors do not require commas between individual entries, because different numbers are positioned one under another.

The concept of a vector (either row or column) is not new; most of us have been using them since childhood. When someone asks the time and we answer, "nine twenty-two," we are using vectors. Effectively, "nine twenty-two" is the row vector [9, 22], where the first number denotes the hour and the second number denotes the minute. Similarly, when someone asks the date and we respond, "three eight seventy-six," we are using vectors again. Here "three eight seventy-six" is the vector [3, 8, 76], where the entries denote month, day, and year, respectively.

One must be careful to use vector notation only when the significance of each entry is clear. If we are discussing dates, [3, 8, 76] is different from [8, 3, 76]. Furthermore [76, 8, 3] makes no sense if the first entry denotes the month.

For certain reports, even vectors may not be convenient, and a more elaborate configuration is useful. As an example, let us return to the stock portfolio just considered. Besides listing the shares actually held, the individual also may wish to list the cost of the stocks. An appropriate listing could be

$$\begin{bmatrix} 100 & 50 & 200 \\ \$8000 & \$15{,}000 & \$12{,}000 \end{bmatrix}.$$

Here the first row pertains to the number of shares held, while the second row pertains to the purchase price of the shares. Columns 1 through 3 pertain to General Motors, IBM, and Exxon, respectively. This report indicates that 200 shares of Exxon cost $12,000 to purchase. A configuration like this is called a matrix.

Definition 4.1 A *matrix* is a rectangular array of elements arranged in horizontal rows and vertical columns enclosed within brackets.

Other examples of matrices are

$$\mathbf{A} = \begin{bmatrix} 2 & 1 & -1 & 0 \\ -3 & 0 & -2 & 4 \end{bmatrix}, \quad \mathbf{B} = \begin{bmatrix} 2 & 4 \\ 0 & \sqrt{5} \\ 1.7 & 1 \end{bmatrix}$$

$$\mathbf{C} = \begin{bmatrix} 1 & 2 & 3 \\ 4 & 5 & 6 \\ 7 & 8 & 9 \end{bmatrix}, \quad \mathbf{D} = \begin{bmatrix} 1 & 0 & 0 & 1 & 0 & 0 \\ 0 & 1 & 1 & 0 & 0 & 1 \\ 0 & 0 & 0 & 0 & 1 & 0 \end{bmatrix}.$$

In this book, we designate matrices by uppercase, boldface, sans serif letters. Since a row vector is simply a matrix having exactly one row, and a column vector is a matrix having exactly one column, we also use uppercase, boldface, sans serif letters to designate vectors.

A matrix per se is nothing more than a table in which the horizontal and vertical lines usually used to separate rows and columns have been deleted. Only the numbers themselves appear. If we have 100 shares of General Motors stock, only the 100 is entered, not the words "General Motors." This notation lends itself well to problem solving by computers. Computers deal strictly with numbers. A computer can accept a table, but only a table of numbers—that is, a matrix. Since only numbers appear, however, we must clearly understand the significance of each entry in a matrix before we use matrix notation.

By convention, the number of rows is given before the number of columns when discussing the size or *order* of a matrix. Matrix **A** above contains two rows and four columns, so we say it has order 2×4 (read "two by four"). Matrix **B** has three rows and two columns, so we say it has order 3×2. The orders of **C** and **D** are 3×3 and 3×6, respectively. A matrix is *square* if it has the same number of rows and columns. Matrix **C** above is square; **A**, **B**, and **D** are not.

The entries in a matrix are called *elements*. Since it is confusing to discuss elements in terms of their values (after all, if the zero element in matrix **D** was referred to, which element would be meant?), we use their positions in the matrix to identify them. We designate these elements by lowercase letters and use subscripts to pinpoint their positions. The first subscript denotes the row, and the second subscript denotes the column. Accordingly, a_{24} represents the element in **A** located in the second row and fourth column, whereas d_{32} designates the element in **D** found in the third row and second column. For the particular matrices (plural of matrix) given above, $a_{24} = 4$ and $d_{32} = 0$. Similarly, $b_{22} = \sqrt{5}$, and c_{37} does not exist since **C** does not have a seventh column.

The *main diagonal* of a matrix is made up of all the elements whose row position equals their column position. Included on the main diagonal are the

1-1 element, the 2-2 element, the 3-3 element, the 4-4 element, and so on, whenever these elements exist. Obviously a 2 × 2 matrix does not have a 3-3 element. Examples of main diagonals, indicated here by the diagonal lines are

$$\begin{bmatrix} 1 & 2 & 3 \\ 4 & 5 & 6 \\ 7 & 8 & 9 \end{bmatrix}, \quad \begin{bmatrix} 1 & 2 & 3 & 4 & 5 \\ 6 & 7 & 8 & 9 & 0 \end{bmatrix}, \quad \text{and} \quad \begin{bmatrix} 1 & 2 & 3 & 4 \\ 0 & 1 & 0 & 1 \\ 2 & 1 & 0 & 2 \\ 3 & 4 & 1 & 1 \\ 7 & 8 & 9 & 1 \end{bmatrix}.$$

As we see later, a particularly important matrix is the identity matrix.

Definition 4.2 An *identity matrix* I is a square matrix having its main diagonal equal to 1 and all other elements equal to 0.

The 3 × 3 and 4 × 4 identity matrices are, respectively,

$$\begin{bmatrix} 1 & 0 & 0 \\ 0 & 1 & 0 \\ 0 & 0 & 1 \end{bmatrix} \quad \text{and} \quad \begin{bmatrix} 1 & 0 & 0 & 0 \\ 0 & 1 & 0 & 0 \\ 0 & 0 & 1 & 0 \\ 0 & 0 & 0 & 1 \end{bmatrix}.$$

4.2 Elementary Operations

In this section, we begin our study of the properties of matrices. For the most part, we deal with the matrix itself and do not concern ourselves with the business situation from which it arises. That will come later. Indeed, all of Chapter 5 is devoted to the single largest area of commercial application: linear programming. Later in this chapter, we also apply matrices to the solution of systems of equations.

Definition 4.3 Two matrices are *equal* if they have the same order and if their corresponding elements are equal.

It follows from this definition that two conditions must be satisfied before two matrices can be called equal: first, they must have exactly the same size and, second, every set of corresponding elements must match. In particular, the matrices

$$\begin{bmatrix} 1 & 2 \\ 3 & 4 \end{bmatrix} \quad \text{and} \quad \begin{bmatrix} 1 & 2 & 0 \\ 3 & 4 & 0 \end{bmatrix}$$

are not equal, since they do not have the same order. The matrices

$$\begin{bmatrix} 1 & 2 \\ 3 & 4 \end{bmatrix} \quad \text{and} \quad \begin{bmatrix} 2 & 3 \\ 1 & 4 \end{bmatrix}$$

have the same order but still are not equal, since their corresponding elements do not match. The 1-1 element in the first matrix is 1, whereas the corresponding 1-1 element in the second matrix is 2. Note that even if two matrices have the same elements they will not be equal unless identical elements appear in the same positions.

Example 1 Find x and y if

$$\begin{bmatrix} 1 & 2 \\ x & 0 \end{bmatrix} = \begin{bmatrix} 1 & 4y \\ 2 & 0 \end{bmatrix}.$$

Solution For two matrices to be equal, all corresponding elements must match. Considering both 2-1 positions, we must have $x = 2$. Considering both 1-2 positions, we require $2 = 4y$ or $y = \frac{1}{2}$.

Definition 4.4 The *sum* of two matrices of the same order is obtained by adding corresponding elements.

Example 2 Find **A** + **B**, **C** + **D**, and **A** + **C** if

$$\mathbf{A} = \begin{bmatrix} 2 & 1 & -1 \\ 0 & 1 & 2 \end{bmatrix}, \quad \mathbf{B} = \begin{bmatrix} -1 & 0 & -1 \\ 0 & 2 & \frac{1}{2} \end{bmatrix},$$

$$\mathbf{C} = \begin{bmatrix} 1 & 2 \\ 3 & 4 \end{bmatrix}, \quad \text{and} \quad \mathbf{D} = \begin{bmatrix} -1 & 2 \\ 0.5 & -2.1 \end{bmatrix}.$$

Solution

$$\mathbf{A} + \mathbf{B} = \begin{bmatrix} 2 & 1 & -1 \\ 0 & 1 & 2 \end{bmatrix} + \begin{bmatrix} -1 & 0 & -1 \\ 0 & 2 & \frac{1}{2} \end{bmatrix}$$

$$= \begin{bmatrix} 2 + (-1) & 1 + 0 & (-1) + (-1) \\ 0 + 0 & 1 + 2 & 2 + \frac{1}{2} \end{bmatrix} = \begin{bmatrix} 1 & 1 & -2 \\ 0 & 3 & 2\frac{1}{2} \end{bmatrix}$$

$$\mathbf{C} + \mathbf{D} = \begin{bmatrix} 1 & 2 \\ 3 & 4 \end{bmatrix} + \begin{bmatrix} -1 & 2 \\ 0.5 & -2.1 \end{bmatrix}$$

$$= \begin{bmatrix} 1 + (-1) & 2 + 2 \\ 3 + 0.5 & 4 + (-2.1) \end{bmatrix} = \begin{bmatrix} 0 & 4 \\ 3.5 & 1.9 \end{bmatrix}.$$

A + **C** is not defined, since the matrices have different orders.

Matrix subtraction is defined similarly to matrix addition; the matrices must be of the same order, and the subtraction is performed on the corresponding elements. For the matrices given in Example 2,

$$\mathbf{C} - \mathbf{D} = \begin{bmatrix} 1 & 2 \\ 3 & 4 \end{bmatrix} - \begin{bmatrix} -1 & 2 \\ 0.5 & -2.1 \end{bmatrix}$$

$$= \begin{bmatrix} 1 - (-1) & 2 - 2 \\ 3 - 0.5 & 4 - (-2.1) \end{bmatrix} = \begin{bmatrix} 2 & 0 \\ 2.5 & 6.1 \end{bmatrix}.$$

We define a *zero matrix* **0** as any matrix having all its elements equal to 0. In particular,

$$\begin{bmatrix} 0 & 0 \\ 0 & 0 \end{bmatrix} \quad \text{and} \quad [0, 0, 0, 0, 0]$$

are the 2 × 2 zero matrix and the 1 × 5 zero row vector, respectively. It follows immediately from Definition 4.4 that, if **A** and **0** have the same order,

A + 0 = A.

Still another simple matrix operation is scalar multiplication.

Definition 4.5 The product of a number c with a matrix **A** is obtained by multiplying each element of **A** by c.

Thus

$$3 \begin{bmatrix} 2 & 1 \\ -1 & 3 \\ 4 & 5 \end{bmatrix} = \begin{bmatrix} 6 & 3 \\ -3 & 9 \\ 12 & 15 \end{bmatrix} \quad \text{and} \quad -\tfrac{1}{2} \begin{bmatrix} 1 & -2 \\ -3 & 4 \end{bmatrix} = \begin{bmatrix} -\tfrac{1}{2} & 1 \\ \tfrac{3}{2} & -2 \end{bmatrix}.$$

The three matrix operations of addition, subtraction, and scalar multiplication are often combined. Such combinations, however, pose no additional difficulties, since the operations themselves are performed one at a time in accordance with the rules just developed.

Example 3 Find $2\mathbf{C} - 3\mathbf{D}$ for the matrices given in Example 2.

Solution

$$2\mathbf{C} - 3\mathbf{D} = 2 \begin{bmatrix} 1 & 2 \\ 3 & 4 \end{bmatrix} - 3 \begin{bmatrix} -1 & 2 \\ 0.5 & -2.1 \end{bmatrix}$$

$$= \begin{bmatrix} 2 & 4 \\ 6 & 8 \end{bmatrix} - \begin{bmatrix} -3 & 6 \\ 1.5 & -6.3 \end{bmatrix}$$

$$= \begin{bmatrix} 2 - (-3) & 4 - 6 \\ 6 - 1.5 & 8 - (-6.3) \end{bmatrix} = \begin{bmatrix} 5 & -2 \\ 4.5 & 14.3 \end{bmatrix}.$$

These elementary operations also can be used to solve equations involving matrices for various unknowns of interest. The algebra is identical to the algebra for equations with numbers.

Example 4 Find **A** if $2\mathbf{A} + 3\mathbf{B} = \mathbf{C}$, where

$$\mathbf{B} = \begin{bmatrix} 1 & 2 \\ 0 & 4 \\ 4 & 0 \end{bmatrix} \quad \text{and} \quad \mathbf{C} = \begin{bmatrix} 3 & -1 \\ 2 & 0 \\ -1 & -1 \end{bmatrix}.$$

Solution Since $2\mathbf{A} + 3\mathbf{B} = \mathbf{C}$, it follows that $2\mathbf{A} = \mathbf{C} - 3\mathbf{B}$ and $\mathbf{A} = \frac{1}{2}\mathbf{C} - \frac{3}{2}\mathbf{B}$. Then,

$$\mathbf{A} = \frac{1}{2}\begin{bmatrix} 3 & -1 \\ 2 & 0 \\ -1 & -1 \end{bmatrix} - \frac{3}{2}\begin{bmatrix} 1 & 2 \\ 0 & 4 \\ 4 & 0 \end{bmatrix} = \begin{bmatrix} \frac{3}{2} & -\frac{1}{2} \\ 1 & 0 \\ -\frac{1}{2} & -\frac{1}{2} \end{bmatrix} - \begin{bmatrix} \frac{3}{2} & 3 \\ 0 & 6 \\ 6 & 0 \end{bmatrix} = \begin{bmatrix} 0 & -\frac{7}{2} \\ 1 & -6 \\ -\frac{13}{2} & -\frac{1}{2} \end{bmatrix}.$$

Example 5 Find x and y if $2\mathbf{A} + \frac{1}{2}\mathbf{B} = \mathbf{C}$, where

$$\mathbf{A} = \begin{bmatrix} x \\ 2 \end{bmatrix}, \quad \mathbf{B} = \begin{bmatrix} 4 \\ y \end{bmatrix}, \quad \text{and} \quad \mathbf{C} = \begin{bmatrix} 1 \\ 0 \end{bmatrix}.$$

Solution Since $2\mathbf{A} + \frac{1}{2}\mathbf{B} = \mathbf{C}$, it follows that

$$2\begin{bmatrix} x \\ 2 \end{bmatrix} + \frac{1}{2}\begin{bmatrix} 4 \\ y \end{bmatrix} = \begin{bmatrix} 1 \\ 0 \end{bmatrix}.$$

Using Definitions 4.5 and 4.4, we obtain

$$\begin{bmatrix} 2x \\ 4 \end{bmatrix} + \begin{bmatrix} 2 \\ \frac{1}{2}y \end{bmatrix} = \begin{bmatrix} 1 \\ 0 \end{bmatrix} \quad \text{and then} \quad \begin{bmatrix} 2x + 2 \\ 4 + \frac{1}{2}y \end{bmatrix} = \begin{bmatrix} 1 \\ 0 \end{bmatrix}.$$

Since matrix (vector) equality implies that corresponding elements are equal, it must be the case that

$$2x + 2 = 1 \quad \text{and} \quad 4 + \tfrac{1}{2}y = 0.$$

Solving each equation separately, we find $x = -\frac{1}{2}$ and $y = -8$.

The last elementary operation we consider in this section is *transposition*, the process of forming a transpose. Formally, the *transpose* of a matrix **A** is a new matrix $\mathbf{A}^T$ obtained from the original matrix by converting all the rows of **A** to the columns of $\mathbf{A}^T$ while preserving order. That is, the first row of **A** becomes the first column of $\mathbf{A}^T$, the second row of **A** becomes the second column of $\mathbf{A}^T$, and so on.

Two examples of matrices and their transposes are

$$A = \begin{bmatrix} 1 & 2 & 3 \\ 4 & 5 & 6 \\ 7 & 8 & 9 \end{bmatrix}, \quad A^T = \begin{bmatrix} 1 & 4 & 7 \\ 2 & 5 & 8 \\ 3 & 6 & 9 \end{bmatrix}$$

and

$$B = \begin{bmatrix} 1 & 0 & -14 & 7 \\ 2 & 1 & 1 & 4 \end{bmatrix}, \quad B^T = \begin{bmatrix} 1 & 2 \\ 0 & 1 \\ -14 & 1 \\ 7 & 4 \end{bmatrix}.$$

Exercises

1. Determine the orders of the following matrices:

$$A = \begin{bmatrix} 2 & -1 \\ 0 & 3 \end{bmatrix}, \quad B = [1,\ 2,\ 0,\ 0,\ 1], \quad C = \begin{bmatrix} 2 & 1 & 3 \\ -4 & 5 & 6 \end{bmatrix},$$

$$D = \begin{bmatrix} 3 & 2 & 1 \\ 4 & 1 & 7 \\ 0 & 0 & 0 \\ 0 & 0 & 0 \end{bmatrix}, \quad E = \begin{bmatrix} -1 \\ 1 \\ -1 \end{bmatrix}.$$

2. Find the (a) 1-2 element and (b) 2-2 element, if they exist, for the matrices in Exercise 1.
3. Find $A + B$ and $A - B$, if they exist, for

(a) $A = \begin{bmatrix} 1 & 3 \\ 0 & -1 \end{bmatrix}, \quad B = \begin{bmatrix} 1 & -2 \\ 0 & 1 \end{bmatrix}$

(b) $A = \begin{bmatrix} 1 \\ 2 \\ -1 \end{bmatrix}, \quad B = \begin{bmatrix} 0 \\ 1 \\ 2 \end{bmatrix}$

(c) $A = \begin{bmatrix} 1 & 0 & 3 \\ 2 & 1 & 1 \end{bmatrix}, \quad B = \begin{bmatrix} 2 & 1 \\ 4 & -1 \end{bmatrix}$

(d) $A = \begin{bmatrix} 1 & 2 & 3 \\ 4 & 5 & 6 \\ 7 & 8 & 9 \end{bmatrix}, \quad B = \begin{bmatrix} 9 & 8 & 7 \\ 6 & 5 & 4 \\ 3 & 2 & 1 \end{bmatrix}$

(e) $A = \begin{bmatrix} 1 \\ 3 \\ 5 \\ 7 \end{bmatrix}, \quad B = [2,\ 4,\ 6,\ 8]$

(f) $A = \begin{bmatrix} 4 & -1 \\ 2 & 7 \\ -1 & 0 \\ 3 & -2 \\ 1 & 1 \end{bmatrix}, \quad B = \begin{bmatrix} 5 & -3 \\ 0 & 2 \\ 1 & 4 \\ -2 & 1 \\ 3 & 0 \end{bmatrix}.$

4. Find $5\mathbf{A} + 2\mathbf{B}$ if

$$\mathbf{A} = \begin{bmatrix} 1 & 3 \\ 0 & -1 \end{bmatrix} \quad \text{and} \quad \mathbf{B} = \begin{bmatrix} -2 & 3 \\ 7 & 4 \end{bmatrix}.$$

5. Find $\mathbf{A} - 3\mathbf{B}$ if

$$\mathbf{A} = \begin{bmatrix} 1 & 2 \\ -1.1 & 7.3 \\ 4 & 0 \end{bmatrix} \quad \text{and} \quad \mathbf{B} = \begin{bmatrix} 2 & -3 \\ 1.7 & 2.4 \\ -4.2 & 0 \end{bmatrix}.$$

6. Find a, b, c, and d if

$$\begin{bmatrix} 1 & a & 3 & b \\ 2 & 1 & 0 & -1 \\ c & 1 & 4 & d \end{bmatrix} = \begin{bmatrix} 1 & 2 & 3 & 4 \\ 2 & 1 & 0 & -1 \\ 4 & 1 & 4 & -5 \end{bmatrix}.$$

7. Do values of p and q exist for which

$$\begin{bmatrix} p & 3 \\ 4 & q \end{bmatrix} = \begin{bmatrix} 4 & 3 \\ 5 & 1 \end{bmatrix} ?$$

8. Find x and y if

(a) $\begin{bmatrix} 2x + 3 \\ y - 7 \end{bmatrix} = \begin{bmatrix} 1 \\ 0 \end{bmatrix}$ (b) $\begin{bmatrix} 2x - y \\ x + y \end{bmatrix} = \begin{bmatrix} 3 \\ 0 \end{bmatrix}.$

9. Find x, y, and z if

$$2 \begin{bmatrix} x \\ 3 \\ 4 \end{bmatrix} + 3 \begin{bmatrix} -1 \\ y \\ 0 \end{bmatrix} = 4 \begin{bmatrix} 0 \\ 4 \\ z \end{bmatrix}.$$

10. Find $\mathbf{B}$ if $-\mathbf{A} + 2\mathbf{B} = 6\mathbf{C}$, where

$$\mathbf{A} = \begin{bmatrix} 1 & 2 & -1 \\ 3 & 0 & 1 \\ 1 & 1 & 1 \end{bmatrix} \quad \text{and} \quad \mathbf{C} = \begin{bmatrix} 2 & -1 & 7 \\ 3 & 0 & 0 \\ 4 & -1 & -5 \end{bmatrix}.$$

11. Find the transposes of the matrices given in Exercise 3.
12. Let $\mathbf{A}$ and $\mathbf{B}$ have the same order. (a) Does it necessarily follow that $\mathbf{A} + \mathbf{B} = \mathbf{B} + \mathbf{A}$? (b) Must $\mathbf{A} - \mathbf{B} = \mathbf{B} - \mathbf{A}$?
13. Let $\mathbf{A}$ and $\mathbf{B}$ have the same order. (a) Does $(\mathbf{A} + \mathbf{B})^T = \mathbf{A}^T + \mathbf{B}^T$? (b) Does $(\mathbf{A} - \mathbf{B})^T = \mathbf{A}^T - \mathbf{B}^T$?
14. The inventory of the Village Appliance Store can be given by a 1×5 vector in which the first entry represents the number of television sets, the second entry represents the number of air conditioners, the third entry represents the number of refrigerators, the fourth entry represents the number of stoves, and the fifth entry represents the number of dishwashers.
    (a) Determine the inventory given on January 1 by [25, 5, 7, 10, 5].
    (b) January sales are given by [5, 0, 1, 2, 1]. What is the inventory on February 1 if no new appliances are added to the stock?
    (c) February sales are given by [4, 1, 2, 1, 1], and new stock added in February is given by [6, 8, 0, 0, 3]. What is the inventory on March 1?
15. The daily gasoline supply of a local service station is given by a 1×3 vector

in which the first entry represents gallons of premium, the second entry represents gallons of regular, and the third entry represents gallons of lead-free gasoline.
(a) Determine the supply given at the close of business on Monday by [6000, 10,000, 2000].
(b) Tuesday's sales are given by [2000, 2500, 1000]. What is the supply on Wednesday morning if no gasoline has been delivered since the previous week?
(c) On Wednesday, the station receives its weekly delivery of gas given by [15,000, 20,000, 6000]. Determine the supply at the end of the day if Wednesday's sales are given by [1500, 1700, 2500].
(d) Thursday's and Friday's sales are reported to the owner as [1700, 2000, 3200] and [2500, 3200, 1800], respectively. Why should the owner be upset?

16. The number of damaged cases delivered by the Homestead Chocolate Company from its various plants during the past year is given by the matrix

$$\begin{bmatrix} 50 & 30 & 40 \\ 150 & 70 & 80 \\ 30 & 10 & 80 \end{bmatrix}.$$

The rows pertain to its three plants in New York, Delaware, and Georgia, respectively. The columns pertain to chocolate bars, 1-pound boxes, and novelty items, respectively. The company's goal for next year is to reduce the number of damaged 1-pound boxes by 10% at each plant, to reduce the number of damaged chocolate bars shipped from its New York plant by 30%, and to reduce the number of damaged novelty items shipped from its Georgia plant by 50% while keeping all other entries the same as this year. What will next year's matrix be if all the goals are reached?

4.3 Matrix Multiplication

As we saw in Section 4.2, the operations of matrix addition, subtraction, and scalar multiplication are simple extensions of the analogous operations for the real numbers. The situation changes drastically when we consider matrix multiplication. No longer is the operation performed on corresponding elements. Not all matrices of the same order can be multiplied together, and even those that can often have a product with a different order. Other differences, some startling, will become apparent as we proceed.

At first, matrix multiplication appears complicated and unmotivated. It is neither. It is precisely this operation that makes matrices useful as tools in solving business problems. The motivation comes from the applications we investigate in later sections. As a means of uncomplicating the operation itself, we present matrix multiplication as a two-step process: (1) determining those matrices that can be multiplied, and (2) giving the rule for performing the multiplication. The first step simplifies the second.

A simple method for determining whether or not two matrices can be multiplied is first to write their respective orders next to each other. For example, consider the product **AB** if

$$\mathbf{A} = \begin{bmatrix} 0 & 1 & 0 \\ -3 & 4 & 2 \end{bmatrix} \quad \text{and} \quad \mathbf{B} = \begin{bmatrix} 6 \\ 7 \\ 8 \end{bmatrix}.$$

Matrix **A** has order 2 × 3, and **B** has order 3 × 1. We write

$(2 \times 3)(3 \times 1).$ (1)

Next consider the adjacent numbers, indicated in expression (1) by the curved arrow. If the adjacent numbers are the same, the multiplication can be performed. If the adjacent numbers are not the same, the multiplication cannot be performed. In expression (1), the adjacent numbers are both 3, so the product **AB** is defined.

The order of the product (when the multiplication is defined) is obtained by canceling the adjacent numbers. In expression (1), we cancel the adjacent 3s, leaving 2 × 1 which is the order of **A** times **B**.

Example 1 Determine whether the products **CD**, **DC**, and **DE** are defined for

$$\mathbf{C} = \begin{bmatrix} 0 & 1 & 2 \\ 3 & 4 & 5 \end{bmatrix}, \quad \mathbf{D} = \begin{bmatrix} 6 & 7 \\ 8 & 9 \\ -1 & -2 \end{bmatrix}, \quad \text{and} \quad \mathbf{E} = \begin{bmatrix} 1 & 0 \\ 2 & -1 \\ 3 & -2 \end{bmatrix}.$$

Solution The orders of **C**, **D**, and **E** are 2 × 3, 3 × 2, and 3 × 2, respectively. For the product **CD**, we write

$(2 \times 3)(3 \times 2).$

Since the adjacent numbers are both 3, the multiplication is defined and the product will have order 2 × 2.

For the product **DC**, we write

$(3 \times 2)(2 \times 3).$

Here the adjacent numbers are both 2, so again the multiplication is defined, but now the product has order 3 × 3.

For the product **DE**, we write

$(3 \times 2)(3 \times 2).$

Now, the adjacent numbers are not equal, and the multiplication cannot be performed.

Note in Example 1 that $\mathbf{CD} \neq \mathbf{DC}$; the order of $\mathbf{CD}$ is 2×2, and the order of $\mathbf{DC}$ is 3×3. In general, the *product of two matrices is not commutative.* That is, interchanging the sequence of the matrices being multiplied usually changes the answer. As such, we must be extremely careful to write the orders of matrices in the correct sequence when we use the method given above.

Knowledge of the size of a product is useful in computing the product. From Example 1, we know that

$$\mathbf{CD} = \begin{bmatrix} 0 & 1 & 2 \\ 3 & 4 & 5 \end{bmatrix} \begin{bmatrix} 6 & 7 \\ 8 & 9 \\ -1 & -2 \end{bmatrix} = \begin{bmatrix} - & - \\ - & - \end{bmatrix}. \tag{2}$$

The product $\mathbf{CD}$ will have a 1-1 element, a 1-2 element, a 2-1 element, and a 2-2 element.

Definition 4.6 The *i-j* element (where *i* is the row and *j* is the column) of a product is obtained by multiplying the elements in the *i*th row of the first matrix by the corresponding elements (the first with the first, the second with the second, and so on) in the *j*th column of the second matrix and adding the result.

Consider the product $\mathbf{CD}$ in Eq. (2). The 1-1 element ($i = 1, j = 1$) is obtained by multiplying the elements in the first row of $\mathbf{C}$ by the corresponding elements in the first column of $\mathbf{D}$ and adding the results. Thus the 1-1 element is

$$[0, 1, 2] \begin{bmatrix} 6 \\ 8 \\ -1 \end{bmatrix} = (0)(6) + (1)(8) + (2)(-1) = 0 + 8 - 2 = 6.$$

The 1-2 element ($i = 1, j = 2$) in $\mathbf{CD}$ is obtained by multiplying the elements in the first row of $\mathbf{C}$ by the corresponding elements in the second column of $\mathbf{D}$ and adding. It is

$$[0, 1, 2] \begin{bmatrix} 7 \\ 9 \\ -2 \end{bmatrix} = (0)(7) + (1)(9) + (2)(-2) = 0 + 9 - 4 = 5.$$

The 2-1 element ($i = 2, j = 1$) in $\mathbf{CD}$ is calculated by multiplying the elements in the second row of $\mathbf{C}$ by the corresponding elements in the first column of $\mathbf{D}$ and summing. It is

$$[3, 4, 5] \begin{bmatrix} 6 \\ 8 \\ -1 \end{bmatrix} = (3)(6) + (4)(8) + (5)(-1) = 18 + 32 - 5 = 45.$$

Finally, the 2-2 element in **CD** is obtained by multiplying the elements in the second row of **C** by the corresponding elements in the second column of **D** and summing. It is

$$[3, 4, 5]\begin{bmatrix} 7 \\ 9 \\ -2 \end{bmatrix} = (3)(7) + (4)(9) + (5)(-2) = 21 + 36 - 10 = 47.$$

Filling the blanks in Eq. (2), we have

$$\mathbf{CD} = \begin{bmatrix} 6 & 5 \\ 45 & 47 \end{bmatrix}.$$

Example 2 Find **AB** and **BA** for

$$\mathbf{A} = \begin{bmatrix} 1 & 0 \\ -3 & 4 \end{bmatrix} \quad \text{and} \quad \mathbf{B} = \begin{bmatrix} 2 & 6 \\ -1 & 5 \end{bmatrix}.$$

Solution

$$\mathbf{AB} = \begin{bmatrix} 1 & 0 \\ -3 & 4 \end{bmatrix}\begin{bmatrix} 2 & 6 \\ -1 & 5 \end{bmatrix} = \begin{bmatrix} (1)(2) + (0)(-1) & (1)(6) + (0)(5) \\ (-3)(2) + (4)(-1) & (-3)(6) + (4)(5) \end{bmatrix}$$

$$= \begin{bmatrix} 2 + 0 & 6 + 0 \\ -6 + (-4) & -18 + 20 \end{bmatrix} = \begin{bmatrix} 2 & 6 \\ -10 & 2 \end{bmatrix}$$

$$\mathbf{BA} = \begin{bmatrix} 2 & 6 \\ -1 & 5 \end{bmatrix}\begin{bmatrix} 1 & 0 \\ -3 & 4 \end{bmatrix} = \begin{bmatrix} (2)(1) + (6)(-3) & (2)(0) + (6)(4) \\ (-1)(1) + (5)(-3) & (-1)(0) + (5)(4) \end{bmatrix}$$

$$= \begin{bmatrix} 2 + (-18) & 0 + 24 \\ -1 + (-15) & 0 + 20 \end{bmatrix} = \begin{bmatrix} -16 & 24 \\ -16 & 20 \end{bmatrix}.$$

Again observe that $\mathbf{AB} \neq \mathbf{BA}$.

Example 3 Find **AB** and **BA** if

$$\mathbf{A} = \begin{bmatrix} 0 & 1 & 0 \\ -3 & 4 & 2 \end{bmatrix} \quad \text{and} \quad \mathbf{B} = \begin{bmatrix} 6 \\ 7 \\ 8 \end{bmatrix}.$$

Solution

$$\mathbf{AB} = \begin{bmatrix} 0 & 1 & 0 \\ -3 & 4 & 2 \end{bmatrix}\begin{bmatrix} 6 \\ 7 \\ 8 \end{bmatrix} = \begin{bmatrix} (0)(6) + (1)(7) + (0)(8) \\ (-3)(6) + (4)(7) + (2)(8) \end{bmatrix}$$

$$= \begin{bmatrix} 0 + 7 + 0 \\ -18 + 28 + 16 \end{bmatrix} = \begin{bmatrix} 7 \\ 26 \end{bmatrix}.$$

BA is not defined, since the adjacent numbers of their orders, $(3 \times 1)(2 \times 3)$ do not match.

Example 4 Find **AB** for

$$A = \begin{bmatrix} 2 & 4 & -1 \\ -4 & -8 & 2 \\ -2 & -4 & 1 \end{bmatrix} \quad \text{and} \quad B = \begin{bmatrix} 4 & 1 & -1 \\ 1 & 1 & 1 \\ 12 & 6 & 2 \end{bmatrix}.$$

Solution

$$AB = \begin{bmatrix} 2 & 4 & -1 \\ -4 & -8 & 2 \\ -2 & -4 & 1 \end{bmatrix} \begin{bmatrix} 4 & 1 & -1 \\ 1 & 1 & 1 \\ 12 & 6 & 2 \end{bmatrix}$$

$$= \begin{bmatrix} (2)(4) + (4)(1) & (2)(1) + (4)(1) & (2)(-1) + (4)(1) \\ + (-1)(12) & + (-1)(6) & + (-1)(2) \\ (-4)(4) + (-8)(1) & (-4)(1) + (-8)(1) & (-4)(-1) + (-8)(1) \\ + (2)(12) & + (2)(6) & + (2)(2) \\ (-2)(4) + (-4)(1) & (-2)(1) + (-4)(1) & (-2)(-1) + (-4)(1) \\ + (1)(12) & + (1)(6) & + (1)(2) \end{bmatrix}$$

$$= \begin{bmatrix} 8 + 4 - 12 & 2 + 4 - 6 & -2 + 4 - 2 \\ -16 - 8 + 24 & -4 - 8 + 12 & 4 - 8 + 4 \\ -8 - 4 + 12 & -2 - 4 + 6 & 2 - 4 + 2 \end{bmatrix} = \begin{bmatrix} 0 & 0 & 0 \\ 0 & 0 & 0 \\ 0 & 0 & 0 \end{bmatrix}.$$

Note that the product of two matrices can equal the zero matrix without either one of the original matrices being zero.

As a commercial application of matrix multiplication, consider the production schedule of a wood cabinet manufacturer given by the matrix

$$P = \begin{bmatrix} 2 & 1 & 5 \\ \frac{1}{2} & 2 & 3 \\ \frac{1}{4} & \frac{1}{4} & \frac{1}{2} \end{bmatrix}.$$

The three columns of **P** pertain, respectively, to the production of lamp bases, television cabinets, and wooden frames for grandfather clocks. The first row denotes the work-hours required to cut and assemble each product, the second row represents the work-hours required to sand and paint each item, while the third row denotes the work-hours required to inspect and crate each item. The hourly wages of a carpenter to cut and assemble, of a decorator, and of an inspector-packer are given, respectively, by the entries of $W = [7.50, 5.20, 3.60]$. The product

$$WP = [7.50, 5.20, 3.60] \begin{bmatrix} 2 & 1 & 5 \\ \frac{1}{2} & 2 & 3 \\ \frac{1}{4} & \frac{1}{4} & \frac{1}{2} \end{bmatrix}$$

$$= [(7.50)(2) + (5.20)(\tfrac{1}{2}) + (3.60)(\tfrac{1}{4}), (7.50)(1) + (5.20)(2) + (3.60)(\tfrac{1}{4}),$$
$$\quad (7.50)(5) + (5.20)(3) + (3.60)(\tfrac{1}{2})]$$
$$= [18.50, 18.80, 54.90]$$

represents the labor costs for manufacturing each lamp base, television cabinet, and grandfather clock frame, respectively.

Exercises

1. The order of **A** is 3 × 4, the order of **B** is 4 × 1, the order of **C** is 1 × 3, the order of **D** is 4 × 3, and the order of **E** is 3 × 3. Determine the order of
   (a) **AB**　　　(b) **BA**　　　(c) **CA**　　　(d) **AD**
   (e) **DA**　　　(f) **BCE**　　　(g) **CEA**　　　(h) **ABCD**
   (i) **DABCE**.

2. Find **AB** and **BA**, if possible, for
$$\mathbf{A} = \begin{bmatrix} 1 & 2 \\ -3 & 1 \end{bmatrix} \quad \text{and} \quad \mathbf{B} = \begin{bmatrix} 11 & -1 \\ 4 & 3 \end{bmatrix}.$$

3. Find **AB** and **BA**, if possible, for
$$\mathbf{A} = \begin{bmatrix} 2 & 0 & -1 & 4 \\ 3 & 6 & 4 & -1 \end{bmatrix} \quad \text{and} \quad \mathbf{B} = \begin{bmatrix} 0 & 5 \\ 2 & 1 \\ -1 & 3 \\ 4 & 1 \end{bmatrix}.$$

4. Find **AB** and **BA**, if possible, for
$$\mathbf{A} = \begin{bmatrix} 2 & 1 & 3 \\ 4 & 1 & 7 \end{bmatrix} \quad \text{and} \quad \mathbf{B} = \begin{bmatrix} -1 & 0 & 4 \\ 3 & 1 & 1 \end{bmatrix}.$$

5. Find **XZ**, **ZX**, and **XY**, if possible, for
$$\mathbf{X} = \begin{bmatrix} 2 \\ 1 \\ 3 \end{bmatrix}, \quad \mathbf{Y} = \begin{bmatrix} 3 \\ 1 \\ 0 \end{bmatrix}, \quad \mathbf{Z} = [2, 1, 4].$$

6. Verify that **AB** = **AC**, but that **B** ≠ **C** for
$$\mathbf{A} = \begin{bmatrix} 2 & 4 \\ 1 & 2 \end{bmatrix}, \quad \mathbf{B} = \begin{bmatrix} 1 & 2 \\ 1 & -1 \end{bmatrix}, \quad \text{and} \quad \mathbf{C} = \begin{bmatrix} 5 & 6 \\ -1 & -3 \end{bmatrix}.$$
Thus the cancelation law is not valid in matrix multiplication.

7. Verify the associative law for the matrices in Exercise 6 by showing that (**AB**)**C** = **A**(**BC**).

8. Verify the distributive law for the matrices in Exercise 6 by showing that **A**(**B** + **C**) = **AB** + **AC**.

9. Find two matrices **A** and **B** such that **AB** = **BA**.

10. The price schedule for a New York to Miami flight is given by **P** = [115, 70, 47], where the vector entries pertain, respectively, to first-class tickets, coach tickets, and student discount tickets. The number of tickets purchased in each category for a particular flight is given by
$$\mathbf{N} = \begin{bmatrix} 8 \\ 111 \\ 18 \end{bmatrix}.$$
What is the significance of the product **PN**?

11. The closing prices of Mr. Dolin's stock portfolio during the past week are given by the matrix

$$\mathbf{P} = \begin{bmatrix} 20 & 20\tfrac{1}{2} & 21\tfrac{1}{4} & 21\tfrac{1}{4} & 21\tfrac{3}{4} \\ 52\tfrac{1}{2} & 52\tfrac{5}{8} & 52\tfrac{3}{4} & 52\tfrac{3}{8} & 52\tfrac{5}{8} \\ 7 & 7 & 6\tfrac{3}{4} & 6\tfrac{1}{2} & 6\tfrac{3}{8} \end{bmatrix},$$

where the columns pertain to the days of the week, Monday through Friday, and the rows pertain to the prices of Breslin Tool and Die, American Citrus Corporation, and Eagle Mining Enterprises, respectively. Mr. Dolin's holdings in each of these companies are given by the vector $\mathbf{H} = [100, 200, 100]$. What is the significance of (a) $\mathbf{HP}$ and (b) $\mathbf{PH}$?

12. Referring to the data for the wood cabinet manufacturer given on page 130, suppose that cutter-assemblers receive a 10% wage increase, decorators receive a 5% increase, and inspector-packers receive a 20% increase.
    (a) Find the new wage vector $\mathbf{W}$.
    (b) Find the new product $\mathbf{WP}$.
    (c) Determine the percent increase in labor cost for each item as a result of the new salary scales.

13. Find $\mathbf{AI}$ and $\mathbf{IA}$ for

$$\mathbf{A} = \begin{bmatrix} 35 & -17 \\ 9 & 23 \end{bmatrix} \quad \text{and} \quad \mathbf{I} = \begin{bmatrix} 1 & 0 \\ 0 & 1 \end{bmatrix}.$$

14. Find $\mathbf{AI}$ and $\mathbf{IA}$ for

$$\mathbf{A} = \begin{bmatrix} 44 & 59 & -6 \\ 87 & 3 & 12 \\ -2 & 21 & 19 \end{bmatrix} \quad \text{and} \quad \mathbf{I} = \begin{bmatrix} 1 & 0 & 0 \\ 0 & 1 & 0 \\ 0 & 0 & 1 \end{bmatrix}.$$

15. Can you generalize the results from Exercises 13 and 14 to the products $\mathbf{AI}$ and $\mathbf{IA}$ for any matrix $\mathbf{A}$?

4.4 Simultaneous Linear Equations

A *system of simultaneous equations* is a set of equations in two or more variables. Examples of systems are

$$\begin{aligned} x + 2y &= 5 \\ 5x - y &= 3, \end{aligned} \tag{3}$$

$$\begin{aligned} 3p - 2q &= 1 \\ -p + 4q &= 3 \\ 2p + q &= 3, \end{aligned} \tag{4}$$

$$\begin{aligned} r + s + 2t &= 0 \\ 2r - 2s + t &= 0. \end{aligned} \tag{5}$$

System (3) is a set of two equations in the two unknowns, x and y. System (4) is a set of three equations in two unknowns, p and q, while system (5) is a set of two equations in three unknowns, r, s, and t.

A system of equations is *linear* if each unknown variable is raised only to the first power and multiplied only by a known number. Systems (3), (4), and (5) are all linear. An example of a system of equations that is not linear is

$$\begin{aligned} x + 3y^2 + 4z &= 2 \\ x - \sqrt{y} + 2z &= 1 \\ 3x + 2y + xz &= 4. \end{aligned} \qquad (6)$$

In the first equation, y appears squared and, in the second equation, it appears to the one-half power (recall that $\sqrt{y} = y^{1/2}$); to be linear each variable must be raised only to the first power. In the third equation, z is multiplied by x; to be linear each variable can be multiplied only by a number. A system is *nonlinear* if any one of its equations is nonlinear. Effectively, system (6) is threefold nonlinear.

For commercial purposes, linear systems of equations are important because they model a large number of business situations. Exercises 8 through 10 are but a few examples. From a mathematical viewpoint, linear systems are important because they can be solved.

Whenever one deals with systems of equations, the primary objective is to find a solution. In system (3), we seek values of x and y that satisfy both equations. In system (4), we seek values of p and q that satisfy all three equations. In general, a *solution* to a system of equations is a set of numbers for the variables that together satisfy *all* the equations in the system.

A solution to system (3) is $x = 1$ and $y = 2$. These values satisfy both equations. A solution to system (4) is $p = 1$ and $q = 1$, since these values satisfy *all* three equations of the system. In contrast, the values $r = -1$, $s = -1$, and $t = 1$ are not a solution to system (5). Although these values satisfy the first equation, they do not satisfy the second equation. A solution must satisfy *all* the equations of the system.

Some systems of equations have many different solutions, while other systems have none. In particular, system (5) has many solutions, of which two are $r = 0, s = 0, t = 0$ and $r = 5, s = 3, t = -4$. System (7) has no solutions, since there are no values of x and y that simultaneously sum to 1 and also to 2:

$$\begin{aligned} x + y &= 1 \\ x + y &= 2. \end{aligned} \qquad (7)$$

In general a system of simultaneous linear equations has either one, infinitely many, or no solutions. In Sections 4.5 and 4.6, we develop a straightforward matrix method for determining which systems have solutions and then a method for obtaining those solutions. The first step is to transform the given equations into matrix notation.

Let us see how we can put system (3),

$$x + 2y = 5 \\ 5x - y = 3, \qquad [3]$$

into matrix notation. We could collect all the unknowns, here x and y, and place them together in the same vector and then collect the coefficients of these unknowns and place them together in their own matrix. We also could combine the right sides of each equation in system (3) into one vector. If we did, we would have

$$\begin{bmatrix} 1 & 2 \\ 5 & -1 \end{bmatrix} \begin{bmatrix} x \\ y \end{bmatrix} = \begin{bmatrix} 5 \\ 3 \end{bmatrix}. \qquad (8)$$

Now, Eq. (8) is certainly in matrix notation, but is it equivalent to system (3)? We can easily verify that it is. If we perform the indicated matrix-vector multiplication, we obtain

$$\begin{bmatrix} 1 & 2 \\ 5 & -1 \end{bmatrix} \begin{bmatrix} x \\ y \end{bmatrix} = \begin{bmatrix} (1x + 2y) \\ (5x + (-1)y) \end{bmatrix},$$

so Eq. (8) becomes

$$\begin{bmatrix} (x + 2y) \\ (5x - y) \end{bmatrix} = \begin{bmatrix} 5 \\ 3 \end{bmatrix}.$$

It then follows from our definition of matrix equality that this last matrix (vector) equation is identical to system (3).

The same procedure can be used to convert any system of simultaneous linear equations to one matrix equation. Collect all the numerical coefficients of the unknown variables into one matrix denoted by **A**. Then collect all the unknown variables into one vector denoted by **X**. Finally, collect all the numbers on the right side of the equations into a vector denoted by **B**. Then the system of equations will be equivalent to the one matrix equation

AX = B. $\qquad (9)$

For Eq. (8),

$$\mathbf{A} = \begin{bmatrix} 1 & 2 \\ 5 & -1 \end{bmatrix}, \quad \mathbf{X} = \begin{bmatrix} x \\ y \end{bmatrix}, \quad \text{and} \quad \mathbf{B} = \begin{bmatrix} 5 \\ 3 \end{bmatrix}.$$

It is common to call **A** the *coefficient matrix*, **X** the *unknown vector*, and **B** the *known vector*.

Example 1 Write the following system in the matrix form of Eq. (9).

$$2x + 3y + 4z = 11$$
$$5x + 6y = 12$$
$$7x - 8y - 9z = 13.$$

Solution Define

$$\mathbf{A} = \begin{bmatrix} 2 & 3 & 4 \\ 5 & 6 & 0 \\ 7 & -8 & -9 \end{bmatrix}, \quad \mathbf{X} = \begin{bmatrix} x \\ y \\ z \end{bmatrix}, \quad \text{and} \quad \mathbf{B} = \begin{bmatrix} 11 \\ 12 \\ 13 \end{bmatrix}.$$

Note that the 2-3 element in **A** is zero, since the coefficient of z in the second equation is zero. The system is equivalent to the matrix equation **AX = B**, or

$$\begin{bmatrix} 2 & 3 & 4 \\ 5 & 6 & 0 \\ 7 & -8 & -9 \end{bmatrix} \begin{bmatrix} x \\ y \\ z \end{bmatrix} = \begin{bmatrix} 11 \\ 12 \\ 13 \end{bmatrix}.$$

Warning: Whenever we convert a system of equations to the form of Eq. (9), we must be careful to place only the coefficients of the first variable in **X** in the first column of **A**. Similarly, only the coefficients of the second variable in **X** should appear in the second column of **A**, and only the coefficients of the last variable in **X** should appear in the last column of **A**. In Example 1, the elements in **X** were ordered x first, y second, and z third. Therefore the first column of **A** contained the coefficients of x, the second column of **A** consisted of the coefficients of y, and the third column of **A** contained the coefficients of z.

Example 2 Write system (4) in matrix form.

Solution Define

$$\mathbf{A} = \begin{bmatrix} 3 & -2 \\ -1 & 4 \\ 2 & 1 \end{bmatrix}, \quad \mathbf{X} = \begin{bmatrix} p \\ q \end{bmatrix}, \quad \text{and} \quad \mathbf{B} = \begin{bmatrix} 1 \\ 3 \\ 3 \end{bmatrix}.$$

System (5) is then equivalent to the matrix equation **AX = B**.

Although Eq. (9) is in matrix notation, it still is not the most convenient form for our purposes. We indicated in Section 4.1 that the most useful matrices are those that involve numbers—just numbers, no words and no letters. Equation (9) contains the vector **X** which consists of letters. If we combine the matrix **A** with the vector **B**, we obtain the *augmented matrix* for Eq. (9). In general, the augmented matrix for any system of equations given in the matrix form **AX =**

B is [**A** | **B**]. In particular, the augmented matrix for Example 1 is

$$\begin{bmatrix} 2 & 3 & 4 & | & 11 \\ 5 & 6 & 0 & | & 12 \\ 7 & -8 & -9 & | & 13 \end{bmatrix}.$$

The augmented matrix for Example 2 is

$$\begin{bmatrix} 3 & -2 & | & 1 \\ -1 & 4 & | & 3 \\ 2 & 1 & | & 3 \end{bmatrix}.$$

The vertical line between **A** and **B** is for convenience only and is used to emphasize which parts of the augmented matrix correspond to which sides of the original equations.

Example 3 Find the augmented matrix for system (5).

Solution For this system,

$$\mathbf{A} = \begin{bmatrix} 1 & 1 & 2 \\ 2 & -2 & 1 \end{bmatrix} \quad \text{and} \quad \mathbf{B} = \begin{bmatrix} 0 \\ 0 \end{bmatrix},$$

hence

$$[\mathbf{A} | \mathbf{B}] = \begin{bmatrix} 1 & 1 & 2 & | & 0 \\ 2 & -2 & 1 & | & 0 \end{bmatrix}.$$

Note that the augmented matrix can be written immediately from the form of equations, provided that the same variables are aligned under each other. We simply write all the coefficients in their natural order and replace the equality signs by a straight line. The reverse process is equally straightforward.

Example 4 Find the system of equations in the variables a, b, c, and d given by the augmented matrix

$$[\mathbf{A} | \mathbf{B}] = \begin{bmatrix} 2 & 0 & -1 & 4 & | & 0 \\ 3 & -2 & 1 & 5 & | & 6 \end{bmatrix}.$$

Solution

$2a + 0b - c + 4d = 0$
$3a - 2b + c + 5d = 6.$

Example 5 Find the system of equations in the variables l, n, and p given by the augmented matrix

$$[A \mid B] = \begin{bmatrix} 1 & -1 & 2 & | & 4 \\ -1 & 0 & 2 & | & 3 \\ 5 & 6 & 7 & | & 0 \end{bmatrix}.$$

Solution

$$\begin{aligned} l - n + 2p &= 4 \\ -l \phantom{{}+0n} + 2p &= 3 \\ 5l + 6n + 7p &= 0. \end{aligned}$$

Exercises

1. Determine which of the following systems of equations are linear:
   (a) $\quad x - 2y + 3z - w = 4$
   $2x \phantom{{}-2y+3z} - z + 4w = 8$
   (b) $\quad 2a - 3b + 4c = 2$
   $a + b - 2c = 0$
   $-a + 2b + 5c = 1$
   (c) $\quad 2a + ab = 3$
   $-a + 2b = 1$
   (d) $\quad 2^3 a + b + c = 3$
   $4a + 2b - c = 2$
   (e) $\quad 2a^3 + b + c = 3$
   $4a + 2b - c = 2$
   (f) $\quad 2(3)^a + b + c = 3$
   $4a + 2b - c = 2$
   (g) $\quad r - 2s = 1$
   $r + 2s = -1$
   $2r - s = 4$
   $r - 2s = 1.$

2. Determine whether or not the proposed values are solutions of the system
   $$\begin{aligned} x - 2y + 2z &= 0 \\ 3y - 2z &= 0 \\ -8x + y + 2z &= 0; \end{aligned}$$
   (a) $x = 1, y = 2, z = 3$
   (b) $x = 1, y = 0, z = -1$
   (c) $x = 0, y = 0, z = 0.$

3. Determine whether the proposed values of z, y, z, and w are solutions of the system
   $$\begin{aligned} 2x + y - z + w &= 0 \\ 3x - 2y + z + 2w &= 2 \\ x - y - z - w &= 1 \\ 4x + 5y + z + 3w &= 2; \end{aligned}$$
   (a) $x = 0, y = -\tfrac{1}{2}, z = 0, w = \tfrac{1}{2}$

(b) $x = 0, y = -\frac{7}{8}, z = -\frac{4}{8}, w = \frac{3}{8}$
(c) $x = 1, y = 0, z = 1, w = -1$.

4. Put systems (a), (b), (d), and (g) in Exercise 1 into the matrix form $\mathbf{AX} = \mathbf{B}$.
5. Find the augmented matrix for the systems of equations given in parts (a), (b), (d), and (g) of Exercise 1.
6. Find the systems of equations in the variables x and y given by the augmented matrix:

(a) $\begin{bmatrix} 2 & 1 & | & 4 \\ 3 & -1 & | & 0 \end{bmatrix}$
(b) $\begin{bmatrix} 1 & 3 & | & 2 \\ 1 & -1 & | & 1 \\ 4 & 1 & | & 2 \end{bmatrix}$.

7. Find the systems of equations in the variables r, s, and t given by the augmented matrix:

(a) $\begin{bmatrix} 1 & 1 & -3 & | & 0 \\ 2 & 4 & 5 & | & 6 \end{bmatrix}$
(b) $\begin{bmatrix} 1 & 0 & 2 & | & 4 \\ 0 & 1 & 2 & | & 3 \\ 2 & 0 & 1 & | & 1 \end{bmatrix}$.

8. A cabinet manufacturer produces lamp bases and television cabinets. Each lamp base b requires 2 hours to cut and assemble, $\frac{1}{2}$ hour to decorate, and 15 minutes to inspect and crate, while each television cabinet c requires 1 hour to cut and assemble, 2 hours to decorate, and 15 minutes to inspect and crate. Each day the manufacturer has available 220 hours for cutting and assembling, 200 hours for decorating, and 35 hours for inspecting and crating. The problem is to determine how many lamp bases and how many television cabinets should be produced in order to use all the available workpower. Show that this problem is equivalent to solving three equations in the two unknowns b and c.

9. The cost C of manufacturing beer is \$80,000 annually in fixed overhead plus \$5 per barrel B. The net sales S is computed on the wholesale price of \$8 per barrel.
    (a) Show that C, B, and S are related by two simultaneous equations.
    (b) Show that the problem of determining how many barrels must be produced for the manufacturer to break even, that is, for the net sales to equal cost, is equivalent to solving three linear equations in the variables C, B, and S.

10. The end-of-the-year employee bonus at Breslin Tool and Die is 10% of the net profit after city and state taxes have been paid. The city tax is 3% of taxable income, while the state tax is 8% of taxable income with credit allowed for the city tax as a pretax deduction. This year, taxable income was \$800,000. Show that b, c, and s, which denote the bonus, city tax, and state tax, respectively, are related by three simultaneous equations.

4.5 Elementary Row Operations

The method we use to solve a given system of linear equations involves transforming the given system into a second system which is easier to solve. Of course, it does us no good to solve the second system of equations unless we can be assured that it has the same solution as the original system. Therefore we must be careful when transforming systems to use only operations that do not affect the original solution. Three such operations are:

(O1) Interchanging the order of two equations.
(O2) Multiplying an equation by a nonzero number.
(O3) Adding to one equation a constant times another equation.

To illustrate these operations, consider the system of equations

$$\begin{aligned} x + y - 2z &= -3 \\ -x - y + z &= 0 \\ 2x + 3y - 2z &= 2, \end{aligned} \qquad (10)$$

which has the solution $x = 1$, $y = 2$, and $z = 3$. If we apply (O1) to the first and third equations, we obtain the new system

$$\begin{aligned} 2x + 3y - 2z &= 2 \\ -x - y + z &= 0 \\ x + y - 2z &= -3, \end{aligned} \qquad (11)$$

which obviously has the same solution as system (10). If we apply (O2) to system (10) by multiplying the first equation by 5, we obtain the system

$$\begin{aligned} 5x + 5y - 10z &= -15 \\ -x - y + z &= 0 \\ 2x + 3y - 2z &= 2. \end{aligned} \qquad (12)$$

System (12) is different from system (10), but it still has the same solution, $x = 1$, $y = 2$, and $z = 3$. Finally, if we apply (O3) to system (10) by adding to the second equation 3 times the third equation (note that we are not changing the third equation, only the second equation), the second equation becomes

$$(-x + 6x) + (-y + 9y) + (z - 6z) = (0 + 6)$$

or

$$5x + 8y - 5z = 6,$$

and the new system is

$$\begin{aligned} x + y - 2z &= -3 \\ 5x + 8y - 5z &= 6 \\ 2x + 3y - 2z &= 2. \end{aligned} \qquad (13)$$

System (13) again has the same solution as system (10).

It should be apparent that operations (O1) through (O3) do not affect the variables in the equations but only the coefficients. Therefore we can save a good deal of writing if we delete the variables and just direct our attention to the coefficients. In other words, we should consider the augmented matrix. Rephrasing operations (O1) through (O3) in words appropriate to the augmented matrix, we obtain the following *elementary row operations:*

(E1) Interchanging the order of two rows.
(E2) Multiplying one row by a nonzero number.
(E3) Adding to one row a constant times another row.

To illustrate these operations, consider the augmented matrix for system (10):

$$\begin{bmatrix} 1 & 1 & -2 & | & -3 \\ -1 & -1 & 1 & | & 0 \\ 2 & 3 & -2 & | & 2 \end{bmatrix}. \tag{14}$$

Applying (E1) to matrix (14) by interchanging the first and third rows, we obtain the new matrix

$$\begin{bmatrix} 2 & 3 & -2 & | & 2 \\ -1 & -1 & 1 & | & 0 \\ 1 & 1 & -2 & | & -3 \end{bmatrix}.$$

This is the augmented matrix for system (11), which verifies that operation (O1) is equivalent to operation (E1). Applying (E2) to matrix (14) by multiplying the first row by 5, we obtain

$$\begin{bmatrix} 5 & 5 & -10 & | & -15 \\ -1 & -1 & 1 & | & 0 \\ 2 & 3 & -2 & | & 2 \end{bmatrix}.$$

This is the augmented matrix for system (12), which verifies the equivalence between operations (O2) and (E2). Finally, applying (E3) to matrix (14) by adding to the second row 3 times the third row, we obtain

$$\begin{bmatrix} 1 & 1 & -2 & | & -3 \\ 5 & 8 & -5 & | & 6 \\ 2 & 3 & -2 & | & 2 \end{bmatrix}.$$

This is the augmented matrix for system (13), which verifies the equivalence between operations (O3) and (E3).

Our objective is still the same: Use operations (O1) through (O3) to transform a given set of equations to another set which is easier to solve. In terms of matrices, we wish to transform one augmented matrix to another simpler augmented matrix by using elementary row operations. In particular, we strive to transform the given augmented matrix into one in upper triangular form.

Definition 4.7 A matrix is in *upper triangular* form if all the elements below the main diagonal (defined in Section 4.1) are zero.

Examples of upper triangular matrices are

$$\begin{bmatrix} 1 & 2 & 3 \\ 0 & 5 & 6 \\ 0 & 0 & 7 \end{bmatrix}, \quad \begin{bmatrix} 1 & 2 \\ 0 & 3 \\ 0 & 0 \\ 0 & 0 \end{bmatrix}, \quad \text{and} \quad \begin{bmatrix} 1 & 2 & 0 & 4 \\ 0 & 3 & 1 & 2 \\ 0 & 0 & 5 & -3 \end{bmatrix}.$$

Again we distinguish the main diagonal by the diagonal line. Note that Definition 4.7 does not place any restrictions on the elements on or above the main diagonal; they can be anything, even zero. The elements below the main diagonal must be zero. The zero matrix and the identity matrix are other upper triangular matrices.

Example 1 Use elementary row operations to transform matrix (14) to upper triangular form.

Solution

$$\begin{bmatrix} 1 & 1 & -2 & | & -3 \\ -1 & -1 & 1 & | & 0 \\ 2 & 3 & -2 & | & 2 \end{bmatrix} \rightarrow \begin{bmatrix} 1 & 1 & -2 & | & -3 \\ 0 & 0 & -1 & | & -3 \\ 2 & 3 & -2 & | & 2 \end{bmatrix}$$
(by adding to the second row 1 times the first row)

$$\rightarrow \begin{bmatrix} 1 & 1 & -2 & | & -3 \\ 0 & 0 & -1 & | & -3 \\ 0 & 1 & 2 & | & 8 \end{bmatrix}$$
(by adding to the third row -2 times the first row)

$$\rightarrow \begin{bmatrix} 1 & 1 & -2 & | & -3 \\ 0 & 1 & 2 & | & 8 \\ 0 & 0 & -1 & | & -3 \end{bmatrix}$$
(by interchanging the second and third rows).

Whenever one element is used to cancel a second element to zero by an elementary row operation, the first element is called the *pivot element*. In Example 1, we first used the 1-1 element to cancel the 2-1 element and then to cancel the 3-1 element. For these operations, the 1-1 element was the pivot.

The process of reducing a matrix to upper triangular form can be simplified if the following three rules are observed:

1. Completely transform one column to the required form before attacking another column. In Example 1, we first converted all the necessary elements in the first column to zero before we considered the second column.
2. Use (E2) to transform the pivot element to 1 *before* using that element to cancel other elements in the same column to 0.
3. Never use an operation if it will change the value of any zero in a previously transformed column.

In addition, it is usually advisable to use (E1) whenever it will result in additional zeros below the main diagonal.

Example 2 Use the elementary row operations to transform to upper triangular form the augmented matrix

$$\begin{bmatrix} 2 & 1 & | & 1 \\ 3 & -1 & | & 4 \\ 2 & 2 & | & 0 \end{bmatrix}.$$

Solution

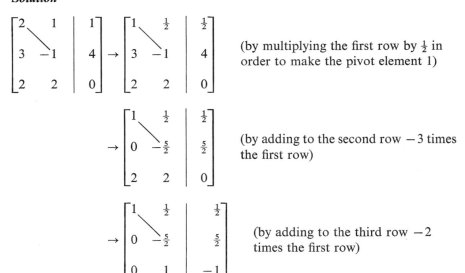

(by multiplying the first row by $\frac{1}{2}$ in order to make the pivot element 1)

(by adding to the second row -3 times the first row)

(by adding to the third row -2 times the first row)

$$\rightarrow \begin{bmatrix} 1 & \frac{1}{2} & & \frac{1}{2} \\ 0 & 1 & & -1 \\ 0 & 1 & & -1 \end{bmatrix}$$

(by multiplying the second row by $-\frac{2}{5}$ in order to make the new pivot element 1)

$$\rightarrow \begin{bmatrix} 1 & \frac{1}{2} & & \frac{1}{2} \\ 0 & 1 & & -1 \\ 0 & 0 & & 0 \end{bmatrix}$$

(by adding to the third row -1 times the second row).

Example 3 Transform to upper triangular form the augmented matrix

$$\begin{bmatrix} 5 & 10 & 30 & | & 122 \\ 10 & 30 & 100 & | & 340 \\ 30 & 100 & 354 & | & 1156 \end{bmatrix}.$$

Solution To avoid fractions, we convert all numbers to decimals. With the aid of a calculator, we compute

$$\begin{bmatrix} 5 & 10 & 30 & | & 122 \\ 10 & 30 & 100 & | & 340 \\ 30 & 100 & 354 & | & 1156 \end{bmatrix} \rightarrow \begin{bmatrix} 1 & 2 & 6 & | & 24.4 \\ 10 & 30 & 100 & | & 340 \\ 30 & 100 & 354 & | & 1156 \end{bmatrix}$$

(by multiplying the first row by $\frac{1}{5} = 0.2$)

$$\rightarrow \begin{bmatrix} 1 & 2 & 6 & | & 24.4 \\ 0 & 10 & 40 & | & 96 \\ 30 & 100 & 354 & | & 1156 \end{bmatrix}$$

(by adding to the second row -10 times the first row)

$$\rightarrow \begin{bmatrix} 1 & 2 & 6 & | & 24.4 \\ 0 & 10 & 40 & | & 96 \\ 0 & 40 & 174 & | & 424 \end{bmatrix}$$

(by adding to the third row -30 times the first row)

$$\rightarrow \begin{bmatrix} 1 & 2 & 6 & | & 24.4 \\ 0 & 1 & 4 & | & 9.6 \\ 0 & 40 & 174 & | & 424 \end{bmatrix}$$

(by multiplying the second row by $\frac{1}{10} = 0.1$)

$$\rightarrow \begin{bmatrix} 1 & 2 & 6 & | & 24.4 \\ 0 & 1 & 4 & | & 9.6 \\ 0 & 0 & 14 & | & 40 \end{bmatrix} \quad \text{(by adding to the third row} \\ -40 \text{ times the second row).}$$

Once the augmented matrix of a given set of equations has been transformed into upper triangular form, the solution can be obtained easily. We show how in Section 4.6.

Exercises

Use elementary row operations to transform the augmented matrices in Exercises 1 through 7 into upper triangular form.

1. $\begin{bmatrix} 1 & 3 & | & -1 \\ 2 & 5 & | & 4 \\ 3 & 5 & | & 0 \end{bmatrix}$

2. $\begin{bmatrix} 1 & 2 & 3 & | & 1 \\ -1 & -3 & 4 & | & 2 \\ 3 & 1 & 0 & | & 1 \end{bmatrix}$

3. $\begin{bmatrix} 0 & 1 & 0 & 2 & | & 1 \\ 1 & 3 & 4 & 1 & | & 2 \\ 1 & 4 & 2 & 1 & | & 7 \end{bmatrix}$

4. $\begin{bmatrix} 2 & 1 & 1 & | & 3 \\ 3 & 0 & 4 & | & 2 \\ 4 & -1 & 3 & | & 0 \end{bmatrix}$

5. $\begin{bmatrix} 0 & 3 & | & 5 \\ 2 & 1 & | & 4 \end{bmatrix}$

6. $\begin{bmatrix} 1.1 & 3 & | & 2.7 \\ 4.9 & -2.4 & | & 7.4 \end{bmatrix}$

7. $\begin{bmatrix} 2 & 1 & 0 & | & 7 \\ 0 & 3 & 1 & | & 1 \\ 5 & 0 & 1 & | & 2 \end{bmatrix}$

8. Transform the augmented matrices found in Exercise 5 in Section 4.4 into upper triangular form.

4.6 Solutions of Simultaneous Linear Equations

Having mastered the material in Section 4.5, we now are able to solve systems of simultaneous equations. The method is quite simple. First, write the augmented matrix for the system of equations. Second, use the elementary row operations to transform this matrix into upper triangular form. Third, write the system of equations that corresponds to the new upper triangular augmented matrix. Fourth, solve the new system. The last step should be easy and will have the same solutions as the original set of equations.

Example 1 Solve system (10) in Section 4.5 for x, y, and z.

Solution The augmented matrix for this system is given by expression (14). Using the results from Example 1 in Section 4.5, we know that matrix (14) can be transformed into

$$\begin{bmatrix} 1 & 1 & -2 & | & -3 \\ 0 & 1 & 2 & | & 8 \\ 0 & 0 & -1 & | & -3 \end{bmatrix}.$$

The corresponding system of equations is

$$\begin{aligned} x + y - 2z &= -3 \\ y + 2z &= 8 \\ -z &= -3. \end{aligned}$$

From the last equation, we have $z = 3$. Substituting this value into the second equation, we obtain $y + (2)(3) = 8$ or $y = 2$. Substituting $z = 3$ and $y = 2$ into the first equation, we calculate $x + 2 - (2)(3) = -3$ or $x = 1$. The solution to system (10) is $x = 1$, $y = 2$, and $z = 3$.

Example 2 Solve the system

$$\begin{aligned} 2p + q &= 1 \\ 3p - q &= 4 \\ 2p + 2q &= 0. \end{aligned}$$

Solution The augmented matrix for this system is

$$\begin{bmatrix} 2 & 1 & | & 1 \\ 3 & -1 & | & 4 \\ 2 & 2 & | & 0 \end{bmatrix}.$$

Using the results from Example 2 in Section 4.5, we know that this matrix can be transformed into the upper triangular matrix

$$\begin{bmatrix} 1 & \tfrac{1}{2} & | & \tfrac{1}{2} \\ 0 & 1 & | & -1 \\ 0 & 0 & | & 0 \end{bmatrix}.$$

The corresponding system of equations is

$$\begin{aligned} p + \tfrac{1}{2}q &= \tfrac{1}{2} \\ q &= -1 \\ 0 &= 0. \end{aligned}$$

Clearly, $q = -1$. Substituting this value into the first equation, we obtain $p + (\tfrac{1}{2})(-1) = \tfrac{1}{2}$ or $p = 1$. The complete solution is $p = 1$ and $q = -1$.

Example 3 Solve the system

$$5d + 10b + 30a = 122$$
$$10d + 30b + 100a = 340$$
$$30d + 100b + 354a = 1156.$$

Solution The augmented matrix for this system is

$$\begin{bmatrix} 5 & 10 & 30 & | & 122 \\ 10 & 30 & 100 & | & 340 \\ 30 & 100 & 354 & | & 1156 \end{bmatrix}.$$

Using the results from Example 3 in Section 4.5, we know that this matrix is transformable into

$$\begin{bmatrix} 1 & 2 & 6 & | & 24.4 \\ 0 & 1 & 4 & | & 9.6 \\ 0 & 0 & 14 & | & 40 \end{bmatrix}.$$

The corresponding system of equations is

$$d + 2b + 6a = 24.4$$
$$b + 4a = 9.6$$
$$14a = 40.$$

By rounding to three decimal places, it follows from the last equation that $a = 40/14 = 2.857$. Substituting this value into the second equation, we have $b + (4)(2.857) = 9.6$ or $b = -1.828$. Finally, from the first equation, we calculate $d + (2)(-1.828) + (6)(2.857) = 24.4$ or $d = 10.914$. The solution is $d = 10.914$, $b = -1.828$, and $a = 2.857$.

Examples 1 through 3 all dealt with systems having a unique (only one) solution. As we know, this is not always the case. How, then, does the method work on a system that has no solutions? As an answer, we consider the system

$$x + y = 1$$
$$x + y = 2,$$

which was determined in Section 4.4 to have no solution. The corresponding augmented matrix and its transformation to upper triangular form is

$$\begin{bmatrix} 1 & 1 & | & 1 \\ 1 & 1 & | & 2 \end{bmatrix} \rightarrow \begin{bmatrix} 1 & 1 & | & 1 \\ 0 & 0 & | & 1 \end{bmatrix} \quad \text{(by adding to the second row } -1 \text{ times the first row).}$$

The equations corresponding to the second matrix are

$$x + y = 1$$
$$0 = 1.$$

The second equation is clearly absurd. We conclude that a system of equations has no solution if and only if the new system, corresponding to the upper triangular matrix, contains an absurd equation similar to $0 = 1$.

The last possibility is a system of equations that has more than one solution. To analyze this situation, first reconsider the solutions of the final set of equations obtained in Examples 1 through 3. Note that in every case we effectively solved each equation for the first variable in that equation. In Example 1, we used the first equation to find x, the second equation to find y, and the third equation to find z. In Example 2, we used the first equation to solve for p, and the second equation to solve for q. This procedure works even when the equations have many solutions. Always solve for the first variable in each equation. The only difference is that the variables will then be in terms of other variables. The variables on the right side of the resulting equations will be arbitrary and can be assigned any number we choose.

Example 4 Solve the system

$$x + 2y - z = 3$$
$$2x + 5y + 2z = 4$$
$$3x + 7y + z = 7.$$

Solution Transforming the augmented matrix for this system into upper triangular form, we obtain

$$\begin{bmatrix} 1 & 2 & -1 & | & 3 \\ 2 & 5 & 2 & | & 4 \\ 3 & 7 & 1 & | & 7 \end{bmatrix} \rightarrow \begin{bmatrix} 1 & 2 & -1 & | & 3 \\ 0 & 1 & 4 & | & -2 \\ 3 & 7 & 1 & | & 7 \end{bmatrix}$$ (by adding to the second row -2 times the first row)

$$\rightarrow \begin{bmatrix} 1 & 2 & -1 & | & 3 \\ 0 & 1 & 4 & | & -2 \\ 0 & 1 & 4 & | & -2 \end{bmatrix}$$ (by adding to the third row -3 times the first row)

$$\rightarrow \begin{bmatrix} 1 & 2 & -1 & | & 3 \\ 0 & 1 & 4 & | & -2 \\ 0 & 0 & 0 & | & 0 \end{bmatrix}$$ (by adding to the third row -1 times the second row).

The equations corresponding to this last matrix are

$$x + 2y - z = 3$$
$$y + 4z = -2$$
$$0 = 0.$$

Using the second equation to solve for y, we obtain $y = -2 - 4z$. Substituting this value into the first equation and solving for x, we calculate

$$x + 2(-2 - 4z) - z = 3, \quad \text{or} \quad x = 7 + 9z.$$

The solution is $x = 7 + 9z$ and $y = -2 - 4z$ with z arbitrary.

If we set $z = 1$, then $x = 16$ and $y = -6$, which is a particular solution. Setting $z = -1$, we obtain $x = -2$ and $y = 2$, which is still another solution. In fact, every choice of z leads to a different but valid solution for the original system.

Example 5 Solve the system

$$a + 2b - c + d = 3$$
$$2a + b + c - 3d = 0.$$

Solution Transforming the augmented matrix for this system into upper triangular form, we obtain

$$\begin{bmatrix} 1 & 2 & -1 & 1 & | & 3 \\ 2 & 1 & 1 & -3 & | & 0 \end{bmatrix} \rightarrow \begin{bmatrix} 1 & 2 & -1 & 1 & | & 3 \\ 0 & -3 & 3 & -5 & | & -6 \end{bmatrix}.$$

The new set of equations is

$$a + 2b - c + d = 3$$
$$-3b + 3c - 5d = -6.$$

Solving the second equation for b, we have $b = 2 + c - \frac{5}{3}d$. Substituting this value into the first equation and solving for a, we calculate

$$a + 2(2 + c - \tfrac{5}{3}d) - c + d = 3 \quad \text{or} \quad a = -1 - c + \tfrac{7}{3}d.$$

The final solution is $a = -1 - c + \tfrac{7}{3}d$ and $b = 2 + c - \tfrac{5}{3}d$ with both c and d arbitrary.

Exercises

1. Solve the following systems for their unique solutions:

    (a) $x + 2y = 3$
    $2x + 5y = 7$

    (b) $4a + 2b = 1$
    $7a - 3b = 4$

    (c) $2.1s - 15t = 3.7$
    $4s + 7.4t = 4$

    (d) $x + 2y - z = 1$
    $2x + 3y - 2z = 4$
    $3x - y + 5z = 0$

    (e) $3a + 4b - 4c = 0$
    $a - 2b + 3c = 1$
    $-2a + b - c = 2$

    (f) $u + 2v - w = 1$
    $u - 3v + w = 2$
    $u + v - 4w = -1$

(g) $\quad a + 2b - c + d = 2$
$\quad\quad -2a - 3b - c - d = 4$
$\quad\quad 4a + 2b - 4c + 5d = 0$
$\quad\quad -3a - 3b + 2c - 7d = 14$

(h) $\quad 5p + 7q = 3$
$\quad\quad 4p + 2q = 5$
$\quad\quad 6p + 12q = 1.$

2. Show that the following systems do not have solutions:

(a) $\quad x + y - 2z = 3$
$\quad\quad 2x + 4y - 3z = 4$
$\quad\quad x + 3y - z = 0$

(b) $\quad 5p + 7q = 3$
$\quad\quad 4p + 2q = 5$
$\quad\quad 6p + 12q = 7.$

3. Show that the following systems have more than one solution and find them:

(a) $\quad x + y - 2z = 3$
$\quad\quad 2x + 4y - 3z = 4$
$\quad\quad x + 3y - z = 1$

(b) $\quad 2a + 3b - c + 2d = 1$
$\quad\quad 4a + 6b - 2c + 4d = 2$
$\quad\quad 3a + b - 2c + d = 2$
$\quad\quad 5a + 4b - 3c + d = 3$

(c) $\quad u - v + 2w = 0$
$\quad\quad 2u + 3v - w = 0$
$\quad\quad 3u + 2v + w = 0$

(d) $\quad u - v + 2w = 0$
$\quad\quad 2u - 2v + 4w = 0$
$\quad\quad 3u - 3v + 6w = 0.$

4.7 Matrix Inversion

The one operation we have not considered is matrix division, but with good reason—this operation is not defined. The closest matrix operation to division is *inversion*, the process of calculating an inverse.

Definition 4.8 The *inverse* of a square matrix **A** is another matrix $\mathbf{A}^{-1}$ satisfying the property

$$\mathbf{A}\mathbf{A}^{-1} = \mathbf{A}^{-1}\mathbf{A} = \mathbf{I}. \tag{15}$$

One matrix is the inverse of another if and only if their product is the identity matrix. Therefore, once we have a candidate for $\mathbf{A}^{-1}$, we can use Definition 4.8 to determine whether or not it really is the inverse. If Eq. (15) is satisfied, $\mathbf{A}^{-1}$ is the inverse of **A**; if Eq. (15) is not satisfied, $\mathbf{A}^{-1}$ is not the inverse of **A**.

Definition 4.8 can be compressed a bit. Since **A** must be square, it can be shown, although not considered here, that if $\mathbf{A}\mathbf{A}^{-1} = \mathbf{I}$, then $\mathbf{A}^{-1}\mathbf{A} = \mathbf{I}$ automatically, and vice versa. Therefore we do not have to check all parts of Eq. (15), but only whether or not $\mathbf{A}\mathbf{A}^{-1} = \mathbf{I}$.

Example 1 Determine whether or not $\mathbf{A}^{-1}$ is the inverse of **A** for the following matrices:

$$\mathbf{A} = \begin{bmatrix} 3 & 5 \\ 1 & 2 \end{bmatrix} \quad \text{and} \quad \mathbf{A}^{-1} = \begin{bmatrix} 2 & -5 \\ -1 & 3 \end{bmatrix}.$$

Solution We form the product $\mathbf{AA}^{-1}$ and check whether or not it is the identity matrix. Here

$$\mathbf{AA}^{-1} = \begin{bmatrix} 3 & 5 \\ 1 & 2 \end{bmatrix} \begin{bmatrix} 2 & -5 \\ -1 & 3 \end{bmatrix} = \begin{bmatrix} 1 & 0 \\ 0 & 1 \end{bmatrix} = \mathbf{I}$$

Therefore $\mathbf{A}^{-1}$ is the inverse of $\mathbf{A}$.

Example 2 Determine whether or not $\mathbf{A}^{-1}$ is the inverse of $\mathbf{A}$ for the following matrices:

$$\mathbf{A} = \begin{bmatrix} 1 & 3 & 2 \\ 0 & 1 & 0 \\ 2 & -1 & -3 \end{bmatrix} \quad \text{and} \quad \mathbf{A}^{-1} = \begin{bmatrix} 3 & -1 & 3 \\ 0 & 1 & 0 \\ 2 & -1 & 1 \end{bmatrix}.$$

Solution Here

$$\mathbf{AA}^{-1} = \begin{bmatrix} 1 & 3 & 2 \\ 0 & 1 & 0 \\ 2 & -1 & -3 \end{bmatrix} \begin{bmatrix} 3 & -1 & 3 \\ 0 & 1 & 0 \\ 2 & -1 & 1 \end{bmatrix} = \begin{bmatrix} 7 & 0 & 5 \\ 0 & 1 & 0 \\ 0 & 0 & 3 \end{bmatrix} \neq \mathbf{I}.$$

Since $\mathbf{AA}^{-1} \neq \mathbf{I}$, $\mathbf{A}^{-1}$ is not the inverse of $\mathbf{A}$.

Although Definition 4.8 provides a test for determining whether or not one matrix $\mathbf{A}^{-1}$ is the inverse of a given matrix $\mathbf{A}$, it does not provide a method for calculating the inverse. The following procedure is such a method. Let $\mathbf{A}$ be the matrix we wish to invert. Augment onto $\mathbf{A}$ the identity matrix $\mathbf{I}$ having the same order as $\mathbf{A}$. Use elementary row operations to transform $\mathbf{A}$ into the identity matrix. Each time an operation is performed, however, perform it on the entire augmented matrix. As $\mathbf{A}$ is transformed into the identity matrix, the part of the augmented matrix that initially corresponded to the identity will be transformed into $\mathbf{A}^{-1}$.

Example 3 Find the inverse of $\mathbf{A} = \begin{bmatrix} 3 & 5 \\ 1 & 2 \end{bmatrix}$.

Solution We first augment onto $\mathbf{A}$ the 2×2 identity, obtaining

$$\begin{bmatrix} 3 & 5 & | & 1 & 0 \\ 1 & 2 & | & 0 & 1 \end{bmatrix}.$$

Next, we use elementary row operations to convert the first part of this matrix to the identity matrix. Each operation, however, is applied to the entire augmented matrix. In particular,

$$\begin{bmatrix} 3 & 5 & | & 1 & 0 \\ 1 & 2 & | & 0 & 1 \end{bmatrix} \rightarrow \begin{bmatrix} 1 & 2 & | & 0 & 1 \\ 3 & 5 & | & 1 & 0 \end{bmatrix} \quad \text{(by interchanging both rows)}$$

$$\rightarrow \begin{bmatrix} 1 & 2 & | & 0 & 1 \\ 0 & -1 & | & 1 & -3 \end{bmatrix} \quad \begin{array}{l}\text{(by adding to the second row } -3 \\ \text{times the first row)}\end{array}$$

$$\rightarrow \begin{bmatrix} 1 & 2 & | & 0 & 1 \\ 0 & 1 & | & -1 & 3 \end{bmatrix} \quad \begin{array}{l}\text{(by multiplying the second row} \\ \text{by } -1\text{)}\end{array}$$

$$\rightarrow \begin{bmatrix} 1 & 0 & | & 2 & -5 \\ 0 & 1 & | & -1 & 3 \end{bmatrix} \quad \begin{array}{l}\text{(by adding to the first row } -2 \\ \text{times the second row).}\end{array}$$

The part of the augmented matrix corresponding to **A** has now been converted to the identity matrix. The other part that initially corresponded to **I** is the inverse of **A**. That is,

$$\mathbf{A}^{-1} = \begin{bmatrix} 2 & -5 \\ -1 & 3 \end{bmatrix}.$$

Example 4 Find the inverse of

$$\mathbf{A} = \begin{bmatrix} 1 & 3 & 2 \\ 0 & 1 & 0 \\ 2 & -1 & -3 \end{bmatrix}.$$

Solution We augment the 3 × 3 identity matrix to **A** and then use elementary row operations to convert the part corresponding to **A** to the identity matrix. The operations, however, are applied to the entire augmented matrix. Here

$$\begin{bmatrix} 1 & 3 & 2 & | & 1 & 0 & 0 \\ 0 & 1 & 0 & | & 0 & 1 & 0 \\ 2 & -1 & -3 & | & 0 & 0 & 1 \end{bmatrix}$$

$$\rightarrow \begin{bmatrix} 1 & 3 & 2 & | & 1 & 0 & 0 \\ 0 & 1 & 0 & | & 0 & 1 & 0 \\ 0 & -7 & -7 & | & -2 & 0 & 1 \end{bmatrix} \quad \begin{array}{l}\text{(by adding to the third row } -2 \\ \text{times the first row)}\end{array}$$

$$\rightarrow \begin{bmatrix} 1 & 0 & 2 & | & 1 & -3 & 0 \\ 0 & 1 & 0 & | & 0 & 1 & 0 \\ 0 & -7 & -7 & | & -2 & 0 & 1 \end{bmatrix} \quad \begin{array}{l}\text{(by adding to the first row } -3 \\ \text{times the second row)}\end{array}$$

$$\rightarrow \begin{bmatrix} 1 & 0 & 2 & | & 1 & -3 & 0 \\ 0 & 1 & 0 & | & 0 & 1 & 0 \\ 0 & 0 & -7 & | & -2 & 7 & 1 \end{bmatrix} \quad \text{(by adding to the third row 7 times the second row)}$$

$$\rightarrow \begin{bmatrix} 1 & 0 & 2 & | & 1 & -3 & 0 \\ 0 & 1 & 0 & | & 0 & 1 & 0 \\ 0 & 0 & 1 & | & \frac{2}{7} & -1 & -\frac{1}{7} \end{bmatrix} \quad \text{(by multiplying the third row by } -\frac{1}{7}\text{)}$$

$$\rightarrow \begin{bmatrix} 1 & 0 & 0 & | & \frac{3}{7} & -1 & \frac{2}{7} \\ 0 & 1 & 0 & | & 0 & 1 & 0 \\ 0 & 0 & 1 & | & \frac{2}{7} & -1 & -\frac{1}{7} \end{bmatrix} \quad \text{(by adding to the first row } -2 \text{ times the third row).}$$

The inverse of **A** is

$$\mathbf{A}^{-1} = \begin{bmatrix} \frac{3}{7} & -1 & \frac{2}{7} \\ 0 & 1 & 0 \\ \frac{2}{7} & -1 & -\frac{1}{7} \end{bmatrix}.$$

The only question still unanswered is whether or not every matrix can be reduced to an identity matrix by elementary row operations. The answer is "no." Matrices that cannot be so reduced, however, do not have inverses. One example is the 2 × 2 zero matrix. A second example is the following.

Example 5 Find the inverse of

$$\mathbf{A} = \begin{bmatrix} 1 & 2 \\ 2 & 4 \end{bmatrix}.$$

Solution Proceeding in the usual manner, we calculate

$$\begin{bmatrix} 1 & 2 & | & 1 & 0 \\ 2 & 4 & | & 0 & 1 \end{bmatrix} \rightarrow \begin{bmatrix} 1 & 2 & | & 1 & 0 \\ 0 & 0 & | & -2 & 1 \end{bmatrix} \quad \text{(by adding to the second row } -2 \text{ times the first row).}$$

But now we are stuck. We are unable to use elementary row operations to reduce the first part of the augmented matrix to the identity matrix. Since the original matrix cannot be so reduced, it does not have an inverse.

One use of the inverse is in solving certain systems of simultaneous linear equations. We showed in Section 4.4 that every system can be written in the matrix form

AX = B. [9]

If **A** is a square matrix and if it has an inverse, we can multiply Eq. (9) by $\mathbf{A}^{-1}$ to obtain

$\mathbf{A}^{-1}$AX = $\mathbf{A}^{-1}$B.

4.7 Matrix Inversion

Since $A^{-1}A = I$ and $IX = X$, we find

$$\boxed{X = A^{-1}B.} \qquad (16)$$

Equation (16) is a matrix representation for the solution of any linear system of simultaneous equations providing A^{-1} exists. The solution vector is the product of A^{-1} and the known vector B.

Example 6 Solve the system

$$3x + 5y = 1$$
$$x + 2y = -6.$$

Solution Using the technique developed in Section 4.4, we first write this system in the matrix form $AX = B$, where

$$A = \begin{bmatrix} 3 & 5 \\ 1 & 2 \end{bmatrix}, \quad X = \begin{bmatrix} x \\ y \end{bmatrix}, \quad \text{and} \quad B = \begin{bmatrix} 1 \\ -6 \end{bmatrix}.$$

We found the inverse of A in Example 3. Therefore it follows from Eq. (16) that

$$\begin{bmatrix} x \\ y \end{bmatrix} = X = A^{-1}B = \begin{bmatrix} 2 & -5 \\ -1 & 3 \end{bmatrix} \begin{bmatrix} 1 \\ -6 \end{bmatrix} = \begin{bmatrix} 32 \\ -19 \end{bmatrix}.$$

Two vectors are equal if and only if their components are equal, hence $x = 32$ and $y = -19$ is the solution to the original system.

Example 7 Solve the system

$$a + 3b + 2c = 1$$
$$b = 4$$
$$2a - b - 3c = 2.$$

Solution This system has the matrix form $AX = B$ with

$$A = \begin{bmatrix} 1 & 3 & 2 \\ 0 & 1 & 0 \\ 2 & -1 & -3 \end{bmatrix}, \quad X = \begin{bmatrix} a \\ b \\ c \end{bmatrix}, \quad \text{and} \quad B = \begin{bmatrix} 1 \\ 4 \\ 2 \end{bmatrix}.$$

We found the inverse of A in Example 4. Using Eq. (16), we now find

$$\begin{bmatrix} a \\ b \\ c \end{bmatrix} = X = A^{-1}B = \begin{bmatrix} \frac{3}{7} & -1 & \frac{2}{7} \\ 0 & 1 & 0 \\ \frac{2}{7} & -1 & -\frac{1}{7} \end{bmatrix} \begin{bmatrix} 1 \\ 4 \\ 2 \end{bmatrix} = \begin{bmatrix} -3 \\ 4 \\ -4 \end{bmatrix}.$$

The solution is $a = -3$, $b = 4$, $c = -4$.

We have two methods for solving systems of linear equations, one by elementary row operations and a second by matrix inversion. If the solution is to be calculated by hand, the method based on elementary row operations is best. It involves fewer computations, and it always works. The method based on matrix inversion does not always work, in particular, when $\mathbf{A}^{-1}$ does not exist. Nonetheless, Eq. (16) provides a representation for the solution of certain systems, which has great theoretical value. Furthermore, the form of Eq. (16) lends itself to solution via a computer if the computer is programmed to do matrix operations.

Exercises

Find the inverses of the matrices given in Exercises 1 through 7.

1. $\begin{bmatrix} 2 & 3 \\ 1 & 2 \end{bmatrix}$
2. $\begin{bmatrix} 0 & 1 \\ 1 & 0 \end{bmatrix}$
3. $\begin{bmatrix} 1 & 2 \\ 3 & 4 \end{bmatrix}$

4. $\begin{bmatrix} 2 & 1 \\ 3 & 5 \end{bmatrix}$
5. $\begin{bmatrix} 1 & 1 & 3 \\ 1 & 2 & -1 \\ 1 & 1 & 1 \end{bmatrix}$
6. $\begin{bmatrix} 1 & 2 & 3 \\ 0 & 1 & 1 \\ 2 & 3 & 4 \end{bmatrix}$

7. $\begin{bmatrix} 1 & 3 & -1 & 1 \\ 2 & 7 & 2 & 1 \\ 0 & 0 & 1 & 3 \\ 0 & 0 & 1 & 1 \end{bmatrix}$

8. Show that $\mathbf{A} = \begin{bmatrix} 1 & -1 \\ -1 & 1 \end{bmatrix}$ does not have an inverse.

9. Show that $\mathbf{A} = \begin{bmatrix} 1 & 2 & 3 \\ 4 & 5 & 6 \\ 7 & 8 & 9 \end{bmatrix}$ does not have an inverse.

10. Show that, if $\mathbf{AA}^{-1} = \mathbf{A}^{-1}\mathbf{A}$, then $\mathbf{A}^{-1}$ must have the same order as $\mathbf{A}$. *Hint:* Let $\mathbf{A}$ have order $m \times n$ and $\mathbf{A}^{-1}$ have order $p \times r$. Show that $n = p$ for $\mathbf{AA}^{-1}$ to be defined and $r = m$ for $\mathbf{A}^{-1}\mathbf{A}$ to be defined. Then show that $n = m$ if $\mathbf{AA}^{-1} = \mathbf{A}^{-1}\mathbf{A}$.

Use inversion to solve the systems given in Exercises 11 through 14.

11. $x + 2y = 3$
    $2x + 5y = 7$.

12. $4a + 2b = 1$
    $7a - 3b = 4$.

13. $3a + 4b - 4c = 0$
    $a - 2b + 3c = 1$
    $-2a + b - c = 2$.

14. $u + 2v - w = 1$
    $u - 3v + w = 2$
    $u + v - 4w = -1$.

Chapter

Linear Programming

All business decisions ultimately deal with maximizing profit or minimizing cost. The total effect of all such decisions determines whether a company makes or loses money. Obviously, analytical tools that provide the means for determining optimal decisions are to be cherished.

In this chapter, we develop such tools for a large class of problems known as linear programming problems. One method is geometric, a second is algebraic. Both depend heavily on the matrix operations developed in Chapter 4, as well as on concepts dealing with inequalities.

5.1 Inequalities

Any two real numbers a and b are related by one of three possibilities: a is less than b, written $a < b$; a equals b, written $a = b$; or a is greater than b, written $a > b$. In particular, $5 < 7$, $-3 < -2$, $4 = \frac{16}{4}$, $8 > 5$, and $0 > -3$. Like equalities, inequalities also satisfy arithmetic properties which are important in problem solving.

Property 5.1 If $a < b$ and if q is any *positive* number, then $aq < bq$. Similarly, if $y > x$, then $yq > xq$.

That is, multiplying both sides of an inequality by a positive number does not change the sense of the inequality; less than remains less than, and greater than remains greater than. Since $5 < 7$, it follows that $5(4) < 7(4)$. Similarly, $20 > 3$, hence $20(10) > 3(10)$.

Property 5.2 If $a < b$ and if q is any *negative* number, then $aq > bq$. Similarly, if $y > x$, then $yq < xq$.

Multiplying both sides of an inequality by a negative number *does* change the sense of the inequality; less than becomes greater than, and greater than becomes less than. Note that $5 < 7$, but $5(-1) > 7(-1)$. Also, $20 > 3$, but $20(-4) < 3(-4)$.

Property 5.3 If $a < b$ and q is any real number, then $a \pm q < b \pm q$. Similarly, if $y > x$, then $y \pm q > x \pm q$.

That is, adding or subtracting the same number from both sides of an inequality does not change the sense of the inequality. Since $5 < 7$, it follows that $5 + 9 < 7 + 9$. Since $20 > 3$, we have $20 - 30 > 3 - 30$.

Property 5.4 If $a < b$ and if $b < c$, then $a < c$. Similarly, if $y > x$ and $x > w$, then $y > w$.

In particular, $5 < 7$ and $7 < 11$, thus $5 < 11$.

The inequalities $a < b$ and $y > x$ are called *strong inequalities*. As useful in business are the *weak inequalities* $a \leq b$ and $y \geq x$, read, respectively, "a is less than *or equal* to b" and "y is greater than *or equal* to x." The word "or" before "equal" in both definitions is important: The statement $a \leq b$ is valid if either a is less than b or a equals b. Obviously both conditions cannot be met simultaneously. It is as correct to write $7 \leq 8$ as it is to write $7 < 8$, although the latter statement is stronger. Another valid expression is $4 \geq \frac{16}{4}$.

Properties 5.1 through 5.4 remain true if the strong inequalities are replaced by weak inequalities.

Using Properties 5.1 through 5.4, one can solve inequalities involving one unknown (for example, finding x if $15 - 2x \leq 31$). The procedure is always the same and is analogous to the techniques described in Section 1.2 for equalities: Use the relevant properties to isolate the unknown on one side of the inequality.

Example 1 Find x if $15 - 2x \leq 31$.

Solution If we subtract 15 from both sides of the inequality, it follows from Property 5.3 that

$$(15 - 2x) - 15 \leq 31 - 15$$
$$-2x \leq 16.$$

Multiplying both sides of this last inequality by $-\frac{1}{2}$, a negative number, we obtain (see Property 5.2 and note the change in the sense of the inequality),

$$(-\tfrac{1}{2})(-2x) \geq (-\tfrac{1}{2})(16)$$
$$x \geq -8.$$

Therefore any value of x greater than or equal to -8 satisfies the original inequality.

Example 2 Find x if $-10 + 5x < 7 + 2x$.

Solution We first subtract $2x$ from both sides of the inequality to group all x terms together. Here

$$(-10 + 5x) - 2x < (7 + 2x) - 2x \quad \text{(Property 5.3)}$$
$$-10 + 3x < 7.$$

To isolate the x terms, we add 10 to both sides of the inequality, resulting in

$$(-10 + 3x) + 10 < 7 + 10 \quad \text{(Property 5.3)}$$
$$3x < 17.$$

Finally, multiplying both sides of this inequality by $\frac{1}{3}$, we find x as

$$(\tfrac{1}{3})(3x) < \tfrac{1}{3}(17) \quad \text{(Property 5.1)}$$
$$x < \tfrac{17}{3}.$$

Any value of x less than $\frac{17}{3}$ satisfies the original inequality.

One linear inequality with two unknowns is slightly harder to solve. For example, find all values of x and y that satisfy the inequality $2x - y > 0$. Such an inequality is *linear* if the associated equation obtained by replacing the inequality by an equality is itself a linear equation as defined in Section 2.1. Consider the inequality $2x - y > 0$. The associated equation is $2x - y = 0$, which is linear, hence the original inequality $2x - y > 0$ is also linear. In contrast, the inequality $5x^2 - 3y \leq 9$ is not a linear inequality, since its associated equation $5x^2 - 3y = 9$ is not a linear equation. From Section 2.3, we recognize it to be quadratic. Surprisingly, we can solve linear inequalities reasonably quickly by extending our knowledge of straight lines.

Every straight line divides a plane into two regions: the region above the line and the region below the line or, depending on the slope of the line, the region to the left of the line and the region to the right of the line. (See Figures 5.1 and 5.2). From Definition 2.1, we know that a straight line satisfies the equation

$$cx + dy = e. \tag{1}$$

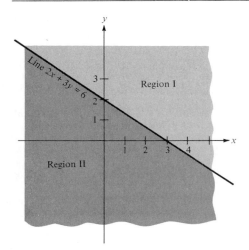

FIGURE 5.1

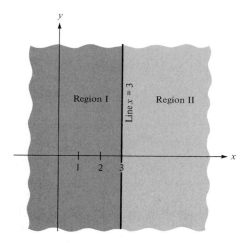

FIGURE 5.2

The two regions on either side of this line satisfy similar relationships. One region satisfies the inequality $cx + dy < e$, and the other region satisfies the inequality $cx + dy > e$.

To determine which region satisfies which inequality, we pick one point, not on the straight line (and therefore in one of the regions), substitute its x- and y-coordinates into the left side of Eq. (1), and simplify. Then the left side of Eq. (1) is either greater than or less than the right side. Whichever inequality is appropriate remains appropriate for the entire region containing the selected point. The other region satisfies the reverse inequality.

As an example, consider the line in Figure 5.1, defined by

$$2x + 3y = 6. \tag{2}$$

We first pick a point not on the line (the origin is usually the easiest) and substitute its coordinates (for the origin $x = 0, y = 0$) into the left side of Eq. (2). Here $2(0) + 3(0) = 0 + 0 = 0$ which is less than the right side of Eq. (2). Therefore every point in region II in Figure 5.1, the region that contains the origin, satisfies $2x + 3y < 6$. Every point in the other region, region I, satisfies $2x + 3y > 6$.

Using this knowledge of straight lines, we give a method for solving linear inequalities involving two unknowns, for example, $cx + dy > e$. First graph the straight line given by the associated equality $cx + dy = e$. Then use the procedures detailed above to locate the appropriate region corresponding to $cx + dy > e$. The x- and y-coordinates of every point in this region satisfy the given inequality.

Example 3 Graph the region $2x - y > 0$.

Solution We first plot the line

$$2x - y = 0, \tag{3}$$

as in Figure 5.3. To determine which side of the line corresponds to the given inequality, we then pick a point not on the line. Since the origin is on the line, we cannot use it. Instead we pick $(0, 2)$, although any other point not on the line would do equally well. Substituting the coordinates of this point into the left side of Eq. (3), we obtain $2(0) - (2) = 0 - 2 = -2$ which is less than

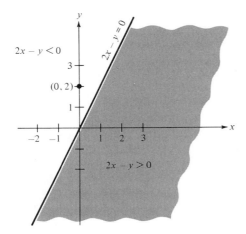

FIGURE 5.3

the right side. The region containing (0, 2) also satisfies the inequality $2x - y < 0$. The opposite area, which appears shaded in Figure 5.3, satisfies the reverse inequality $2x - y > 0$ and is the one of interest.

Since points on a line satisfy an equality, they cannot satisfy a corresponding inequality. In Example 3, the region $2x - y > 0$ does not include the straight line $2x - y = 0$. We emphasize this fact graphically by not letting the shaded region touch the line in Figure 5.3. Most often, however, we are interested in regions that satisfy weak inequalities of the form $2x - y \geq 0$. This region includes the area $2x - y > 0$ *and* the straight line $2x - y = 0$. In such cases, we emphasize the inclusion of the line by having the shaded region touch the line.

Example 4 Graph the region $x + y \leq 6$.

Solution The line

$$x + y = 6 \tag{4}$$

is plotted in Figure 5.4. Since the origin is not on the line, we substitute its coordinates into the left side of Eq. (4) and obtain $0 + 0 = 0$ which is less than 6. The region containing the origin, which appears shaded in Figure 5.4, satisfies the required inequality. Since the desired inequality is weak, the points on the line also satisfy it, hence the shaded region in Figure 5.4 touches the line.

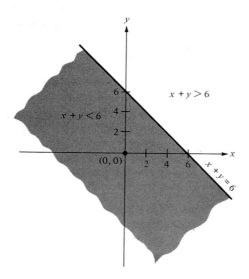

FIGURE 5.4

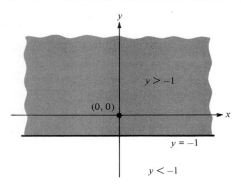

FIGURE 5.5

Example 5 Graph the region $y \geq -1$.

Solution The line
$$y = -1 \tag{5}$$
is plotted in Figure 5.5. When we substitute the coordinates of the origin ($x = 0, y = 0$) into Eq. (5), the left side of the equation becomes zero which is greater than the right side. The region containing the origin also satisfies $y > -1$ and, along with the line itself, represents the area of interest, which appears shaded in Figure 5.5.

Exercises

1. Determine which of the following statements are *not* true:
   - (a) $5 < 9$
   - (b) $5 \leq 9$
   - (c) $1 \leq 5 - 4$
   - (d) $1 \geq 5 - 4$
   - (e) $1 < 5 - 4$
   - (f) $3 + 2 \leq 4 - 1$
   - (g) $-2 \geq -3$
   - (h) $-2 > -3$
   - (i) $-2 \leq -3$.

2. Solve the following inequalities for x:
   - (a) $\frac{1}{2}x - 19 < 3$
   - (b) $2x + 14 \geq 5$
   - (c) $-2x + 14 \geq 5$
   - (d) $x + 5 \leq 2x - 7$
   - (e) $3x - 19 < -5x + 7$
   - (f) $-\frac{1}{2}x + 2 \geq -\frac{1}{3}x + \frac{1}{2}$.

3. Graph the following inequalities:
   - (a) $2x + 5y \leq 10$
   - (b) $2x - 5y \leq 10$
   - (c) $-2x + 5y \leq 10$
   - (d) $-2x - 5y \leq 10$
   - (e) $2x + 5y \leq -10$
   - (f) $x - 3y \geq -6$
   - (g) $y \geq 0$
   - (h) $2x - 3y > 12$
   - (i) $2x + 5y < 0$
   - (j) $x + 2y \leq 0$
   - (k) $x \geq 0$
   - (l) $y < 3$.

5.2 Systems of Linear Inequalities

In all linear programming problems, we are faced with not one but a set of linear inequalities, and we need to find the region that simultaneously satisfies all the prescribed conditions. That is, we seek the points whose coordinates satisfy all the given inequalities simultaneously.

Example 1 Find all points that satisfy the system

$2x - y > 0$
$x + y \le 6$.

Solution We first find the regions that satisfy the individual inequalities. This has already been done in Examples 3 and 4 in Section 5.1. Since these regions are not the final answer, we do not shade them in totally but rather indicate, as in Figure 5.6, which sides of the individual lines satisfy the individual inequalities. We are seeking the region that satisfies *both* inequalities. That is, we want the area in Figure 5.6 that is to the right of the line $2x - y = 0$ *and also* to the left of the line $x + y = 6$. The appropriate area appears shaded in Figure 5.7.

Example 2 Find all points that satisfy the system

$x + y \le 2$
$2x + y \ge -1$
$y \ge -3$.

Solution Using the procedure detailed in Section 5.1, we first find the regions that satisfy the individual inequalities. We then indicate each region on the same graph, as in Figure 5.8. The region we want must satisfy all the inequalities simultaneously. As such, it must be to the left of the line $x + y = 2$, to the right of the line $2x + y = -1$, and above the line $y = -3$. The appropriate area appears shaded in Figure 5.9.

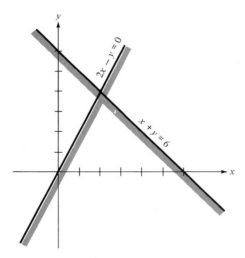

FIGURE 5.6

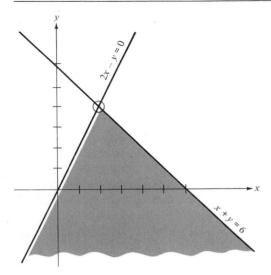

FIGURE 5.7

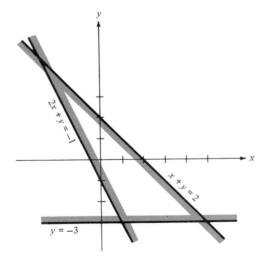

FIGURE 5.8

Example 3 Find all points that satisfy the system

$$3x + 10y \leq 33{,}000$$
$$5x + 8y \leq 42{,}000$$
$$x \geq 0$$
$$y \geq 0.$$

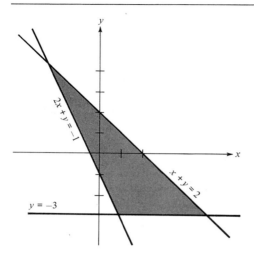

FIGURE 5.9

Solution The regions that satisfy the individual inequalities are indicated in Figure 5.10. The region we seek must lie below the line $3x + 10y = 33,000$, below the line $5x + 8y = 42,000$, to the right of the line $x = 0$, and above the line $y = 0$. The appropriate area is shaded in Figure 5.11.

For our purposes, the most important points of a region specified by a set of inequalities are the *cornerpoints* of that region. For Examples 1 through 3, these cornerpoints are nothing more than points of intersection of lines bounding

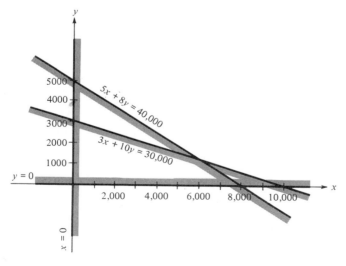

FIGURE 5.10

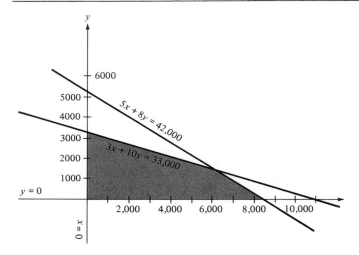

FIGURE 5.11

the region. Consider the shaded region in Figure 5.7 bounded by the two lines $x + y = 6$ and $2x - y = 0$. The one cornerpoint for this region is circled on the graph.

To find the coordinates of the cornerpoint in Figure 5.7, we note that the point lies on both straight lines. Since this cornerpoint lies on the line $x + y = 6$, its x- and y-coordinates must satisfy the equation of this line. Since the cornerpoint also lies on the line $2x - y = 0$, its x- and y-coordinates must satisfy the equation of that line too. Therefore the x- and y-coordinates must satisfy simultaneously the equations

$$x + y = 6$$

and

$$2x - y = 0. \tag{6}$$

We can find these coordinates algebraically by solving system (6) for x and y with the techniques of Section 4.6. Doing so, we find $y = 4$ and $x = 2$, hence the cornerpoint of Figure 5.7 is (2, 4).

The general procedure for finding the coordinates of a cornerpoint is first to locate the lines that give rise to the cornerpoint and then to solve the equations of those lines for the points of intersection.

Example 4 Find all cornerpoints of the shaded region in Figure 5.9.

Solution This region has three cornerpoints which are circled in Figure 5.12. Starting with the highest point and working in a clockwise direction, we find

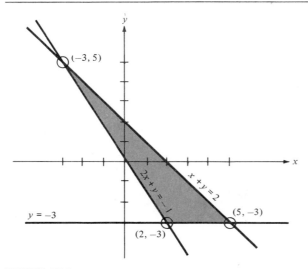

FIGURE 5.12

that the first cornerpoint lies on the lines $x + y = 2$ and $2x + y = -1$, so its coordinates must satisfy the system

$$x + y = 2$$
$$2x + y = -1.$$

Solving this system by the method given in Section 4.6, we compute $y = 5$ and $x = -3$. The cornerpoint is $(-3, 5)$.

The second cornerpoint lies on the lines defined by the equations $x + y = 2$ and $y = -3$, so its coordinates must satisfy the system

$$x + y = 2$$
$$y = -3.$$

The augmented matrix for this system is already in upper triangular form, hence we solve directly for $y = -3$ and $x = 5$. The second cornerpoint is $(5, -3)$.

The last cornerpoint lies on the lines defined by the equations $2x + y = 1$ and $y = -3$, so its coordinates must satisfy the system

$$2x + y = 1$$
$$y = -3.$$

Solving this system directly, we obtain $y = -3$ and $x = 2$. The last cornerpoint is $(2, -3)$.

Example 5 Find all cornerpoints of the shaded region in Figure 5.11.

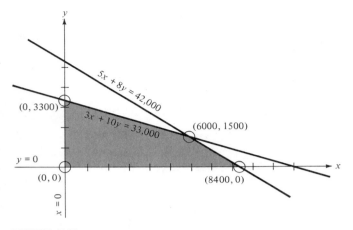

FIGURE 5.13

Solution This region has four cornerpoints which are circled in Figure 5.13. Starting with the highest point and working in a clockwise direction, we find that the coordinates of the first cornerpoint must satisfy the system

$$x = 0$$
$$3x + 10y = 33{,}000.$$

The solution is $x = 0$ and $y = 3300$.

The coordinates of the second cornerpoint must satisfy the system

$$3x + 10y = 33{,}000$$
$$5x + 8y = 42{,}000.$$

Solving this system for x and y, we find that the cornerpoint is (6000, 1500).

The coordinates of the third cornerpoint must satisfy the system

$$5x + 8y = 42{,}000$$
$$y = 0.$$

Solving, we calculate $x = 8400$ and $y = 0$. The last cornerpoint is obviously (0, 0).

Exercises

In each of the following problems, graph the regions defined by the given inequalities and find the cornerpoints of the regions.

1. $x + 2y \geq 4$
   $2x - y \leq 3.$

2. $x + 2y \geq 4$
   $2x - y \geq 3.$

3. $x - 4y \leq 4$
   $3x + 4y \leq 12$
   $x \geq 0.$

4. $x + 4y \leq 1$
   $3x + 4y \leq 12$
   $x \geq 6.$

5. $3x + y \geq -3$
   $x - y \leq 4$
   $y \leq 2.$

6. $10x + 8y \leq 80$
   $x - 4y \leq 12$
   $x \geq 0.$

7. $10x + 8y \leq 80$
   $x - 4y \leq 12$
   $x \leq 0.$

8. $10x + 8y \leq 80$
   $x - 4y \leq 12$
   $y \geq -3.$

9. $5x + 7y \leq 35$
   $7x - 5y \leq 35$
   $-x - 5y \leq 10$
   $x - y \leq 5.$

10. $5x + 7y \leq 35$
    $7x - 5y \geq 35$
    $-x - 5y \leq 10$
    $x - y \leq 5.$

11. $2x + y \leq 10$
    $5x + 8y \leq 40$
    $x \geq 0$
    $y \geq 0.$

12. $7x + 4y \leq 140$
    $x + 2y \leq 40$
    $x \geq 0$
    $y \geq 0.$

5.3 Optimizing a Functional: A Geometric Approach

Definition 5.1 A *linear functional* in the variables x and y is an expression of the form

$$c_1 x + c_2 y, \qquad (7)$$

where c_1 and c_2 are known numbers.

A linear functional is simply a number times the first variable x plus a number times the second variable y. It has no other form. The expression $3x - 7y$ is a linear functional ($c_1 = 5$, and $c_2 = -7$), whereas the expressions

$$xy + 2x \qquad (8)$$

and

$$3x - 4y^2 \qquad (9)$$

are not. In expression (8), the term xy is a variable times a variable, while in expression (9) the variable y appears squared. For an expression to be a linear functional, only known numbers can multiply variables and the variables must appear only to the first power. Also note that a linear functional is *not* an equation; there is no equality sign in expression (7). It is simply an expression in a special form.

We can generalize Definition 5.1 to expressions including three or more variables. An expression in the three variables x, y, and z is a linear functional if it has the form

$$c_1 x + c_2 y + c_3 z, \tag{10}$$

where c_1, c_2, and c_3 are known numbers. An expression in the four variables r, s, t, and u is a linear functional if it has the form

$$c_1 r + c_2 s + c_3 t + c_4 u, \tag{11}$$

where c_1, c_2, c_3, and c_4 are known numbers. In this section, we deal only with linear functionals in two variables; we return to the other cases in Section 5.5.

A linear programming model consists of two parts: (1) a linear functional, and (2) a set of linear inequalities similar to those considered in Section 5.2. The linear functional is called the *objective function*, and it generally represents profit or cost in dollars. The problem is to optimize (either maximize or minimize) the objective function while simultaneously satisfying all the inequalities. For example, if the objective function denotes profit, we will seek those values of the variables, say x and y, that maximize the profit and also satisfy some given set of inequalities. If the objective function represents cost, we will want those values of the variables that minimize the cost but still satisfy the given inequalities. A representative linear programming problem is

Maximize $\quad 7x + 22y$,

subject to $\quad \begin{aligned} 3x + 10y &\leq 33{,}000 \\ 5x + 8y &\leq 42{,}000 \\ x &\geq 0 \\ y &\geq 0. \end{aligned}$ $\tag{12}$

Here we seek the values of x and y that simultaneously satisfy all the inequalities and maximize $7x + 22y$. The first part is easy. In Example 3 in Section 5.2, we located all the points that satisfied the inequalities; they appear shaded in Figure 5.11. But which one of these points maximizes $7x + 22y$? To answer this question, and thereby solve the linear programming problem, we need the concept of a bounded region.

Definition 5.2 A region is *bounded* if it can be totally included within a (very large) circle.

The region shaded in Figure 5.11 is bounded, since it can be included within a circle as shown in Figure 5.14. The region shaded in Figure 5.7 is not bounded, since no circle, no matter how large, can contain the entire area.

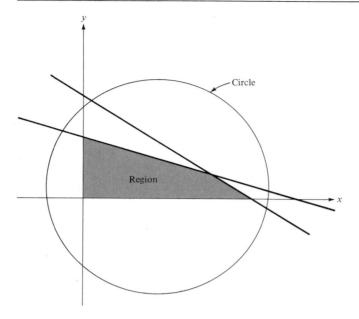

FIGURE 5.14

It is a well-known result in linear programming that any linear functional subject to a set of linear inequalities takes on its maximum (and minimum) at a cornerpoint of the region defined by the inequalities, providing the region is bounded. If the region is not bounded, this result is not valid. Consequently, the following four-step procedure is recommended for geometrically solving a linear programming problem in two variables:

1. Find the region that satisfies the set of given inequalities.
2. Check whether or not the region is bounded. If not, do not continue, since this method is not applicable.
3. If the region is bounded, find all the cornerpoints.
4. Substitute each cornerpoint into the objective function and determine which one optimizes it.

In this book, we consider only regions that are bounded. Occasionally, it is possible to transform unbounded regions into bounded ones, and we consider such procedures briefly in Section 5.4.

Example 1 Maximize $7x + 22y$,

subject to $\quad 3x + 10y \leq 33{,}000$
$\qquad\qquad 5x + 8y \leq 42{,}000$
$\qquad\qquad\quad\ x \geq 0$
$\qquad\qquad\quad\ y \geq 0.$

Solution Following the recommended procedure, we first find the region that satisfies all the inequalities. This was done in Example 3 in Section 5.2, and the region appears shaded in Figure 5.11. From Figure 5.14, we note that the region is bounded. The cornerpoints for this region were found in Example 5 in Section 5.2. Finally, substituting the coordinates of these cornerpoints into the objective function of this problem, $7x + 22y$, we complete Table 5.1. It is now clear that the maximum value of $7x + 22y$ is 75,000, which occurs when $x = 6000$ and $y = 1500$.

TABLE 5.1

| Cornerpoints | | Objective function |
x	y	$7x + 22y$
0	0	$7(0) + 22(0) = 0$
0	3300	$7(0) + 22(3300) = 72,600$
6000	1500	$7(6000) + 22(1500) = 75,000$††
8400	0	$7(8400) + 22(0) = 58,800$

†† Maximum.

Example 2 Maximize $0.75x + 1.10y$,

subject to $3x + 4y \leq 7400$
$7x + 6y \leq 14,000$
$x \geq 0$
$y \geq 0.$

Solution Using the techniques developed in Section 5.2, we first find the region that satisfies all the given inequalities. It appears shaded in Figure 5.15. Note that this region is bounded. We then find the cornerpoints (see Section 5.2) which also are displayed in Figure 5.15. Substituting the coordinates of these cornerpoints into the objective function, $0.75x + 1.10y$, we compute the entries in Table 5.2. The maximum value of the objective function is 2035 with $x = 0$ and $y = 1850$.

TABLE 5.2

| Cornerpoints | | Objective function |
x	y	$0.75x + 1.10y$
0	0	$0.75(0) + 1.10(0) = 0$
0	1850	$0.75(0) + 1.10(1850) = 2035$††
1200	950	$0.75(1200) + 1.10(950) = 1945$
2014.29	0	$0.75(2014.29) + 1.10(0) = 1510.72$

†† Maximum.

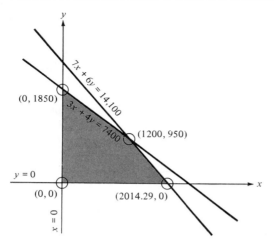

FIGURE 5.15

Example 3 Maximize $0.85x + 1.00y$,

subject to $\quad 3x + 4y \leq 7400$
$ 7x + 6y \leq 14{,}100$
$ x \geq 0$
$ y \geq 0.$

Solution The only difference between this problem and the previous one is the objective function. Since the inequalities are identical, the region of interest and the cornerpoints remain the same as those displayed in Figure 5.15. Substituting these cornerpoints into the new objective function, we complete Table 5.3. It follows that the maximum value of this objective function is 1970 which is assumed when $x = 1200$ and $y = 950$.

TABLE 5.3

Cornerpoints		Objective function
x	y	$0.85x + 1.00y$
0	0	$0.85(0) + 1.00(0) = 0$
0	1850	$0.85(0) + 1.00(1850) = 1.850$
1200	950	$0.85(1200) + 1.00(950) = 1970$††
2014.29	0	$0.85(2014.29) + 1.00(0) = 1712.15$

†† Maximum.

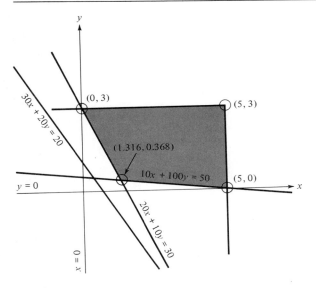

FIGURE 5.16

Example 4 Minimize $0.03x + 0.02y$,

subject to
$$20x + 10y \geq 30$$
$$10x + 100y \geq 50$$
$$30x + 20y \geq 20$$
$$x \geq 0, \quad x \leq 5$$
$$y \geq 0, \quad y \leq 3.$$

Solution The region that satisfies all the given inequalities is displayed in Figure 5.16, as are the appropriate cornerpoints. Note that the conditions $x \geq 0$, $y \geq 0$, and $30x + 20y \geq 20$ are not necessary; if they were deleted from the problem, the region would remain the same. Such inequalities are called *superfluous*. Substituting the coordinates of the cornerpoints into the objective function, we complete Table 5.4. The minimum value of $0.03x + 0.02y$ is 0.047 with $x = 1.316$ and $y = 0.368$.

TABLE 5.4

| Cornerpoints | | Objective function |
x	y	$0.03x + 0.02y$
0	3	$0.03(0) + 0.02(3) = 0.06$
5	3	$0.03(5) + 0.02(3) = 0.21$
5	0	$0.03(5) + 0.02(0) = 0.15$
1.316	0.368	$0.03(1.316) + 0.02(0.368) = 0.047$†

† Minimum.

It is possible for an objective function to take on its maximum or minimum at more than one cornerpoint. It can be shown (although not considered here) that, whenever two cornerpoints optimize a linear functional, every point on the line segment connecting these cornerpoints also optimizes the functional.

Example 5 Minimize $8x + 5y$,

$$\text{subject to} \quad \begin{aligned} x + y &\leq 2 \\ 2x + y &\geq -1 \\ y &\geq -3. \end{aligned}$$

Solution The bounded region that satisfies all the inequalities is displayed in Figure 5.12 in Section 5.2. The appropriate cornerpoints are plotted there as well. Substituting the coordinates of these points into the given objective function, we calculate the entries in Table 5.5. The minimum occurs at two cornerpoints, $x = -3, y = 5$ *and* $x = 2, y = -3$. Therefore the minimum is taken on by *all* points on the line segment between the points $(-3, 5)$ and $(2, -3)$.

TABLE 5.5

Cornerpoints		Objective function
x	y	$8x + 5y$
-3	5	$8(-3) + 5(5) = 1$†
5	-3	$8(5) + 5(-3) = 25$
2	-3	$8(2) + 5(-3) = 1$†

† Minimum.

Exercises

1. Maximize $3x + 2y$,

    $$\text{subject to} \quad \begin{aligned} 60x + 10y &\leq 240{,}000 \\ 2x + 5y &\leq 47{,}200 \\ x \geq 0 \quad y &\geq 0. \end{aligned}$$

2. Minimize $8000x + 7000y$,

    $$\text{subject to} \quad \begin{aligned} 4x + 3y &\geq 160 \\ 6x + 10y &\geq 350 \\ x \geq 0, \quad x &\leq 31 \\ y \geq 0, \quad y &\leq 31. \end{aligned}$$

    Are any of these inequalities superfluous?

3. Maximize $15x + 30y$,

    $$\text{subject to} \quad \begin{aligned} 0.4x + 0.8y &\leq 2240 \\ 0.6x + 0.4y &\leq 1440 \\ x \geq 0 \quad y &\geq 0. \end{aligned}$$

    Hint: Multiply the first two equations by 10.

4. Maximize $40x + 60y$,

   subject to $30x + 50y \leq 80{,}000$
   $10x + 5y \leq 14{,}000$
   $4x + 4y \leq 7200$
   $x \geq 0$
   $y \geq 0.$

5. Minimize $21{,}600 - 270x - 180y$,

   subject to $x + \tfrac{1}{2}y \leq 90$
   $5x + 3y \leq 225$
   $2x + \tfrac{1}{2}y \leq 76$
   $x \geq 0, \quad x \leq 40$
   $y \geq 0, \quad y \leq 60.$

Are any of these inequalities superfluous? Note that the objective function is *not* of the form given in Definition 5.1. Nonetheless, the recommended procedure given on page 170 is still valid. Why?

6. Minimize $1.00x + 0.25y$,

   subject to $0.60x + 0.05y \geq 100$
   $0.40x + 0.10y \geq 100$
   $0.08y \geq 20$
   $x \geq 0, \quad x \leq 1200$
   $y \geq 0, \quad y \leq 2000.$

7. Maximize $2x + 3y$,

   subject to $12x + 14y \leq 14{,}952$
   $x + 4y \leq 1892$
   $0.03x + 0.12y \leq 72$
   $x \geq 0$
   $y \geq 0.$

5.4 Linear Programming Problems

We are ready to attack several linear programming problems. Unfortunately, the initial form in which these problems appears does not lend itself to immediate mathematical analysis. Problems are stated in words, whereas our method is applicable to inequalities. Our first task therefore is to convert the words into mathematical relationships or models. Once this is done, the solution to the problem is obtained easily by the method developed in Section 5.3.

The following four-step procedure is recommended for converting linear programming problems to suitable formats:

Step 1 Determine the quantity to be maximized or minimized and write this quantity in equation form.

The objective of any linear programming problem is the optimization of a particular quantity, and this goal is stated clearly in each problem. Putting this quantity in equation form serves to define the unknown variables that appear in the remainder of the problem.

Step 2 Locate and isolate all physical requirements and restrictions inherent in the problem.

Often, Step 2 is accomplished best by completing a table similar to Table 5.6.

TABLE 5.6

	First requirement	Second requirement	Third requirement
First product			
Second product			
Third product			
Fourth product			
Constraints			

Step 3 Write the requirements and constraints found in Step 2 as inequalities. In general, each column of Table 5.6 requires one inequality.

Step 4 Establish any hidden constraints.

Hidden constraints are those that are not given explicitly in the statement of the problem but which nonetheless exist. For example, if two products are being manufactured, negative amounts of each will not be produced. Let x and y represent the amount of each product being manufactured. Two hidden, but obvious, constraints are $x \geq 0$ and $y \geq 0$.

We illustrate this four-step procedure for mathematically modeling linear programming problems with the following examples.

Example 1 A wood cabinet manufacturer produces cabinets for television consoles and frames for grandfather clocks, both of which must be assembled and decorated. Every television cabinet requires 3 hours to assemble and 5 hours to decorate. In contrast, each grandfather clock frame requires 10 hours to assemble and 8 hours to decorate. The profit on each cabinet and clock frame is $7 and $22, respectively. The manufacturer has 33,000 hours available each week for assembling these products (825 assemblers working 40 hours per week) and 42,000 hours available each week for decorating (1050 decorators working 40 hours per week). How many units of each product should the manufacturer produce weekly to maximize profit? Assume that all units produced can be sold.

Solution The objective is to maximize profit. Since the profit on each cabinet is $7 and the profit on each clock frame is $22, the total profit P will be

P = $7 times the number of cabinets produced plus $22 times the number of frames produced.

But how many of each product will be produced? This is precisely what we are asked to find, so at this point these values are unknowns. We set

x = number of television cabinets to be produced
y = number of grandfather clock frames to be produced.

Then, $P = 7x + 22y$.

Each cabinet and each clock frame requires a certain amount of time to assemble and decorate. These times are the physical requirements of the problem. The total labor available represents the constraints. Since there are two requirements, one for assembling and one for decorating, and two products, cabinets and clock frames, for this problem Table 5.6 becomes Table 5.7.

To complete Step 3 of the general procedure, we begin with the column in Table 5.7 listing assembly time. Each cabinet requires 3 hours to assemble. Since the plant will manufacture x cabinets, it must allocate 3 times x hours for assembling all its cabinets. Each clock frame requires 10 hours to assemble. Since the plant will manufacture y frames, it must allocate 10 times y hours for assembling all its clock frames. Therefore the manufacturer must allocate $3x + 10y$ hours each week for assembly operations. The hours used for assembling all units must be less than or equal to the hours available for assembling, hence $3x + 10y \leq 33{,}000$.

The last column in Table 5.7 deals with decorating time. Each cabinet requires 5 hours to decorate. Since the plant will manufacture x cabinets, it must allocate 5 times x hours for decorating all its cabinets. Each clock frame

TABLE 5.7

	Assembly time (hours)	Decorating time (hours)
Each television cabinet	3	5
Each grandfather clock frame	10	8
Available time	33,000	42,000

requires 8 hours to decorate. Since the plant will manufacture y frames, it must allocate 8 times y hours for decorating all its clock frames. Therefore each week the manufacturer must allocate $5x + 8y$ hours for decorating operations. The hours used for decorating the units must be less than or equal to the hours available for decorating, hence $5x + 8y \leq 42{,}000$.

Since the manufacturer will not produce a negative number of cabinets or clock frames, we also have the hidden conditions $x \geq 0$ and $y \geq 0$.

Collecting our results, we find that the given problem is modeled by the system

Maximize $P = 7x + 22y$,

subject to $\quad 3x + 10y \leq 33{,}000$
$\qquad\qquad\; 5x + 8y \leq 42{,}000$
$\qquad\qquad\qquad\quad\; x \geq 0$
$\qquad\qquad\qquad\quad\; y \geq 0.$

The solution to this problem is given in Example 1 in Section 5.3 as $x = 6000$ and $y = 1500$. That is, the company should produce 6000 television cabinets and 1500 grandfather clock frames weekly for a maximum profit of $75,000.

Example 2 The Lawn Care Company produces two different lawn fertilizers, Green Power Regular and Green Power Delux. The profit on each bag of Regular is 75¢, whereas the profit on each bag of Delux is $1.10. Each bag of Regular contains 3 pounds of active ingredients (a combination of nitrogen, phosphoric acid, and potash) and 7 pounds of inert substances. In contrast, each bag of Delux contains 4 pounds of active ingredients and 6 pounds of inert substances. Because of limited warehouse facilities, the company can stock only 7400 pounds of active ingredients and 14,100 pounds of inert substances. How many bags of each product should the Lawn Care Company produce daily if its objective is to maximize profit?

Solution The objective is to maximize profit. Since the profit on each bag of Regular is 75¢ and the profit on each bag of Delux is $1.10, the total profit P (in dollars) is

$P = 0.75$ times the number of bags of Regular produced plus 1.10 times the number of bags of Delux produced.

But how many of each product will be produced? This is precisely what we are asked to find, so these quantities are presently unknown. As such, we set

$x =$ number of bags of Regular to be produced
$y =$ number of bags of Delux to be produced.

Then $P = 0.75x + 1.10y$.

TABLE 5.8

	Amounts of active ingredients	Amounts of inert substances
Each bag of Regular	3	7
Each bag of Delux	4	6
Available quantities	7400	14,100

Every bag of fertilizer requires a specified amount of active ingredients and inert substances. These represent the physical requirements of the problem. The total quantities available represent the constraints. Both are displayed in Table 5.8.

Considering the second column of Table 5.8, we note that each bag of Regular contains 3 pounds of active ingredients. Since the company will produce x bags, it will consume 3 times x pounds of active ingredients to produce all its Regular. Each bag of Delux contains 4 pounds of active ingredients. Since the company will produce y bags, it will consume 4 times y pounds of active ingredients to produce all its Delux. Therefore the company will consume $3x + 4y$ pounds of active ingredients daily. Since the Lawn Care Company cannot consume more active ingredients than it has, it follows that $3x + 4y \leq 7400$.

Turning to the third column of Table 5.8, we see that each bag of Regular contains 7 pounds of inert substances. Since the company will produce x bags, it will consume 7 times x pounds of inert substances to produce all its Regular. Each bag of Delux contains 6 pounds of inert substances. Since the company will produce y bags, it will consume 6 times y pounds of inert substances to produce all its Delux. Therefore the company will consume $7x + 6y$ pounds of inert substances. Again, Lawn Care Company cannot consume more than it has, so $7x + 6y \leq 14{,}100$.

Hidden conditions are $x \geq 0$ and $y \geq 0$. Collecting all results, we find that the original problem is modeled by the system

Maximize $P = 0.75x + 1.10y$,

subject to $\quad 3x + 4y \leq 7400$
$\phantom{\text{subject to }\quad} 7x + 6y \leq 14{,}100$
$\phantom{\text{subject to }\quad\quad\quad\;\;} x \geq 0$
$\phantom{\text{subject to }\quad\quad\quad\;\;} y \geq 0.$

The solution to this problem is given in Example 2 in Section 5.3 as $x = 0$ and $y = 1850$. That is, the company should produce no Regular and 1850 bags of Delux for a maximum daily profit of $2035.

If the profit for Regular were 85¢ per bag, and that for Delux $1 per bag, it follows from Example 3 in Section 5.3 that the maximum profit would be

$1970 with a daily production schedule of 1200 bags of Regular and 950 bags of Delux.

Example 3 The Heartland Cereal Company produces All-Pro, a ready-to-eat breakfast food made from a blend of fortified cereal and dehydrated dairy products. Each serving of All-Pro is guaranteed to contain at least 30 grams of protein, 50 units of calcium, and 20 grams of carbohydrate. Heartland's supplier certifies that every ounce of fortified cereal contains exactly 20 grams of protein, 10 units of calcium, and 30 grams of carbohydrate. Each ounce of the dehydrated dairy product contains exactly 10 grams of protein, 100 units of calcium, and 20 grams of carbohydrate. The fortified cereal and dehydrated milk product cost 3¢ and 2¢, respectively. How much of each ingredient should be used in each serving of All-Pro in order to minimize supply cost while still maintaining the guaranteed nutritional composition?

Solution The objective is to minimize cost, which is 3¢ times the ounces of fortified cereal used plus 2¢ times the ounces of dehydrated milk product used. Since these ounces are unknown, we set

x = ounces of fortified cereal used in each serving of All-Pro
y = ounces of dehydrated milk product used in each serving of All-Pro.

Then, the cost (in dollars) is $C = 0.03x + 0.02y$.

The requirements and constraints for this problem are given in Table 5.9.

Each ounce of cereal provides 20 grams of protein, and each ounce of the milk product provides 10 grams of protein. Since the company uses x ounces of cereal and y ounces of the milk product in each serving of All-Pro, it follows that a serving contains $20x + 10y$ grams of protein. The actual protein content must be greater than or equal to the guaranteed minimum of 30 grams per serving, hence $20x + 10y \geq 30$.

Similarly, the calcium content of each serving is $10x + 100y$. Again this content must be greater than or equal to the guaranteed minimum of 50 grams per serving, hence $10x + 100y \geq 50$.

TABLE 5.9

	Protein content	Calcium content	Carbohydrate content
Each ounce of fortified cereal	20	10	30
Each ounce of dehydrated dairy product	10	100	20
Guaranteed minimum in each serving of All-Pro	30	50	20

5.4 Linear Programming Problems   181

An analysis of the carbohydrate requirement leads to the inequality $30x + 20y \geq 20$.

Since the company cannot use negative quantities of either supply, we also have the hidden conditions $x \geq 0$ and $y \geq 0$. Collecting all results, we obtain the mathematical problem:

Minimize $C = 0.03x + 0.02y$,

subject to
$$\begin{aligned} 20x + 10y &\geq 30 \\ 10x + 100y &\geq 50 \\ 30x + 20y &\geq 20 \\ x &\geq 0 \\ y &\geq 0. \end{aligned}$$

To solve this problem, we follow the procedure described in Section 5.3 and first locate the set of points that simultaneously satisfy all the inequalities. The appropriate region appears shaded in Figure 5.17. Note that the region is not bounded! Therefore the recommended procedure is not valid. Either we must find another procedure for solving this problem, or there are additional hidden conditions which can be used to bound the region. The second alternative is the case here.

The Heartland Cereal Company will not use more ingredients in each serving of All-Pro than is absolutely necessary. What, then, is the most they will use? If All-Pro were made from just fortified cereal, they would need $1\frac{1}{2}$

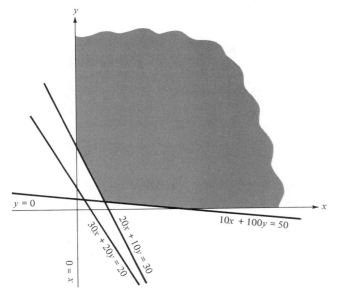

FIGURE 5.17

ounces to meet the protein requirement, 5 ounces to meet the calcium requirement, and $\frac{2}{3}$ ounce to meet the carbohydrate requirement. Therefore 5 ounces would meet all the requirements. Note that each ounce of cereal contains all the necessary ingredients. Five ounces of cereal supplies $5(20) = 100$ grams of protein, $5(10) = 50$ units of calcium, and $5(30) = 150$ grams of carbohydrate. This is too much protein and carbohydrate but exactly the required amount of calcium. Any lesser amount of cereal would not meet the calcium requirement.

Similarly, if All-Pro were made from just dehydrated milk product, Heartland would need 3 ounces to meet the protein requirement, $\frac{1}{2}$ ounce to meet the calcium requirement, and 1 ounce to meet the carbohydrate requirement. It follows that 3 ounces of the milk product would meet all requirements. Note that 3 ounces of the milk product supplies $3(10) = 30$ grams of protein, $3(100) = 300$ units of calcium, and $3(20) = 60$ grams of carbohydrate. This is too much calcium and carbohydrate, but exactly the amount of protein required. Any lesser amount of milk product by itself would not meet the protein requirement.

Since All-Pro will probably be a blend of fortified cereal and dehydrated milk product, it is possible that Heartland will need less than the maximum amount of both supplies to meet the guaranteed minimums. Hence we have the additional hidden conditions $x \leq 5$ and $y \leq 3$.

If we add these last two inequalities to the ones previously obtained, we have Example 4 in Section 5.3 (see Figure 5.16). There we found the solution to be $x = 1.316$ and $y = 0.368$ for a minimum cost of 4.7¢ per serving.

It is interesting to note that this minimum cost blend actually contains 46.84 grams of carbohydrate, which is 26.84 grams more than the guaranteed minimum.

Example 4 Continental Motors manufactures taxis and airport limousines for fleet operators throughout the country at its two plants. Plant A produces 75 taxis and 6 limousines daily, while plant B produces 50 taxis and 2 limousines each day. To fulfill contractual obligations to fleet owners in March, the company must manufacture 3000 taxis and 186 limousines. The costs of operating plants A and B, respectively, are $200,000 and $145,000 per day. How many days should each plant be operated during March to fulfill all contractual obligations at a minimum cost?

Solution We wish to minimize the cost which is $200,000 times the number of days plant A is in operation plus $145,000 times the number of days plant B is in operation. Since these numbers are currently unknown, we set

x = number of days plant A will operate in March
y = number of days plant B will operate in March.

Then, the cost C (in dollars) is $C = 200,000x + 145,000y$.

TABLE 5.10

	Taxis	Limousines
Daily production of plant A	75	6
Daily production of plant B	50	2
Contractual obligation	3000	186

The requirements and constraints for this problem are given in Table 5.10. Considering each column separately, we obtain the inequalities

$75x + 50y \geq 3000$ (from the second column)
$6x + 2y \geq 186$ (from the third column).

The hidden conditions $x \geq 0$ and $y \geq 0$ are also applicable.

Plotting these four inequalities, we find that the region of interest, which appears shaded in Figure 5.18, is not bounded. Thus the procedure recom-

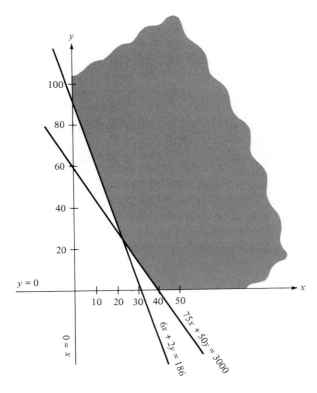

FIGURE 5.18

mended in Section 5.3 is not applicable. We try to remedy this situation by seeking additional hidden constraints which, when combined with our previous inequalities, will result in a bounded region. For this problem, we note that March has 31 days, hence $x \leq 31$ and $y \leq 31$.

The original word problem has been converted into the following mathematical problem:

Minimize $C = 200{,}000x + 145{,}000y$,

subject to
$$75x + 50y \geq 3000$$
$$6x + 2y \geq 186$$
$$x \geq 0, \quad y \geq 0$$
$$x \leq 31, \quad y \leq 31.$$

Using the procedure recommended in Section 5.3 for solving such problems, we find that plant A should operate 22 days and plant B should operate 27 days with a net minimum cost of $8,315,000.

Exercises

1. Redo Example 4 if the daily costs for operating plants A and B, respectively, are $200,000 and $250,000.
2. Redo Example 4 if plant A produces 75 taxis and 2 limousines at a cost of $175,000 per day, while plant B produces 50 taxis and 6 limousines at a cost of $157,000 per day.
3. The Atlas Pipe Company manufactures two types of metal joints, each of which must be molded and threaded. Type-I joints require 60 seconds on the molding machine and 2 seconds on the threading machine, whereas type-II joints require 10 seconds on the molding machine and 5 seconds on the threading machine. Each day the molding machines are operable for 240,000 seconds and the threading machine is operable for 47,200 seconds. The profit on each type-I and type-II joint is 3¢ and 2¢, respectively. The objective is to determine the number of joints of each type to be produced that will maximize profit. (a) Formulate the problem mathematically and solve. (b) Solve the problem if the profit on each type-I joint is 2¢ and the profit on each type-II joint is 4¢.
4. Eagle Mining Enterprises owns two mines in West Virginia. The ore from each mine is separated into two grades before it is shipped. Mine I produces 4 tons of high-grade ore and 6 tons of low-grade ore daily with an operating cost of $8000. In contrast, mine II produces 3 tons of high-grade ore and 10 tons of low-grade ore daily at a cost of $7000 per day. The mining company has contracts which require it to deliver 160 tons of high-grade ore and 350 tons of low-grade ore each month. Determine the number of days each mine should be operated to meet contractual obligations at a minimum cost.
5. The Heaven Rest Mattress Company produces a regular mattress labeled Sleeper and an extra-firm mattress labeled Johnny Firm. Each Sleeper requires 0.4 hours for basic assembling (binding the springs) and 0.6 hours for finishing (adding the padding and covering the unit with material). Each Johnny Firm requires 0.8

hours for assembling and 0.4 hours for finishing. The current labor force provides 2240 hours and 1440 hours per week for assembling and finishing, respectively. The profit on the Sleeper mattress is $15 per mattress; each Johnny Firm returns a profit of $30. How many mattresses of each type should Heaven Rest produce weekly to maximize its profit?

6. The King Sleep Corporation also produces two lines of mattresses, regular and extra firm. Each regular mattress requires 30 springs, 10 pounds of padding, and 4 yards of material, while each extra-firm mattress requires 50 springs, 5 pounds of padding and 4 yards of material. The wholesale price is $40 for each regular mattress and $60 for each extra-firm mattress. The maximum weekly inventory King Sleep can stock is 80,000 springs, 14,000 pounds of padding, and 7200 yards of material. How many mattresses should King Sleep produce weekly to maximize its income? How much of each supply (springs, padding, and material) is consumed with the optimal production schedule?

7. A caterer has in stock 1200 gallons of a premixed fruit juice containing 60% orange juice and 40% grapefruit juice, and 2000 gallons of a premixed fruit drink containing 5% orange juice, 10% grapefruit juice, 8% cranberry juice, and 77% filler (sugar, water, and flavorings). The fruit juice costs $1 per gallon, while the fruit drink costs 25¢ per gallon. From this stock, the caterer is to fulfill a contract for 1000 gallons of a fruit drink for a corporation picnic. This drink must contain at least 10% orange juice, 10% grapefruit juice, and 2% cranberry juice. How much of each stock item should be used to fulfill the contract at a minimum cost?

8. A manufacturer has available 14,952 feet of rope, 1892 feet of lumber, and 72 gallons of varnish with which to produce outdoor swings (wood seat and rope sides) and treehouse ladders (wood steps and rope sides). Each swing requires 12 feet of rope, 1 foot of lumber, and 0.03 gallons of varnish, and returns a profit of $2. Each ladder requires 14 feet of rope, 4 feet of lumber, and 0.12 gallons of varnish, and returns a profit of $3. How many units of each product should the manufacturer produce to maximize profit?

9. The decorating department of a large store has 40 sofas and 60 club chairs which must be reupholstered by the end of the week. First the material for each piece must be cut by a cutter, then the frame is reupholstered by an upholsterer, and finally the pillows are stuffed and finished by a sewer. The department manager has available 90 hours of cutters' time, 450 hours of upholsterers' time, and 76 hours of sewers' time each week. Whatever pieces cannot be completed in this time must be given to an outside contractor who charges the store $270 per sofa and $180 per chair. From experience, the department manager knows that each sofa requires 1 hour of cutting, 10 hours of upholstering, and 2 hours of sewing, while each club chair needs 0.5 hours of cutting, 6 hours of upholstering, and 0.5 hours of sewing. The objective is to determine the number of sofas and chairs that can be finished within the department so as to minimize the cost of using the outside contractor.

(a) Let x denote the number of sofas and y the number of club chairs that can be finished within the department. Show that two hidden conditions are $x \leq 40$ and $y \leq 60$.

(b) Model the department mathematically and solve for the optimal x and y. *Hint:* See Exercise 5 in Section 5.3.

(c) How many hours are required from the cutters in this optimal solution?
(d) Show that the objective is equivalent to maximizing $270x + 180y$.

10. The manager of a supermarket meat department finds that she has 160 pounds of round steak, 600 pounds of chuck steak, and 300 pounds of pork in stock on Saturday morning. From experience, she knows that she can sell half these quantities as straight cuts. The remaining meat will have to be ground and combined into hamburger meat and picnic patties for which there is a large weekend demand which generally cannot be fulfilled with existing supplies. Each pound of hamburger meat must contain at least 20% ground round and 60% ground chuck. Each pound of picnic patties must contain at least 30% ground pork and 50% ground chuck. The remainder of each product can consist of an inexpensive nonmeat filler which the store has in unlimited quantities. How many pounds of each product should be made if the objective is to minimize the pounds of meat that must be stored in the supermarket over Sunday?

In Exercises 11 through 14, set up but do not solve the mathematical relations that model the problems. Each problem contains more than two variables and is solved most easily by the simplex method to be considered in Sections 5.5 through 5.7.

11. A wood cabinet manufacturer produces cabinets for television consoles, frames for grandfather clocks, and lamp bases, all of which must be assembled, decorated, and crated. Each television console requires 3 hours to assemble, 5 hours to decorate, and 0.1 hour to crate, and returns a profit of $7. Each clock frame requires 10 hours to assemble, 8 hours to decorate, and 0.6 hour to crate, and returns a profit of $22. Each lamp base requires 1 hour to assemble, 1 hour to decorate, and 0.1 hour to crate, and returns a profit of $2. The manufacturer has 30,000, 40,000, and 120 hours available weekly for assembling, decorating, and crating, respectively. How many units of each product should be manufactured to maximize profit?

12. Redo Exercise 4 if in addition Eagle Mining also owns a third mine which produces 4 tons of high-grade ore and 6 tons of low-grade ore daily at an operating cost of $7500 per day.

13. A can of dog food is guaranteed to contain at least 10 units of protein, 20 units of mineral matter, and 6 units of fat. The product is composed of a blend of four different ingredients. Ingredient I contains 10 units of protein, 2 units of mineral matter, and $\frac{1}{2}$ unit of fat per ounce. Ingredient II contains 1 unit of protein, 40 units of mineral matter, and 3 units of fat per ounce. Ingredient III contains 1 unit of protein, 1 unit of mineral matter, and 6 units of fat per ounce. Ingredient IV contains 5 units of protein, 10 units of mineral matter, and 3 units of fat per ounce. The cost of each ingredient by the ounce is 2¢, 2¢, 1¢, and 3¢, respectively. How many ounces of each should be used to minimize the cost of the dog food but still meet the guaranteed composition?

14. A distributor with warehouses in New York and Atlanta must supply stores in New Orleans, Omaha, and Cleveland with chocolate for their candy counters. The distributor has 200 pounds of chocolate stored at its New York facility and 350 pounds at its Atlanta facility. Back orders from each of the stores include 100 pounds for New Orleans, 150 pounds for Omaha, and 125 pounds for Cleveland, although each store accepts up to 25 pounds more than ordered. Delivery costs for each store from each warehouse are given in Table 5.11, where all figures are in cents per pound.

TABLE 5.11

	To New Orleans	To Omaha	To Cleveland
From New York	25	20	10
From Atlanta	10	15	30

Determine how shipments should be made to minimize the total delivery cost.

5.5 Slack Variables and Standard Form

As long as linear programming problems involve only two unknowns, geometric methods provide a straightforward procedure for obtaining solutions. The situation changes drastically when more than two variables are involved. Problems with three variables can still be done geometrically, but the graphs are three-dimensional and difficult to draw. More variables require correspondingly more dimensions which do not exist geometrically. Fortunately there are nongeometric methods which work well regardless of the number of unknowns. The most useful one is the simplex method.

One requirement for the simplex method is that all variables be nonnegative. Hidden conditions of the form $x \geq 0$ and $y \geq 0$ are always assumed.

A second requirement is that the inequalities be in standard form. A maximum linear programming problem, one in which the objective function is to be maximized, is in *standard form* if all the inequalities are less than or equal to ($\leq$). The only exceptions are the variables themselves, which are assumed nonnegative. An example of such a problem is the following:

Maximize $4x + 10y + 6z$,

subject to $\quad 2x + 4y + z \leq 12$
$\phantom{\text{subject to }\quad}6x + 2y + z \leq 26$
$\phantom{\text{subject to }\quad}5x + y + 2z \leq 80,$

assuming $x \geq 0$, $y \geq 0$, and $z \geq 0$.

In contrast, a minimum linear programming problem, one in which the objective function is to be minimized, is in *standard form* if all the inequalities are greater than or equal to ($\geq$). An example of such a problem is the following:

Minimize $4x + 3y + 3z$,

subject to $\quad x + 2y \phantom{{}+z} \geq 2$
$\phantom{\text{subject to }\quad}3x + y + z \geq 4,$

assuming x, y, and z are all nonnegative.

Any linear programming problem not already in standard form can be converted to it. Such procedures are beyond the scope of this book and for our purposes unnecessary. All commercial problems of interest to us will be in standard form, as was the case for all the examples in Section 5.4.

If it is assumed that the basic requirements are met (all the variables are nonnegative and the problem is in standard form), the first step in the simplex method is to convert all inequalities to equalities. For maximum linear programming problems this is done by adding new quantities called *slack variables* to the inequalities. To see how, we again consider Example 1 in Section 5.4, dealing with the production of cabinets and clock frames by assemblers and decorators. The constraint inequalities are

$$3x + 10y \le 33{,}000 \tag{13}$$

$$5x + 8y \le 42{,}000. \tag{14}$$

From Eq. (13), we know that $3x + 10y$ is either less than 33,000 or equal to it. If $3x + 10y$ is actually less than 33,000, there must exist a positive number, call it s_1, such that

$$3x + 10y + s_1 = 33{,}000. \tag{15}$$

In particular, if $3x + 10y = 20{,}000$, then $s_1 = 13{,}000$. If, however, $3x + 10y = 28{,}000$, then $s_1 = 5000$. If $3x + 10y = 33{,}000$, then Eq. (15) is still valid, providing $s_1 = 0$. Thus Eq. (15) is equivalent to Eq. (13) if $s_1 \ge 0$.

Similarly, we know from Eq. (14) that $5x + 8y$ is either less than 42,000 or equals 42,000. If $5x + 8y$ is actually less than 42,000, then there must exist a positive number, call it s_2, such that

$$5x + 8y + s_2 = 42{,}000. \tag{16}$$

In particular, if $5x + 8y = 30{,}000$, then $s_2 = 12{,}000$. If $5x + 8y = 15{,}000$, then $s_2 = 27{,}000$. If $5x + 8y$ equals 42,000, then Eq. (16) is still valid, providing $s_2 = 0$. In any event, Eq. (16) is equivalent to Eq. (14) if $s_2 \ge 0$.

Variables such as s_1 and s_2, which are added to one side of a linear inequality in a standard maximum linear programming problem to convert that inequality to an equality, are called *slack variables*. These quantities represent the waste involved in a given production schedule. In the preceding problem, s_1 represents the hours the assemblers will be idle, and s_2 represents the hours the decorators will be idle if x cabinets and y consoles are produced. Once values for x and y are specified, we can use Eqs. (15) and (16) to solve for s_1 and s_2. In keeping with the first requirement of the simplex method, slack variables, like all other variables, must be nonnegative.

Slack variables can be incorporated into the objective function by adding them to the objective function with zero coefficients. For the cabinet–clock frame problem just considered, the original objective function was found in Example 1 in Section 5.4 to be $7x + 22y$. This is identical to the new objective function $7x + 22y + 0s_1 + 0s_2$, since $0s_1 = 0s_2 = 0$.

With these changes, the old linear programming problem

Maximize $7x + 22y$,

$$\begin{aligned} \text{subject to} \quad 3x + 10y &\leq 33{,}000 \\ 5x + 8y &\leq 42{,}000 \\ x &\geq 0 \\ y &\geq 0, \end{aligned}$$

is completely equivalent to the new problem

Maximize $7x + 22y + 0s_1 + 0s_2$,

$$\begin{aligned} \text{subject to} \quad 3x + 10y + s_1 &= 33{,}000 \\ 5x + 8y + s_2 &= 42{,}000, \end{aligned}$$

assuming x, y, s_1, and s_2 are all nonnegative.

The second form, however, is the one required for the simplex method.

Example 1 Convert the following problem to an equivalent problem involving slack variables.

Maximize $2a - 4b + c + d$,

$$\begin{aligned} \text{subject to} \quad a + 3b + 4d &\leq 4 \\ 2a + b &\leq 3 \\ b + 4c + d &\leq 3 \\ a \geq 0, \quad b &\geq 0 \\ c \geq 0, \quad d &\geq 0. \end{aligned}$$

Solution Adding the slack variables s_1, s_2, and s_3 to the left sides of the first three inequalities, respectively, and then adding all three slack variables to the objective function with zero coefficients, we obtain

Maximize $2a - 4b + c + d + 0s_1 + 0s_2 + 0s_3$,

$$\begin{aligned} \text{subject to} \quad a + 3b + 4d + s_1 &= 4 \\ 2a + b + s_2 &= 3 \\ b + 4c + d + s_3 &= 3, \end{aligned}$$

assuming a, b, c, d, s_1, s_2, and s_3 are nonnegative.

The procedure we have developed for converting inequalities to equalities is useful only when the inequalities are less than or equal to. It is not appropriate for inequalities of the opposite sense. After all, if $3x + 10y \geq 33{,}000$, no positive number added to the left side of the inequality will result in an equality. Such an addition would make the left side still greater. To change $3x + 10y \geq 33{,}000$ to an equality, we must subtract a nonnegative number from the left side, obtaining $3x + 10y - s_1 = 33{,}000$.

We will not handle minimum problems this way. The simplex method as described in Section 5.6 is for maximum problems in standard form, in which all the inequalities are less than or equal to. Minimum problems are solved by first converting them to other maximum problems. We show how in Section 5.7.

Exercises

In each of the following problems, convert the given systems to equivalent ones with slack variables.

1. Maximize $3x + 2y$,

    subject to $60x + 10y \leq 240{,}000$
    $2x + 5y \leq 47{,}200$
    $x \geq 0$
    $y \geq 0.$

2. Maximize $15x + 30y$,

    subject to $0.4x + 0.8y \leq 2240$
    $0.6x + 0.4y \leq 1440$
    $x \geq 0$
    $y \geq 0.$

3. Minimize $2a + 7b$,

    subject to $a - 3b \geq 2$
    $2a + 4b \geq 1$
    $a \geq 0$
    $b \geq 0.$

4. Maximize $15x + 10y + 14z$,

    subject to $6x + 5y + 3z \leq 26$
    $4x + 2y + 5z \leq 8$
    $x \geq 0, \quad y \geq 0, \quad z \geq 0.$

5. Maximize $2x + 3y + 4z + 6w$,

    subject to $x + y + z + w \leq 15$
    $7x + 5y + 3z + 2w \leq 120$
    $3x + 5y + 10z + 15w \leq 100$
    $x \geq 0, \quad y \geq 0$
    $z \geq 0, \quad w \geq 0.$

6. Maximize $15r + 10s + 14t$,

   subject to $30r + 50s + 40t \leq 80{,}000$
   $10r + 5s + 7t \leq 14{,}000$
   $4r + 4s + 6t \leq 7200$
   $14r + 7s + 10t \leq 21{,}000$
   $r \geq 0, \quad s \geq 0, \quad t \geq 0.$

5.6 Maximizing a Functional: The Simplex Method

The simplex method is a matrix procedure for solving linear programming problems with any number of unknown variables. In this section, we develop the method for maximum problems in standard form. We consider minimum problems in Section 5.7.

To clarify the steps involved, we first apply the simplex method to the linear programming problem given by

Maximize $4x + 10y + 6z$,

subject to $2x + 4y + z \leq 12$
$6x + 2y + z \leq 26$ \hfill (17)
$5x + y + 2z \leq 80,$

assuming x, y, and z are nonnegative.

The method is as follows.

Step 1 Use slack variables to convert the constraint inequalities to equations.

We showed how to add slack variables to a problem in Section 5.5. Performing this operation on Eq. (17), we obtain

Maximize $4x + 10y + 6z + 0s_1 + 0s_2 + 0s_3$,

subject to $2x + 4y + z + s_1 = 12$
$6x + 2y + z + s_2 = 26$ \hfill (18)
$5x + y + 2z + s_3 = 80,$

assuming $x, y, z, s_1, s_2,$ and s_3 are nonnegative.

Step 2 Write the augmented matrix for the constraint equations. Add to this matrix one additional row consisting of the *negatives* of the coefficients of the objective function. Whenever a suitable coefficient is not given (for example, the right side of the objective function), use zero.

For system (18), the appropriate augmented matrix is

$$\left[\begin{array}{cccccc|c} 2 & 4 & 1 & 1 & 0 & 0 & 12 \\ 6 & 2 & 1 & 0 & 1 & 0 & 26 \\ 5 & 1 & 2 & 0 & 0 & 1 & 80 \\ \hline -4 & -10 & -6 & 0 & 0 & 0 & 0 \end{array}\right]. \qquad (19)$$

For convenience, we use a horizontal line to separate the coefficients of the constraint equations from those of the objective function.

Step 3 Locate the most negative number in the last row and designate the column in which this number appears as the *work column*. If more than one equally negative number exists, choose one.

The elements in the last row of matrix (19) are -4, -10, -6, and 0. Here -10 is the most negative, so the second column becomes the work column.

Step 4 Form ratios by dividing each *positive* element of the work column into the element in the same row and last column. Designate the element in the work column that yields the smallest ratio as the pivot element. If more than one element yields the same smallest ratio, choose one.

For matrix (19), the ratios of interest are $12/4 = 3$, $26/2 = 13$, and $80/1 = 80$. Since $12/4$ is the smallest ratio, the element 4 becomes the pivot. Note that we did not consider the ratio $0/-10$ since -10 is not a positive element of the work column.

Step 5 Use elementary row operations with the pivot element to reduce all other elements in the work column to zero. Recall that the first step in such a procedure is to convert the pivot element to 1.

Applying elementary row operations to matrix (19) with the pivot element 4, we obtain

$$\left[\begin{array}{cccccc|c} \frac{1}{2} & 1 & \frac{1}{4} & \frac{1}{4} & 0 & 0 & 3 \\ 5 & 0 & \frac{1}{2} & -\frac{1}{2} & 1 & 0 & 20 \\ \frac{9}{2} & 0 & \frac{7}{4} & -\frac{1}{4} & 0 & 1 & 77 \\ \hline 1 & 0 & -\frac{7}{2} & \frac{5}{2} & 0 & 0 & 30 \end{array}\right]. \qquad (20)$$

Note that, after Step 5 has been completed, the elements of the work column consist of one 1 and all the rest 0s. We refer to any column having this form as a *reduced column*. Matrix (20) has three reduced columns, the second, fifth, and sixth.

Step 6 Repeat Steps 3 through 5 until there are no negative numbers in the last row.

The only negative entry in the last row of matrix (20) is $-\frac{7}{2}$, so column 3 becomes the new work column. Applying Step 4, we obtain the ratios $3/\frac{1}{4} = 12$, $20/\frac{1}{2} = 40$, and $77/\frac{7}{4} = 44$, hence the $\frac{1}{4}$ element becomes the new pivot. The ratio $30/-\frac{7}{2}$ was not considered, since $-\frac{7}{2}$ is not positive. Using the second elementary row operation to convert $\frac{1}{4}$ to 1 and the third elementary row operation to reduce the other elements in the new work column to 0, we obtain

$$\begin{bmatrix} 2 & 4 & 1 & 1 & 0 & 0 & | & 12 \\ 4 & -2 & 0 & -1 & 1 & 0 & | & 14 \\ 1 & -7 & 0 & -2 & 0 & 1 & | & 56 \\ \hline 8 & 14 & 0 & 6 & 0 & 0 & | & 72 \end{bmatrix}. \qquad (21)$$

There are no negative entries in the last row of matrix (21), and Step 6 is complete.

Step 7 The solution of the linear programming problem is obtained by assigning to each variable associated with a reduced column the value appearing in the last column and same row as the 1 element. Assign all other variables (those that do not correspond to a reduced column) the value zero.

The reduced columns in matrix (21) are the third, fifth, and sixth, which represent the coefficients of the variables z, s_2, and s_3. Thus $z = 12$, $s_2 = 14$, and $s_3 = 56$. Furthermore, since the first, second, and fourth columns are not reduced, we assign to their corresponding variables, x, y, and s_1, the value zero. The maximum value of the objective function is obtained either by substituting these values of x, y, and z into it, $4(0) + 10(0) + 2(12) = 72$, or, more directly, by reading the value from the last row and last column of matrix (21).

Example 1 Use the simplex method to solve Example 1 in Section 5.4.

Solution The mathematical model for this problem is

Maximize $7x + 22y$,

subject to $3x + 10y \leq 33{,}000$
$\qquad\qquad 5x + 8y \leq 42{,}000$,

assuming x and y are nonnegative.

Introducing the slack variables s_1 and s_2, as in Eqs. (15) and (16) in Section 5.5, we obtain the same problem in equation form as

Maximize $7x + 22y + 0s_1 + 0s_2$,

subject to $\quad 3x + 10y + s_1 \quad\quad = 33{,}000$
$\quad\quad\quad\quad\, 5x + 8y \quad\quad + s_2 = 42{,}000,$

assuming x, y, s_1, and s_2 are nonnegative.

The associated augmented matrix with the extra row for the objective function is

$$\left[\begin{array}{cccc|c} 3 & 10 & 1 & 0 & 33{,}000 \\ 5 & 8 & 0 & 1 & 42{,}000 \\ \hline -7 & -22 & 0 & 0 & 0 \end{array}\right]. \tag{22}$$

The work column is column 2, and the appropriate ratios are $33{,}000/10 = 3300$ and $42{,}000/8 = 5250$. Therefore the pivot element is 10. Using elementary row operations, we reduce matrix (22) to

$$\left[\begin{array}{cccc|c} 0.3 & 1 & 0.1 & 0 & 3300 \\ 2.6 & 0 & -0.8 & 1 & 15{,}600 \\ \hline -0.4 & 0 & 2.2 & 0 & 72{,}600 \end{array}\right]. \tag{23}$$

Repeating Steps 3 through 5 for matrix (23), we find that the new work column is column 1, the appropriate ratios are $3300/0.3 = 11{,}000$ and $15{,}600/2.6 = 6000$, and the new pivot element is 2.6. Using elementary row operations, we reduce matrix (23) to

$$\left[\begin{array}{cccc|c} 0 & 1 & 0.19 & -0.11 & 1{,}500 \\ 1 & 0 & -0.31 & 0.38 & 6{,}000 \\ \hline 0 & 0 & 2.08 & 0.15 & 75{,}000 \end{array}\right].$$

Step 6 is now complete. The reduced columns are the first and second, which correspond to the variables x and y. The solution is $x = 6000$, $y = 1500$, $s_1 = 0$, and $s_2 = 0$, with a maximum profit of \$75,000.

Example 2 Use the simplex method to solve the linear programming problem given by

Maximize $2a + 4b + c + d$,

subject to $\quad a + 3b \quad\quad + 4d \le 4$
$\quad\quad\quad\quad 2a + b \quad\quad\quad\quad \le 3$
$\quad\quad\quad\quad\quad\quad\; b + 4c + d \le 3,$

assuming a, b, c, and d are nonnegative.

Solution First, adding the three slack variables s_1, s_2, and s_3 to the three inequalities, respectively, as in Example 1 in Section 5.5, and then, forming the augmented matrix, we obtain

$$\left[\begin{array}{ccccccc|c} 1 & 3 & 0 & 4 & 1 & 0 & 0 & 4 \\ 2 & 1 & 0 & 0 & 0 & 1 & 0 & 3 \\ 0 & 1 & 4 & 1 & 0 & 0 & 1 & 3 \\ \hline -2 & -4 & -1 & -1 & 0 & 0 & 0 & 0 \end{array}\right].$$

The work column is column 2, and the pivot element is 3, since it yields the smallest number among the ratios 4/3, 3/1, and 3/1. Multiplying the first row by $\frac{1}{3}$ to make the pivot element 1 and reducing all other elements in the work column to 0 by elementary row operations, we compute

$$\left[\begin{array}{ccccccc|c} \frac{1}{3} & 1 & 0 & \frac{4}{3} & \frac{1}{3} & 0 & 0 & \frac{4}{3} \\ \frac{5}{3} & 0 & 0 & -\frac{4}{3} & -\frac{1}{3} & 1 & 0 & \frac{5}{3} \\ -\frac{1}{3} & 0 & 4 & -\frac{1}{3} & -\frac{1}{3} & 0 & 1 & \frac{5}{3} \\ \hline -\frac{2}{3} & 0 & -1 & \frac{13}{3} & \frac{4}{3} & 0 & 0 & \frac{16}{3} \end{array}\right].$$

The new work column is the third, and the pivot element is 4, since it is the only positive element in that column and therefore the only one with which we can form ratios. Multiplying the third row by $\frac{1}{4}$ to convert the pivot element to 1 and using elementary row operations to convert all other elements in the work column to 0, we obtain

$$\left[\begin{array}{ccccccc|c} \frac{1}{3} & 1 & 0 & \frac{4}{3} & \frac{1}{3} & 0 & 0 & \frac{4}{3} \\ \frac{5}{3} & 0 & 0 & -\frac{4}{3} & -\frac{1}{3} & 1 & 0 & \frac{5}{3} \\ -\frac{1}{12} & 0 & 1 & -\frac{1}{12} & -\frac{1}{12} & 0 & \frac{1}{4} & \frac{5}{12} \\ \hline -\frac{3}{4} & 0 & 0 & \frac{17}{4} & \frac{5}{4} & 0 & \frac{1}{4} & \frac{23}{4} \end{array}\right].$$

The next work column is column 1, and the new pivot element is $\frac{5}{3}$, since it yields the smallest number among the ratios $\frac{4/3}{1/3} = 4$ and $\frac{5/3}{5/3} = 1$. Using elementary row operations to convert $\frac{5}{3}$ to 1 and all other elements in the work column to 0, we calculate

$$\left[\begin{array}{ccccccc|c} 0 & 1 & 0 & \frac{8}{5} & \frac{2}{5} & -\frac{1}{5} & 0 & 1 \\ 1 & 0 & 0 & -\frac{4}{5} & -\frac{1}{5} & \frac{3}{5} & 0 & 1 \\ 0 & 0 & 1 & -\frac{3}{20} & -\frac{1}{10} & \frac{1}{20} & \frac{1}{4} & \frac{1}{2} \\ \hline 0 & 0 & 0 & \frac{73}{20} & \frac{11}{10} & \frac{9}{20} & \frac{1}{4} & \frac{13}{2} \end{array}\right].$$

Step 6 of the simplex method now is complete. The reduced columns are the first three; the solution is $a = 1$, $b = 1$, $c = \frac{1}{2}$, and $d = 0$ (also $s_1 = s_2 = s_3 = 0$), with a maximum value of the objective function equal to $\frac{13}{2}$.

Two modifications are required in Steps 1 through 7 if the simplex method is to work for all maximum problems in standard form.

Modification 1 If all elements in a work column are negative (in which case Step 4 cannot be performed), the linear programming problem will not have a solution.

Modification 2 If more than one reduced column has the 1 element in the same row after Step 6 is completed, the linear programming problem will have infinitely many solutions. Some solutions are obtained by assigning to one of the variables (your choice) the value in the last column and same row as the 1 element and assigning the value 0 to the other corresponding variables. Once two complete solutions are found, every point on the straight-line segment between these points is also a solution.

Example 3 Solve the linear programming problem given by

Maximize $2x + 7y$,

subject to $\quad -x + y \le 3$
$\qquad\qquad -x - y \le -5,$

assuming x and y are nonnegative.

Solution Adding slack variables s_1 and s_2, respectively, to the two constraint inequalities and then forming the augmented matrix for this new system, we obtain

$$\begin{bmatrix} -1 & 1 & 1 & 0 & | & 3 \\ -1 & -1 & 0 & 1 & | & -5 \\ -2 & -7 & 0 & 0 & | & 0 \end{bmatrix}.$$

The first pivot element is the 1 element in the 1-2 position. Applying the elementary row operations, we compute

$$\begin{bmatrix} -1 & 1 & 1 & 0 & | & 3 \\ -2 & 0 & 1 & 1 & | & -2 \\ -9 & 0 & 7 & 0 & | & 21 \end{bmatrix}.$$

The new work column is the first, but now Step 4 cannot be performed since all the elements in that column are negative. It follows from Modification 1 that the original problem does not have a solution.

From the geometry of the problem, the result from Example 3 is expected. The region defined by the inequalities, shaded in Figure 5.19, is unbounded. No matter what values of x and y are chosen, other values of the variables further

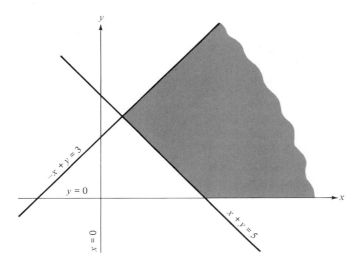

FIGURE 5.19

from the origin always exist which yield an even greater value of the objective function.

Example 4 Solve the linear programming problem defined by

Maximize $3x + 2y + 2z$,

subject to $2x + 2y + z \leq 10$
$x + y + 2z \leq 4$,

assuming x, y, and z are nonnegative.

Solution The augmented matrix for this system is

$$\begin{bmatrix} 2 & 2 & 1 & 1 & 0 & | & 10 \\ 1 & 1 & 2 & 0 & 1 & | & 4 \\ -3 & -3 & -2 & 0 & 0 & | & 0 \end{bmatrix}.$$

Here we have a choice for the first work column; it is immaterial which one we choose. Arbitrarily choosing column 1, we then use the 2-1 element as the pivot. Applying the elementary row operations, we compute

$$\begin{bmatrix} 0 & 0 & -3 & 1 & -2 & | & 2 \\ 1 & 1 & 2 & 0 & 1 & | & 4 \\ 0 & 0 & 4 & 0 & 3 & | & 12 \end{bmatrix}.$$

It follows that $z = 0$, $s_2 = 0$, $s_1 = 2$, and either x or y equals 4. Using Modification 2, we first set $x = 4$ and $y = 0$ to obtain one solution, and then set $x = 0$ and $y = 4$ to obtain a second solution. One complete solution to the original problem is $x = 4$, $y = 0$, and $z = 0$ (also $s_1 = 2$ and $s_2 = 0$), while a second complete solution is $x = 0$, $y = 4$, and $z = 0$ (again with $s_1 = 2$ and $s_2 = 0$). Furthermore, every point on the line segment between (4, 0, 0) and (0, 4, 0) is also a solution. The maximum value of the objective function for any of these points is 12.

Finally, it should be noted that the simplex method does not always lead to all solutions of a linear programming problem when more than one solution exists. As an example, consider the program given by

Maximize $3x + 3y$,

subject to $\quad x + y \leq 5$
$\qquad\qquad 2x + y \leq 8$,

assuming x and y are nonnegative.

The augmented matrix for this problem is

$$\left[\begin{array}{cccc|c} 1 & 1 & 1 & 0 & 5 \\ 2 & 1 & 0 & 1 & 8 \\ \hline -3 & -3 & 0 & 0 & 0 \end{array}\right].$$

We have a choice for the first work column. Selecting the second column (which avoids fractions) and pivoting on the 1-2 element, we compute

$$\left[\begin{array}{cccc|c} 1 & 1 & 1 & 0 & 5 \\ 1 & 0 & -1 & 1 & 3 \\ \hline 0 & 0 & 3 & 0 & 15 \end{array}\right].$$

The solution is $x = 0$ and $y = 5$ (with $s_1 = 0$ and $s_2 = 3$). Solving the same problem with the geometric method developed in Section 5.3, we find that the problem has many solutions, all lying on the line segment between $x = 0$, $y = 5$ and $x = 2$, $y = 3$.

This apparent weakness in the simplex method, not always producing all solutions, is of little consequence as a practical matter. The objective was to maximize a quantity, perhaps profit. The simplex method achieves the objective. In the last example, we found a solution that produced a maximum value for the objective function of 15. Other solutions may exist, but they will not improve the objective.

Exercises

Use the simplex method to solve the following linear programming problems.
1. Exercise 1 in Section 5.3.
2. Exercise 3 in Section 5.3.
3. Exercise 4 in Section 5.3.
4. Exercise 5 in Section 5.3.
5. Maximize $x + 2y + 4z$,

    subject to $\quad x + 3y + 2z \leq 30$
    $\qquad\qquad\quad 2x + y + 3z \leq 12$,

    assuming x, y, and z are nonnegative.
6. Maximize $x + y + z$,

    subject to $\quad x + y + z \leq 12$
    $\qquad\qquad\quad 2x + 2y + z \leq 16$,

    assuming x, y, and z are nonnegative.
7. Maximize $15x + 10y + 14z$,

    subject to $\quad 6x + 5y + 3z \leq 26$
    $\qquad\qquad\quad 4x + 2y + 5z \leq 8$,

    assuming x, y, and z are nonnegative.
8. Maximize $2x + 3y + 4z + 6w$,

    subject to $\quad x + y + z + w \leq 15$
    $\qquad\qquad\quad 7x + 5y + 3z + 2w \leq 120$
    $\qquad\qquad\quad 3x + 5y + 10z + 15w \leq 100$,

    assuming x, y, z, and w are nonnegative.
9. Maximize $5a + 2b + 3c + d$,

    subject to $\quad 2a + b + 5c + d \leq 10$
    $\qquad\qquad\quad 3a + b + 3c + d \leq 12$
    $\qquad\qquad\quad a + 2b - c - 2d \leq 3$.

5.7 Minimizing a Functional: The Simplex Method

Every minimum linear programming problem in standard form can be converted to a maximum problem in standard form through the matrix operation of transposition first introduced in Section 4.2. The new problem is called the *dual* of the original problem.

The first step in converting a minimum problem to a maximum problem is to write an augmented matrix for the given minimum problem, similar to Step 2 of the simplex method but with two exceptions. First, no slack variables are introduced and, second, the coefficients of the objective function are placed in the augmented matrix as they appear and not as their negatives. For the problem defined by

Minimize $12x + 26y + 80z$,

subject to $\quad 2x + 6y + 5z \geq 4$
$\quad\quad\quad\quad 4x + 2y + z \geq 10$ \hfill (24)
$\quad\quad\quad\quad x + y + 2z \geq 6$,

assuming x, y, and z are nonnegative. The augmented matrix is

$$\begin{bmatrix} 2 & 6 & 5 & | & 4 \\ 4 & 2 & 1 & | & 10 \\ 1 & 1 & 2 & | & 6 \\ \hline 12 & 26 & 80 & | & 0 \end{bmatrix}. \tag{25}$$

For the problem defined by

Minimize $4x + 3y + 3z$,

subject to $\quad x + 2y \quad\quad \geq 2$
$\quad\quad\quad\quad 3x + y + z \geq 4$
$\quad\quad\quad\quad\quad\quad\quad 4z \geq 1$ \hfill (26)
$\quad\quad\quad\quad 4x + \quad\quad z \geq 1$,

assuming x, y, and z are nonnegative. The augmented matrix is

$$\begin{bmatrix} 1 & 2 & 0 & | & 2 \\ 3 & 1 & 1 & | & 4 \\ 0 & 0 & 4 & | & 1 \\ 4 & 0 & 1 & | & 1 \\ \hline 4 & 3 & 3 & | & 0 \end{bmatrix}. \tag{27}$$

Taking the transpose of such an augmented matrix yields the augmented matrix (again without any slack variables and the coefficients of the objective function entered as they appear and not as their negatives) for the dual maximum problem. As an illustration, consider system (24). The augmented matrix for this system is matrix (25), having as its transpose

$$\begin{bmatrix} 2 & 4 & 1 & | & 12 \\ 6 & 2 & 1 & | & 26 \\ 5 & 1 & 2 & | & 80 \\ \hline 4 & 10 & 6 & | & 0 \end{bmatrix}. \tag{28}$$

The maximum problem modeled by matrix (28) is

Maximize $4X + 10Y + 6Z$,

subject to
$$\begin{aligned} 2X + 4Y + Z &\leq 12 \\ 6X + 2Y + Z &\leq 26 \\ 5X + Y + 2Z &\leq 80, \end{aligned} \qquad (29)$$

assuming X, Y, and Z are nonnegative. System (29) is called the dual of system (24). For emphasis, we use capital letters for the variables of the dual system.

As another example, consider the minimum problem defined by system (26). The augmented matrix for that system is given by matrix (27) with transpose

$$\begin{bmatrix} 1 & 3 & 0 & 4 & | & 4 \\ 2 & 1 & 0 & 0 & | & 3 \\ 0 & 1 & 4 & 1 & | & 3 \\ \hline 2 & 4 & 1 & 1 & | & 0 \end{bmatrix}. \qquad (30)$$

The maximum problem in standard form modeled by matrix (30) is

Maximize $2X + 4Y + Z + W$,

subject to
$$\begin{aligned} X + 3Y + 4W &\leq 4 \\ 2X + Y &\leq 3 \\ Y + 4Z + W &\leq 3, \end{aligned} \qquad (31)$$

assuming X, Y, Z, and W are nonnegative. System (31) is the dual of system (26).

The procedure for solving a minimum linear programming problem in standard form is as follows.

1. Find the dual maximum problem.
2. Apply the simplex method through Step 6 to the dual maximum problem.
3. The solution to the original minimum problem is found in the last row of the final matrix obtained with the simplex method in the columns associated with the slack variables.

Example 1 Solve system (24).

Solution The dual maximum problem is given by system (29). Applying the simplex method to this system [see expressions (17) through (21) in Section 5.6], we obtain

$$\begin{bmatrix} 2 & 4 & 1 & 1 & 0 & 0 & | & 12 \\ 4 & -2 & 0 & -1 & 1 & 0 & | & 14 \\ 1 & -7 & 0 & -2 & 0 & 1 & | & 56 \\ \hline 8 & 14 & 0 & 6 & 0 & 0 & | & 72 \end{bmatrix}.$$

Columns associated with the slack variables: columns 3, 4, 5 (indicated by overbrace).

Solution to the minimum problem (indicated by underbrace on bottom row).

The solution to the minimum problem is $x = 6$, $y = 0$, and $z = 0$, with a minimum value of the objective function equal to 72.

Example 2 Solve system (26).

Solution The dual maximum problem is given by system (31). Applying the simplex method to (31) (see Example 2 in Section 5.6), we obtain the matrix

$$\begin{bmatrix} 0 & 1 & 0 & \frac{8}{5} & \frac{2}{5} & -\frac{1}{5} & 0 & | & 1 \\ 1 & 0 & 0 & -\frac{4}{5} & -\frac{1}{5} & \frac{3}{5} & 0 & | & 1 \\ 0 & 0 & 1 & -\frac{3}{20} & -\frac{1}{10} & \frac{-1}{20} & \frac{1}{4} & | & \frac{1}{2} \\ \hline 0 & 0 & 0 & \frac{73}{20} & \frac{11}{10} & \frac{9}{20} & \frac{1}{4} & | & \frac{13}{2} \end{bmatrix}.$$

Columns associated with the slack variables.

Solution to the minimum problem.

The solution to the minimum problem is $x = \frac{11}{10}$, $y = \frac{9}{20}$, and $z = \frac{1}{4}$, with a minimum value of the objective function equal to $\frac{13}{2}$.

As long as the dual problem has a solution, so will the original minimum problem. If the dual does not have a solution, that is, if Modification 1 of the simplex method is required, the original minimum problem also will not have a solution.

Exercises

Find the dual maximum problem in standard form for each of the minimum problems given in Exercises 1 through 5.
1. Exercise 3 in Section 5.5.

2. Exercise 2 in Section 5.3.
3. Exercise 6 in Section 5.3.

Use the simplex method to solve the following linear programming problems.
4. Minimize $33{,}000x + 42{,}000y$,

   subject to $\quad 3x + 5y \geq 7$
   $\qquad\qquad\; 10x + 8y \geq 22$,

   assuming x and y are nonnegative.
5. Minimize $80{,}000x + 14{,}000y + 7200z$,

   subject to $\quad 30x + 10y + 4z \geq 40$
   $\qquad\qquad\; 50x + 5y + 4z \geq 60$,

   assuming x, y, and z are nonnegative.
6. Minimize $240{,}000x + 47{,}200y$,

   subject to $\quad 60x + 2y \geq 3$
   $\qquad\qquad\; 10x + 5y \geq 2$,

   assuming x and y are nonnegative.
7. Minimize $22{,}400r + 14{,}400s$,

   subject to $\quad 4r + 6s \geq 15$
   $\qquad\qquad\; 8r + 4s \geq 30$,

   assuming r and s are nonnegative.
8. Minimize $12a + 16b$,

   subject to $\quad a + 2b \geq 1$
   $\qquad\qquad\; 3a + b \geq 2$
   $\qquad\qquad\; 2a + 3b \geq 4$,

   assuming a and b are nonnegative.
9. Minimize $15x + 120y + 100z$,

   subject to $\quad x + 7y + 3z \geq 2$
   $\qquad\qquad\; x + 5y + 5z \geq 3$
   $\qquad\qquad\; x + 3y + 10z \geq 4$
   $\qquad\qquad\; x + 2y + 15z \geq 6$,

   assuming x, y, and z are nonnegative.

CALCULUS

Part 2

Chapter

The Derivative

In this chapter, we begin our study of the derivative. The derivative has many commercial applications in economics, operations research, and production scheduling. It can be used to solve a large class of optimization problems that cannot be handled with the linear programming techniques developed in Chapter 5. We reserve our study of these applications, however, until Chapter 7. Here we direct our attention to a graphical interpretation of the derivative as a rate of change, and to the specific rules for calculating derivatives. Such considerations necessarily begin with the concept of a function.

6.1 Concept of a Function

Table 6.1 illustrates a relationship between the years 1970 through 1975 and the number of cars sold during each year by Village Distributors, a small new car dealer. Each year's sales are arranged under the corresponding year and the relationship between the two quantities, year and number of cars sold, is clear.

TABLE 6.1

Year	1970	1971	1972	1973	1974	1975
Cars sold	160	145	155	102	95	151

The same clarity is not evident in Report 6.1.

Report 6.1

Summary of new car sales:

During the years 1970 through 1975, Village Distributors had a spotty sales record. On three occasions, sales went over the 150 mark (151, 155, and 160), but during other years they fell under 150, twice drastically (95 and 102) and once marginally (145).

Report 6.1 is wordier than Table 6.1, but it is less useful. The correspondence between the individual years and the number of cars sold during each year is missing. Obviously, this correspondence is important.

The notion of two distinct sets of quantities (like years and number of cars sold) and a rule of correspondence between the sets (achieved in Table 6.1 by arranging corresponding entries under each other) is central to the concept of a function. In fact, it describes a function.

Definition 6.1 A *function* is a rule of correspondence between two sets, which assigns to each element in the first set exactly one element (but not necessarily a different one) in the second set.

A function therefore has three components: a first set (perhaps years), a second set (perhaps numbers), and a rule of correspondence between the two sets. This rule must be complete in that an assignment must be made to each and every element of the first set. As an example, take the first set to be all the people in the world, take the second set to be all positive numbers, and use the rule, "Assign to each person his or her exact weight." This is a function. We have two sets and a rule which assigns to every element in the first set (people) exactly one element in the second set (his or her weight).

As a second example, take the first set to all be the cars in the world, take the second set to be all the colors, and use the rule of correspondence, "Assign to each car its color." This is *not* a function. Although we have two valid sets, the rule cannot handle a car with a red body and a white roof. Such a car must be assigned *two* colors, red and white, and this is not valid for function rules.

For a function, the rule of correspondence must assign exactly *one* element of the second set to each element of the first set.

It is often useful to visualize a function as a machine. The machine is programmed to (1) accept elements from the first set, (2) transform these elements according to the rule it has been given, and finally (3) emit the result which is an element of the second set. But the machine is incapable of thought. If it receives an input for which the rule does not apply, it will break down. Functions act similarly. The rule of correspondence must be capable of assigning one element of the second set to each input element of the first set. If the rule cannot do this, it is not a legitimate function.

In the people-weight example above, we feed the machine a person and it emits a number, that person's weight. The machine can handle any person we feed it. In the car-color example, however, we feed the machine a car. If the car has two colors, the machine will have trouble. Since it cannot think, it will not know which color to assign to a red and white car. It will break down.

The people-weight example is interesting in another respect. Although each person is assigned one number, all the assigned numbers are not necessarily different. Two different people can have the same weight. Nonetheless, we still have a function. Definition 6.1 requires only that each element of the first set be assigned one element of the second set, but *not necessarily* a *different one*.

It is too wordy to refer constantly to the two sets under consideration as the first set and the second set. More commonly the first set is called the *domain*, and the second set is called the *range*.

Example 1 Determine whether or not the two sets of numbers displayed in Figure 6.1 and the correspondence illustrated by the arrows is a function.

Solution Since each element of the domain has exactly one element of the

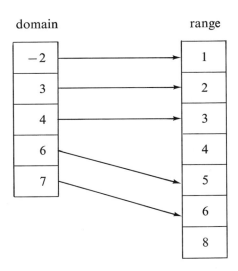

FIGURE 6.1

210 6 The Derivative

range assigned to it, the relationship indicated is a function. Although some elements in the second set are not paired with elements in the first set, this does not matter. Two sets are given, and every element in the first set has assigned to it one element in the second set.

Example 2 Determine whether or not the two sets of numbers displayed in Figure 6.2 and the correspondence illustrated by the arrows is a function.

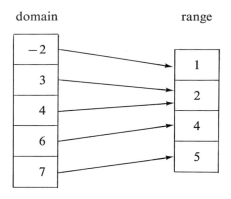

FIGURE 6.2

Solution Since all the conditions specified in Definition 6.1 are satisfied (we have two sets and each element in the domain is assigned one element in the range), this relationship is a function. As in the people-weight example we considered previously, some elements of the range are assigned twice. Again, this is of no consequence.

Example 3 Determine whether or not the relationship illustrated in Figure 6.3 is a function.

Solution This relationship is not a function, since one element in the domain

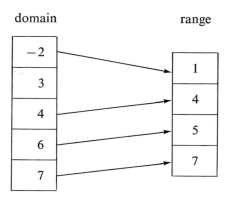

FIGURE 6.3

is not matched with any element in the range. Definition 6.1 requires that each and every element of the first set be assigned some element in the range.

The importance of clearly designating which set is the domain and which is the range cannot be underestimated. To illustrate the pitfalls involved in interchanging their roles, we return to the function defined in Example 2 and reverse the roles of the two sets. Since rules of correspondence act on the elements in the domain, we also reverse the direction of the arrows. The results are illustrated in Figure 6.4. The relationship given in Figure 6.4 is not a function. Here element 2 in the domain has *two* elements in the range assigned to it. Definition 6.1 requires that each element in the domain be assigned *one* element in the range.

The selection of which set is to be called the domain and which set is to be called the range is left to the discretion of the person defining the relationship. In general, however, the decision is based on the context of the application. Only after the selection is made and the rule of correspondence given can a decision be made as to whether or not the components form a function.

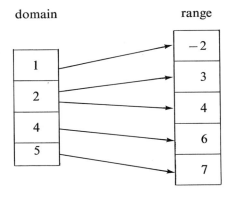

FIGURE 6.4

Exercises

In Exercises 1 through 14, determine whether or not the given relationships are functions.

1.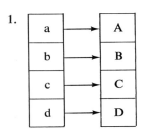

2.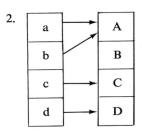

3.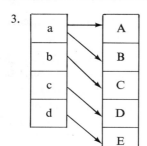

4.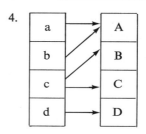

5.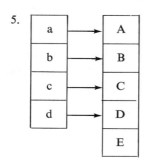

6.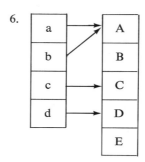

7.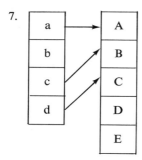

8.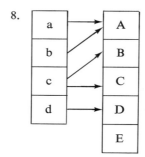

9.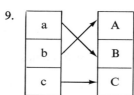

10.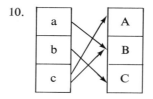

11.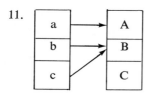

12.

x	1	2	3	4
y	6	7	8	9

13.

x	1	2	3	4
y	6	6	7	8

14.

x	1	1	3	4
y	6	7	8	9

15. Determine whether or not the following correspondences constitute functions.
    (a) The correspondence between students in a class and their heights.
    (b) The correspondence between students in a class and their names.
    (c) The correspondence between names and students in a class.
    (d) The correspondence between stocks listed on the New York Stock Exchange and their closing prices.
    (e) The correspondence between closing stock prices and stocks listed on the New York Stock Exchange.
    (f) The correspondence between all the cars in the United States and their colors.
    (g) The correspondence between the prime interest lending rate and the banks located in San Francisco.

6.2 Mathematical Functions

We know that a function consists of three components: a domain, a range, and a rule. The domain and range can be any two sets (people, cars, colors, numbers, etc.), while the rule can be given in a variety of ways (arrows, tables, words, etc.). In business situations, the primary concern is with sets of numbers (representing price, demand, advertising expenditures, cost, or profit) and rules defined by mathematical equations.

At first glance, it may seem strange to think of an equation as a rule, but it is. Consider two identical sets of real numbers and the equation $y = 15x + 10$, where x represents a number in the domain and y represents a number in the range. The equation is nothing more than the rule, "Multiply each element in the domain by 15 and add 10 to the result." Similarly, the equation $y = x^2 - 7$ is the rule, "Square each element in the domain and then subtract 7 from the result."

Whenever we have two sets of numbers and a rule given by an equation where the variable x denotes an element in the domain and the variable y denotes an element in the range, we simply say that y *is a function of* x and write $y = f(x)$. If the elements in the domain are denoted by some other variable, say P, and if the elements in the range are denoted by D, we write $D = f(P)$, read "D is a function of P." Care must be taken, however, not to interpret the notation $y = f(x)$ as "y equals f times x." Simply put, $y = f(x)$ is shorthand notation for the rather cumbersome statement, "We have two sets of numbers and a rule which satisfies Definition 6.1; the rule is given by a mathematical equation where x and y denote elements in the domain and range, respectively."

Even this shorthand notation often is simplified. Rather than saying "given the function $y = f(x)$," it is common to say "given the function $f(x)$." For example, the function $f(x) = x^2$ denotes two sets of real numbers and the rule that assigns to each element in the domain its square. Here the value $x = 2$ is assigned its square, $2^2 = 4$, the value $x = 3$ is assigned its square, $3^2 = 9$, and the value $x = 7$ is assigned its square, $7^2 = 49$. We can calculate the corresponding value for any x-value simply by replacing x in $f(x) = x^2$ by the particular value of x of interest. Effectively, the variable x is only a placeholder.

More generally, given any function $f(x)$, the value of the function for a particular value of x is found by replacing x with that particular value. Notationally, $f(2)$ represents the value of the function $f(x)$ at $x = 2$, $f(3)$ represents the value of the function $f(x)$ at $x = 3$, and $f(7)$ represents the value of the function $f(x)$ at $x = 7$.

Example 1 Find $f(2)$, $f(0)$, and $f(-1)$ if $f(x) = 2x^2 - 3x + 4$.

Solution This function assigns to each value of x the number obtained by squaring x and multiplying by 2, subtracting from the result x times 3, and finally adding 4. We are interested in values of this function for $x = 2$, $x = 0$, and $x = -1$. In particular,

$$f(2) = 2(2)^2 - 3(2) + 4 = 8 - 6 + 4 = 6$$
$$f(0) = 2(0)^2 - 3(0) + 4 = 0 - 0 + 4 = 4$$

and

$$f(-1) = 2(-1)^2 - 3(-1) + 4 = 2 + 3 + 4 = 9.$$

Since x is only a placeholder, it can be replaced by any other quantity, as long as we are consistent in replacing every x by the new quantity. For example, if the function $f(x) = 8x$ is to be evaluated at $x =$ hot dog, then f(hot dog) $=$ 8 hot dog. Similarly for $x = \Delta x$, where Δx is simply a combination of the characters Δ and x, $f(\Delta x) = 8\Delta x$. Also $f(\$ \ @ \ q) = 8\$ \ @ \ q$. Although these last examples are contrived, they serve to indicate the placeholder quality of the variable x.

Example 2 Find $f(-2)$, $f(z)$, $f(h + 1)$, and $f(\Delta x)$ if $f(P) = P^3 - 2P$.

Solution This rule assigns to each value of P the value obtained by cubing P and subtracting from the result 2 times the value of P. In particular,

$$f(-2) = (-2)^3 - 2(-2) = -8 + 4 = -4$$
$$f(z) = (z)^3 - 2z = z^3 - 2z$$
$$f(h + 1) = (h + 1)^3 - 2(h + 1) = (h^3 + 3h^2 + 3h + 1) - (2h + 2)$$
$$= h^3 + 3h^2 + h - 1$$

and

$$f(\Delta x) = (\Delta x)^3 - 2\Delta x.$$

In any function, one variable always depends on the value of another variable. We select a variable from the domain first, independent of any other quantity, and then use the rule of correspondence to find the value of the second variable in the range. Not surprisingly, therefore, the variable in the domain is called the *independent variable*, and the variable in the range is called the *dependent variable*.

One cannot always tell by looking at an equation which variable is the independent variable and which is the dependent variable. Since the independent variable comes from the domain, while the dependent variable represents an element in the range, the decision is equivalent to determining which set is the domain and which is the range. As mentioned in Section 6.1, this choice is dictated by the physical situation the equation models.

As an example, consider a hypothetical equation which relates coal production to steel production. For the coal company, the amount of coal mined depends on the demand from the steel manufacturers who use the coal to make steel. The amount of steel produced, a decision made independently of the coal company, dictates the amount of coal to be mined. Here steel production is the independent variable, and coal production is the dependent variable. The situation is different, however, from the viewpoint of the steel producer. No coal means no steel; the amount of steel depends heavily on the amount of available coal. To the steel producer, coal production may be the independent variable, especially if coal production cannot meet present demand.

Once the decision has been made as to which quantity is the independent variable and which is the dependent variable, and again we stress that this choice is dictated by the physical situation, we can use our notation to communicate this decision to others. Conventionally, the equation $y = f(x)$ indicates that x is the independent variable and y is the dependent variable. Similarly, $D = f(P)$ indicates that P is the independent variable and D is the dependent variable. Note that $D = f(P)$ is even read "D is a function of (depends on) P."

To this point, we have said very little about the domain and the range other than that they must exist, and for mathematical functions they must be sets of numbers. In many business situations, this is not restrictive enough.

Consider the following demand-price equation for oranges applicable to a particular grocery store:

$$D = P^2 - 34P + 289, \tag{1}$$

which is plotted in Figure 6.5. Recall that a demand-price equation relates the price of a product, denoted here as P, to the demand D. As illustrated in the

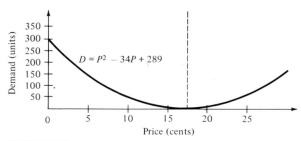

FIGURE 6.5

graph, the demand for oranges decreases as the price increases from 0 to 17¢. Beyond 17¢ however, a strange relationship occurs: The demand for oranges increases with an increase in price. If the store owner blindly uses Eq. (1), he or she will come to the ridiculous conclusion that a price of 15¢ per orange will result in a demand for 4 oranges and a price of $10 (1000¢) per orange will result in a demand for 966,289 oranges.

Obviously something is wrong. Whereas Eq. (1) represents a valid relationship between price and demand for certain prices, specifically prices between 0 and 17¢, it is not a reasonable relationship for other prices, such as $10 per orange. In representing this relationship without some indication of the values for P for which it is valid, we have neglected important information. A more complete description is

$$D = P^2 - 34P + 289 \qquad (P \text{ an integer lying between 0 and 17, inclusive}). \qquad (2)$$

The restriction, "P an integer lying between 0 and 17, inclusive," is a restriction on the domain. That is, only integers between 0 and 17 are considered the domain and are to be matched with elements of the second set of demands.

The domain therefore is nothing more than the set of all *allowable* values from which we can select the independent variable. If the domain is not given explicitly, it is taken to be all real numbers. One exception occurs when it is clear from the given equation that certain numbers cannot be values of the independent variable.

Example 3 Determine the domain for the function $y = 1/(x - 2)$.

Solution An admissible domain for x is all real numbers except 2, since the function is not defined at $x = 2$. A more complete description for this function is

$$y = \frac{1}{x - 2} \qquad (x \text{ any real number except 2}).$$

For mathematical functions, we usually define the range to be the set of all numbers between minus infinity and plus infinity. Typically, not all these values are used by a given equation, but Definition 6.1 does not stipulate that every value in the range must be assigned. In Example 1 in Section 6.1 the value 4 was not used. Note that we can always pinpoint the values of the dependent variable used by applying the rule of correspondence to all the elements in the domain.

Example 4 Determine (a) the independent variable, (b) the dependent variable, (c) the domain, and (d) the range for the function given by $y = f(x) = 30x^2 + 20x + 10, 0 \leq x \leq 10$.

Solution (a) x is the independent variable. (b) y is the dependent variable. (c) The domain of the function consists of all numbers between 0 and 10. (d) The range of the function consists of all numbers between plus and minus infinity. Note, however, that only y-values lying between 10 and 3210 are actually used by this function.

Exercises

1. Consider the function $D = z^2 - 30z + 225, 0 \leq z \leq 10$, z an integer.
   (a) What is the domain of the function? (List the values.)
   (b) What is the range of the function?
   (c) What values of the range are actually taken by D? (List them.)
2. The following equations relate values of x to values of y. For each equation list a possible domain and range to qualify the sets of numbers and the equation as a function.
   (a) $y = 3 + 4x$
   (b) $y = 1/(x - 3)$
   (c) $y = 10 + 18x^2$
   (d) $y = x^2 - 1/x$
   (e) $y = (x + 4)/(x + 2)$
   (f) $f(x) = (x + 2)/(x + 3)(x - 6)$
   (g) $f(x) = x^2 + 3x + 10$
   (h) $f(x) = (x^2 + 3)/(x + 5)$.
3. Determine whether or not the relationship defined by $y = 2 + 3x, 0 \leq x \leq \infty$ is a function. Determine whether the inverse relationship is a function. (*Hint:* Solve for x in terms of y.)
4. (a) Determine whether or not the relationship $y = +\sqrt{x}$ is a function for $0 \leq x < \infty$.
   (b) Determine whether or not the relationship $y = \pm\sqrt{x}$ is a function for $0 \leq x < \infty$.
5. Determine whether or not $z = 2w^2 + 4, 0 \leq w \leq 4$, is a function. Is the inverse relationship a function? (Solve for w in terms of z using the quadratic formula given in Section 1.5.)
6. Given the function $f(x) = x^2 + 2x + 3$, find (a) $f(2)$, (b) $f(0)$, (c) $f(10)$, (d) $f(-1)$, and (e) $f(1/x)$.
7. Given the function $y = f(x) = x^2 + 3x - 6$, find (a) $f(2)$, (b) $f(5)$, (c) $f(0)$, (d) $f(a + b)$, (e) $f(x + \Delta x)$.
8. Given the function $f(x) = x^3 + 6x - 4$, find (a) $f(0)$, (b) $f(1)$, (c) $f(3)$, (d) $f(5)$, (e) $f(a + b)$, (f) $f(x^2)$, (g) $f(x + \Delta x)$.
9. Given the function $f(a) = a + 2a^2 + a^3$, find (a) $f(2)$, (b) $f(d)$, (c) $f(x + y)$, (d) $f(2a)$.

10. A store owner has determined that the demand for a particular brand of specialty shoes is related to price by the equation $D = P^2 - 28P + 196$. Determine a domain for this equation so that the resulting function represents a plausible demand curve.

6.3 Average Rate of Change

An extremely useful measure in business forecasting is the rate at which a quantity changes. For example, it is useful in predicting sales for the month of April to know that sales in March totaled 10,000 units. Even more useful, however, is the additional information that sales have been increasing on the average at the rate of 2000 units per month. Similarly, a company needs information on wage scales for the past year before it can prepare next year's budget. In addition, however, information on the average yearly increase in salaries is also important. In both these situations a knowledge of the rates that quantities change, either changes in sales per month or changes in wages paid per hour, increases one's ability to forecast future requirements accurately.

A rate of change measures the change in one quantity associated with a change in a second quantity. If an individual traveled 500 miles in 10 hours, the average change in distance with respect to time is 50 miles per hour. That is, on the average, each hourly increase in driving time resulted in an increase of 50 miles traveled. We calculated this average rate of change by dividing the total change in miles driven by the total change in hours traveled.

In general, we define the average rate of change in one quantity with respect to a second quantity as the ratio of the two changes. Denoting the first and second quantities by y and x, respectively, we have

$$\text{Average rate of change in } y \text{ with respect to } x = \frac{\text{change in } y}{\text{change in } x}. \tag{3}$$

Example 1 Using the information presented in Figure 6.6, find the average rate of change in sales with respect to months over the 3-month period, April through June.

Solution Reading directly from the graph, we find that total sales as of April 1 were 10,000 units, while total sales at the end of June (July 1) were 44,000. The change in sales is $44,000 - 10,000 = 34,000$, and the change in time is 3 months. Therefore

$$\text{Average rate of change} = \frac{34,000}{3} = 11,333 \text{ units per month.}$$

Example 2 Redo Example 1 for the 3-month period, June through August.

Solution Reading directly from Figure 6.6, we find that total sales as of June 1 were 26,000, while total sales at the end of August (September 1) were 69,000.

6.3 Average Rate of Change **219**

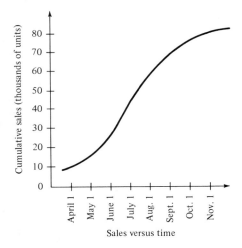

FIGURE 6.6

The change is 69,000 − 26,000 = 43,000, and the change in time is 3 months. Therefore

Average rate of change = $\dfrac{43{,}000}{3}$ = 14,333 units per month.

Examples 1 and 2 together illustrate the point that average rates of changes depend on the intervals under consideration. Different intervals can result in different average rates of change.

Example 3 Using the information presented in Figure 6.7, determine the

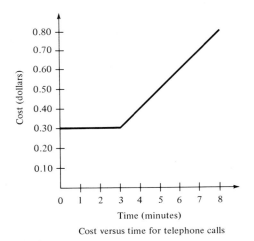

FIGURE 6.7

average rate of change in telephone costs with respect to minutes for (a) the first 3 minutes and (b) the second 3 minutes of a 6-minute call.

Solution (a) Over the first 3 minutes the final cost and the initial cost are both the same, 30¢. The change in cost is therefore $30 - 30 = 0$¢. The time span is 3 minutes, so the average rate of change $= 0/3 = 0$¢ per minute. That is, there is no rate change over the first 3 minutes. Regardless of the amount of time spent (up to 3 minutes) the final cost remains the same, 30¢. (b) Over the second 3-minute interval, the final cost and initial cost are 60¢ and 30¢, respectively; the change is $60 - 30 = 30$¢. Thus the average rate of change $= 30/3 = 10$¢ per minute. Over this interval, each additional minute results in an additional average charge of 10¢.

In computing rates of change, we have tacitly assumed that there was a relationship or rule of correspondence between the two quantities involved. That is, we assumed that the first was a function of the second. In the previous examples, we assumed that distance was a function of time (the more time spent traveling, the greater the distances covered), that sales were a function of months, and that telephone costs were a function of minutes spent making a call. In Eq. (3), we assumed that y was a function of x.

When y is a known function of x, that is, $y = f(x)$, we can rewrite Eq. (3) somewhat more neatly. Suppose we are interested in the average rate of change over the interval $[x_1, x_2]$ where the notation $[x_1, x_2]$ denotes all values of x between x_1 and x_2 inclusive. The change in x is simply $x_2 - x_1$. At x_1 the associated value of y is $y_1 = f(x_1)$, whereas at x_2 the associated value of y is $y_2 = f(x_2)$. The change in y is $y_2 - y_1$ or, equivalently, $f(x_2) - f(x_1)$. Therefore

$$\text{Average rate of change in } y \text{ with respect to } x = \frac{y_2 - y_1}{x_2 - x_1} \qquad (4)$$

or, equivalently,

$$\text{Average rate of change in } y \text{ with respect to } x = \frac{f(x_2) - f(x_1)}{x_2 - x_1}. \qquad (5)$$

Example 4 Determine the average rate of change in the function illustrated in Figure 6.8 between the points $x = 1$ and $x = 4$.

Solution At the points $x_1 = 1$ and $x_2 = 4$ the corresponding y-values are $y_1 = 3$ and $y_2 = 12$. Over the 3-unit interval from $x = 1$ to $x = 4$, the value of the function (y-values) changes 9 units. Using Eq. (4) we obtain

$$\text{Average rate of change} = \frac{12 - 3}{4 - 1} = \frac{9 \text{ units of } y}{3 \text{ units of } x} = 3 \text{ units of } y \text{ per unit of } x.$$

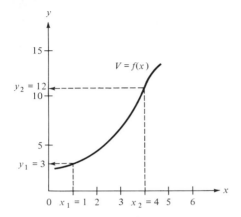

FIGURE 6.8

If the relationship between y and x is given by an algebraic equation rather than a graph, we can use Eq. (5) to calculate the average rate of change directly.

Example 5 Determine the average rates of change in the function $f(x) = x^2 - 2x + 4$ over the interval $x = 2$ to $x = 5$ and then over the interval $x = 2$ to $x = 4$.

Solution For $x_1 = 2$, $f(x_1) = f(2) = (2)^2 - 2(2) + 4 = 4$ and, for $x_2 = 5$, $f(x_2) = f(5) = (5)^2 - 2(5) + 4 = 19$. Using Eq. (5), we compute

$$\text{Average rate of change} = \frac{f(5) - f(2)}{5 - 2} = \frac{19 - 4}{5 - 2} = 5.$$

On the average, over the interval $x = 2$ to $x = 5$, $f(x)$ changes 5 units for every 1-unit change in x.

Over the interval $[2, 4]$, we have

$$\text{Average rate of change} = \frac{f(x_2) - f(x_1)}{x_2 - x_1} = \frac{f(4) - f(2)}{4 - 2} = \frac{12 - 4}{4 - 2} = 4.$$

Thus, over the interval $x = 2$ to $x = 4$, the function changes 4 units on the average for each 1-unit change in x.

The right side of Eq. (4) should be familiar; it was used in Chapter 2 to define the slope of a straight line. Since the physical significance of the slope is a rate of change, our work in this section reinforces our previous results. It also extends these results to all functions. For any function $y = f(x)$, whether it is a linear equation or not, the quantity $[f(x_2) - f(x_1)]/(x_2 - x_1)$ represents an average rate of change.

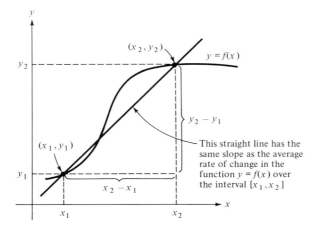

FIGURE 6.9

Average rates of change have geometric significance. If $y_1 = f(x_1)$ and $y_2 = f(x_2)$, the average rate of change over $[x_1, x_2]$ is the slope of the straight line through the points (x_1, y_1) and (x_2, y_2). This relationship is illustrated in Figure 6.9.

Returning to Example 1, we see that we can calculate the average rate of change for April through June by first drawing a straight line through the two points in Figure 6.6 corresponding to April 1 and July 1. This is done in Figure 6.10. The slope of this line is the average rate of change.

To obtain the average rate of change for the months of June through

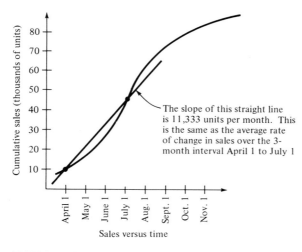

FIGURE 6.10

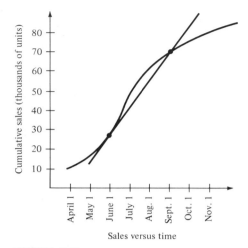

Sales versus time

FIGURE 6.11

August, we first draw a straight line through the points on Figure 6.6 corresponding to June 1 and September 1. This is done in Figure 6.11. The slope of this line is the average rate of change over the months June through August. Note that its slope is different from the one in Figure 6.10.

For linear equations, the slope and the average rate of change are identical. Indeed, we used the latter to define the former. The same is *not* true for other functions. The average rate of change is still $(y_2 - y_1)/(x_2 - x_1)$. The slope of a nonlinear function, however, is defined as the instantaneous rate of change of the function, which is the subject of Section 6.4.

Exercises

1. Find the average rate of change in the function $f(x) = x^2 - 4x + 5$ over the following intervals: (a) [1, 10], (b) [1, 8], (c) [−2, 1], (d) [−3, −1].
2. Find the average rate of change in the function $y = 1/x$ over the following intervals: (a) [4, 6], (b) [3, 7] (c) [2, 4].
3. Determine the average rate of change in the following functions over the interval [1, 5]:
   (a) $f(x) = 3x - 4$
   (b) $f(x) = x^2 + 6x + 2$
   (c) $y = x^3 + 5$
   (d) $S = 2 - 3t^2$.
4. Use Eq. (5) to find a general expression for the average rate of change in the function $f(x) = x^2 - 4x + 5$ over the interval $[x_1, x_2]$. Use this expression to find the average rates of change over the intervals given in Exercise 1.
5. Use Eq. (4) to find a general expression for the average rate of change in the function $y = 1/x$ over the interval $[x_1, x_2]$. Use this expression to find the average rates of change over the intervals given in Exercise 2.

6.4 Instantaneous Rates of Change

Although average rates of change are useful for many decision-making purposes, they are not always sufficient. Sometimes events change so rapidly that weekly averages, daily averages, and even hourly averages are not indicative of the actual situation. In such cases, instantaneous rates of change are needed, that is, rates effective for an instant of time.

Intuitively, instantaneous rates can be obtained from average rates by computing average rates of change over smaller and smaller intervals. The process is clearer when viewed graphically. Figure 6.12 depicts the average rate of change in a function $y = f(x)$ over the interval $[1, 4]$. To find the instantaneous rate of change in this function at $x = 1$, we find average rates of change over smaller and smaller intervals, with each interval beginning at $x = 1$. This is shown in Figure 6.13. We see that the slopes of the lines approach the slope of the tangent line to the curve $y = f(x)$ at $x = 1$. The slope of the tangent line is defined as the instantaneous rate of change in the function at $x = 1$. Geometrically, the slope of the tangent line also is called the *slope of curve $y = f(x)$ at $x = 1$*.

As a numerical example, let us find the instantaneous rate of change in the function $y = x^2$ at $x = 1$. The process involves calculating average rates of change over smaller and smaller intervals, each interval beginning at $x = 1$. Arbitrarily starting with the interval $[1, 2]$, we have $x_1 = 1$ and $x_2 = 2$. The corresponding y-values for the function $y = x^2$ are $y_1 = (x_1)^2 = (1)^2 = 1$

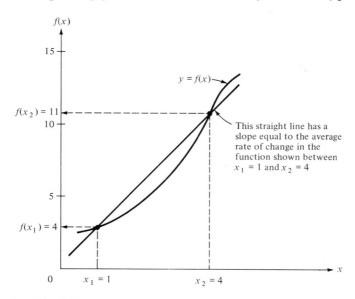

FIGURE 6.12

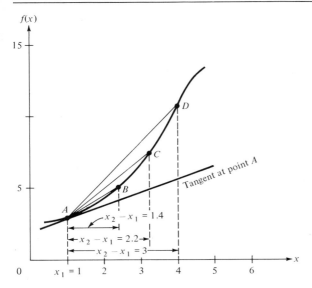

FIGURE 6.13

and $y_2 = (x_2)^2 = (2)^2 = 4$. Thus

$$\text{Average rate of change} = \frac{y_2 - y_1}{x_2 - x_1} = \frac{4 - 1}{2 - 1} = 3.$$

We now shorten the interval to $[1, 1.5]$ (any smaller interval would do) and compute the average rate of change over this smaller interval. For the interval $[1, 1.5]$, $x_1 = 1$ and $x_2 = 1.5$, so $y_1 = (x_1)^2 = (1)^2 = 1$ and $y_2 = (x_2)^2 = (1.5)^2 = 2.25$. Thus

$$\text{Average rate of change} = \frac{y_2 - y_1}{x_2 - x_1} = \frac{2.25 - 1}{1.5 - 1} = 2.5.$$

Continuing in this manner, we generate Table 6.2.

Apparently, the average rates of change approach 2.0 which is the instantaneous rate of change in the function $y = x^2$ at $x = 1$. This value 2.0 is also the slope of the curve $y = x^2$ at $x = 1$. A graphical representation of these rates is given in Figure 6.14.

Example 1 Find the instantaneous rate of change in $y = 2x^2 - 1$ at $x = 2$.

Solution We compute average rates of change over smaller and smaller intervals, each beginning at $x = 2$. These results are tabulated in Table 6.3.

TABLE 6.2

x_1	x_2	y_1	y_2	Average rate $= \dfrac{y_2 - y_1}{x_2 - x_1}$
1	2	1	4	3
1	1.5	1	2.25	2.5
1	1.2	1	1.44	2.2
1	1.1	1	1.21	2.1
1	1.05	1	1.1025	2.05
1	1.01	1	1.0201	2.01
1	1.005	1	1.010025	2.005
1	1.001	1	1.002001	2.001
1	1.0005	1	1.00100025	2.0005
1	1.0001	1	1.00020001	2.0001

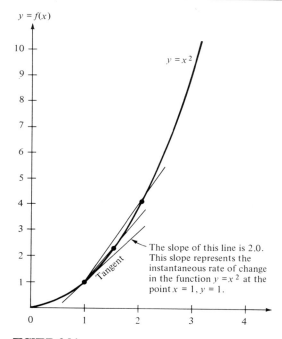

FIGURE 6.14

We see that the average rates of change approach 8.0 which is the instantaneous rate of change in $y = 2x^2 - 1$ at $x = 2$. The value 8.0 also is the slope of the curve $y = 2x^2 - 1$ at $x = 2$.

Figure 6.15 is a plot of the function $f(x) = 2x^2 - 1$ considered in Example 1. Note that every point on the curve has a line that can be drawn tangent to it, similar to the tangent at (1, 1). Accordingly, there is an infinite number of instantaneous rates of change associated with this particular function. The

6.4 Instantaneous Rates of Change

TABLE 6.3

x_1	x_2	y_1	y_2	Average rate = $\dfrac{y_2 - y_1}{x_2 - x_1}$
2	4	7	31	12
2	3	7	17	10
2	2.5	7	11.5	9
2	2.1	7	7.82	8.2
2	2.05	7	7.405	8.1
2	2.01	7	7.0802	8.02
2	2.001	7	7.008002	8.002
2	2.0005	7	7.0040005	8.001
2	2.0001	7	7.00080002	8.0002

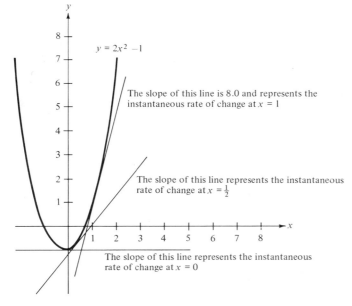

FIGURE 6.15

slope of each tangent line represents the instantaneous rate of change in the function at the x-coordinate of the point under consideration.

Example 2 Find the instantaneous rate of change in the function $y = x^2$ at $x = 2$.

Solution We compute the average rates of change over smaller and smaller intervals, each interval beginning at $x = 2$. These results are tabulated in Table 6.4. The average rates of change approach 4.0 which is the instantaneous

228 6 The Derivative

TABLE 6.4

x_1	x_2	y_1	y_2	Average rate = $\dfrac{y_2 - y_1}{x_2 - x_1}$
2	5	4	25	7
2	4	4	16	6
2	3	4	9	5
2	2.5	4	6.25	4.5
2	2.1	4	4.41	4.1
2	2.01	4	4.0401	4.01
2	2.005	4	4.020025	4.005
2	2.001	4	4.004001	4.001
2	2.0001	4	4.00040001	4.0001

rate of change in $y = x^2$ at $x = 2$. Compare this result with the instantaneous rate of change of the same function at $x = 1$ (Table 6.2).

As an illustration of why instantaneous rates of change are important to the business community, let us consider Figure 6.16 which shows the cost of producing a particular textbook as a function of the number of copies produced. Three tangents are drawn at points A, B, and C, corresponding to production runs of 10,000, 20,000, and 30,000 respectively. The slope of each tangent line represents the instantaneous rate of change (here cost per book) for the associated production run.

The slope of the tangent line at A is greater than the slope of the tangent line at B, which in turn is greater than the slope of the tangent line at C. Accordingly, the instantaneous rate of change, cost per book, is greater at A than at B, and the instantaneous rate of change at B is still greater than at C.

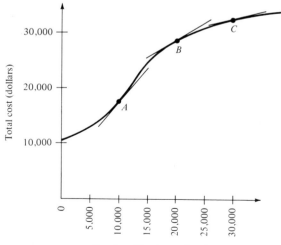

FIGURE 6.16

That is, the additional cost of producing book 10,001 is greater than the additional cost of producing book 20,001, which in turn is greater than the additional cost of producing book 30,001. As more books are produced, it becomes cheaper per book to produce additional books.

We devote all of Chapter 7 to discussing other important commercial applications of instantaneous rates of change. For the remainder of this chapter, we confine ourselves to developing simple methods for computing these rates. Finding instantaneous rates of change by calculating average rates of change over smaller and smaller intervals, as in Examples 1 and 2, is tedious and time-consuming. The work usually can be simplified with an algebraic procedure.

In the previous two examples, we calculated $(y_2 - y_1)/(x_2 - x_1)$ or, equivalently, $[f(x_2) - f(x_1)]/(x_2 - x_1)$ over the interval $[x_2, x_1]$ as x_2 approached x_1. This concept of having one point approach another point, here x_2 approaching x_1, is fundamental to all of calculus. We are interested in knowing what happens as x_2 gets arbitrarily close to x_1; the words *arbitrarily* and *close* are both important. First, we are not interested in what happens when x_2 equals x_1, but only in the result as x_2 gets close to x_1. Just how close is close is another matter, and we use the word *arbitrarily* to signify as close as the mind can conceive. The mathematical notation for this concept is $\lim_{x_2 \to x_1}$, read "the limit as x_2 approaches x_1."

An example may clarify the concept. Imagine a talented bug placed on one end of a yardstick and instructed to move one-half the distance remaining to the other end of the yardstick every second. During the first second, the bug moves $\frac{1}{2}$ yard. There is still $\frac{1}{2}$ yard to go and, since it must move half the remaining distance each second, the bug moves $\frac{1}{4}$ yard during the second second and still has $\frac{1}{4}$ yard to go. During the third second, the bug moves one-half of this distance or $\frac{1}{8}$ yard. It continues this process forever. Does the bug ever reach the other end of the yardstick. Obviously not, since it still has some distance left after each jump. But it does get close—arbitrarily close.

Leaving the bug, we consider the mathematical problem of calculating $\lim_{x_2 \to 5} (2x_2 - 1)$, that is, finding the value of the quantity $2x_2 - 1$ as x_2 approaches 5. As x_2 gets arbitrarily close to 5, $2x_2$ approaches 10 and $2x_2 - 1$ approaches 9. We write $\lim_{x_2 \to 5} (2x_2 - 1) = 9$. The same answer could have been obtained by simply substituting $x_2 = 5$ into $2x_2 - 1$, but such a substitution is procedurally wrong. First, $\lim_{x_2 \to 5}$ is shorthand notation for x_2 approaching 5, not x_2 equaling 5. More importantly, however, direct substitution does not always work, as Example 3 illustrates.

Example 3 Evaluate $\lim_{x_2 \to 3} \dfrac{(x_2)^2 - 9}{x_2 - 3}$.

Solution Simply substituting the value 3 for x_2 in the given expression yields 0/0 which is arithmetically meaningless. However, by applying the tabular method of substituting values of x_2 close to 3, we find that the quotient approaches 6. That is, the limit exists, but cannot be obtained by direct substitution.

Rather than using the tabular approach we can find this limit more directly. Note that the numerator $(x_2)^2 - 9$ can be factored into $(x_2 - 3)(x_2 + 3)$. If we then cancel the common term from both the numerator and the denominator, we have

$$\lim_{x_2 \to 3} \frac{(x_2)^2 - 9}{x_2 - 3} = \lim_{x_2 \to 3} \frac{(x_2 - 3)(x_2 + 3)}{(x_2 - 3)}$$

$$= \lim_{x_2 \to 3} (x_2 + 3)$$

$$= 6.$$

Example 3 is a prototype for many problems involving instantaneous rates of change, in which we must calculate

$$\text{Instantaneous rate of change in } f(x) \text{ at } x_1 = \lim_{x_2 \to x_1} \frac{f(x_2) - f(x_1)}{x_2 - x_1}. \quad (6)$$

If we simply substitute $x_2 = x_1$ into the right side of Eq. (6), we obtain 0/0 which is meaningless. The algebraic approach used in Example 3, however, is often successful. That is, we first factor $x_2 - x_1$ from the numerator and then cancel it and the denominator of Eq. (6) before letting x_2 approach x_1.

Example 4 Find the instantaneous rate of change in $f(x) = x^2$ at $x = 1$ using Eq. (6).

Solution Here $x_1 = 1$ (the point at which we want information about the function) and $f(x) = x^2$, hence $f(x_1) = (x_1)^2 = (1)^2 = 1$ and $f(x_2) = (x_2)^2$. Note that we do not substitute a numerical value for x_2, since it represents a moving point which will later approach $x_1 = 1$. Using Eq. (6), we obtain

$$\lim_{x_2 \to 1} \frac{f(x_2) - f(x_1)}{x_2 - x_1} = \lim_{x_2 \to 1} \frac{(x_2)^2 - 1}{x_2 - 1}$$

$$= \lim_{x_2 \to 1} \frac{(x_2 - 1)(x_2 + 1)}{(x_2 - 1)}$$

$$= \lim_{x_2 \to 1} (x_2 + 1)$$

$$= 2.$$

Compare this approach to the one used in Table 6.2.

Rather than working with Eq. (6) directly, it is sometimes easier first to change notation by setting $h = x_2 - x_1$. Here h represents the distance between

x_2 and x_1. On solving $h = x_2 - x_1$ for x_2, we also have $x_2 = x_1 + h$, and we can replace the requirement "x_2 approaches x_1" with the requirement "h approaches 0." That is, since $x_2 = x_1 + h$, the condition that h approaches 0 is equivalent to the condition that x_2 approaches x_1. Substituting the quantities $h = x_2 - x_1$, $x_2 = x_1 + h$, and $\lim_{x_2 \to x_1} = \lim_{h \to 0}$ into Eq. (6), we obtain

$$\text{Instantaneous rate of change in } f(x) \text{ at } x_1 = \lim_{h \to 0} \frac{f(x_1 + h) - f(x_1)}{h}. \qquad (7)$$

The last expression gives the instantaneous rate of change in a given function $y = f(x)$ at a specific point x_1. Generally, one needs the instantaneous rate of change for a given function at many points. For example, in the textbook case illustrated in Figure 6.16, we found the instantaneous rate of change at three points. To find instantaneous rates at several points, we can proceed in one of two ways. First, we can evaluate Eq. (7) separately at all the points x_1 that we need. A more efficient procedure is first to evaluate Eq. (7) at the arbitrary point $x_1 = x$ and then substitute the required values of x into the result.

Replacing x_1 by x in Eq. (7), we conclude that

$$\text{Instantaneous rate of change in } f(x) \text{ at any point } x = \lim_{h \to 0} \frac{f(x + h) - f(x)}{h}. \qquad (8)$$

We can use Eq. (8) to calculate the instantaneous rate of change in a given function $f(x)$ at any point x. Should we require the instantaneous rate of change at a particular point, we simply evaluate the resulting expression at the required point.

The following four-step process is recommended for evaluating Eq. (8).

Step 1 Find $f(x + h)$ by replacing x with the quantity $x + h$ in the given function $f(x)$.

Step 2 Calculate $f(x + h) - f(x)$ by subtracting $f(x)$ from the expression for $f(x + h)$ found in Step 1. Simplify the resulting difference.

Step 3 Divide by h.

Step 4 Let h approach zero.

Example 5 Determine a general expression for the instantaneous rate of change in the function $f(x) = x^2$ from Eq. (8). Evaluate this expression at the points $x = 1$, $x = 3$, and $x = 5$.

Solution Using the recommended four-step procedure for evaluating Eq. (8) with $f(x) = x^2$, we find

Step 1: $f(x + h) = (x + h)^2$.

Step 2: $\begin{aligned}f(x + h) - f(x) &= (x + h)^2 - x^2 \\ &= (x^2 + 2xh + h^2) - x^2 \\ &= 2xh + h^2.\end{aligned}$

Step 3: $\dfrac{f(x + h) - f(x)}{h} = \dfrac{2xh + h^2}{h} = 2x + h.$

Step 4: As h approaches zero, the expression $2x + h$ approaches $2x$. Therefore

$$\lim_{h \to 0} \frac{f(x + h) - f(x)}{h} = \lim_{h \to 0} (2x + h) = 2x.$$

The instantaneous rate of change in the function $f(x) = x^2$ at any point x is $2x$. In particular, the instantaneous rate of change at $x = 1$ is $2(1) = 2$. The instantaneous rate of change at $x = 3$ is $2(3) = 6$, whereas the instantaneous rate of change at $x = 5$ is $2(5) = 10$.

Example 6 Determine a general expression for the instantaneous rate of change in the function $f(x) = 2x^2 + 3x - 2$.

Solution Following the recommended four-step procedure, we have

Step 1: $f(x + h) = 2(x + h)^2 + 3(x + h) - 2$.

Step 2: $\begin{aligned}f(x + h) - f(x) &= [2(x + h)^2 + 3(x + h) - 2] - [2x^2 + 3x - 2] \\ &= [2x^2 + 4xh + 2h^2 + 3x + 3h - 2] \\ &\quad - [2x^2 + 3x - 2] \\ &= 4xh + 2h^2 + 3h.\end{aligned}$

Step 3: $\dfrac{f(x + h) - f(x)}{h} = \dfrac{4xh + 2h^2 + 3h}{h}$

$= 4x + 2h + 3.$

Step 4: As h approaches zero, the expression $4x + 2h + 3$ approaches $4x + 3$. Therefore

$$\lim_{h \to 0} \frac{f(x + h) - f(x)}{h} = \lim_{h \to 0} (4x + 2h + 3) = 4x + 3,$$

which is a general expression for the instantaneous rate of change in $f(x) = 2x^2 + 3x - 2$ at any point x.

We conclude this section with a few observations. First, the general expression for the instantaneous rate of change in a given function at an arbitrary point x is itself a function of x. In Example 5 we found the instantaneous rate of change in $f(x) = x^2$ to be $2x$, while in Example 6 we found the instantaneous rate of change in $f(x) = 2x^2 + 3x - 2$ to be $4x + 3$. This is not at all surprising. Since most functions are not linear equations, it follows that the graphs of such functions are not straight lines. Therefore most curves have different tangents at different points on the curve, and these different tangents have different slopes. (See Figures 6.15 and 6.16.) Since the slopes of the tangent lines are the instantaneous rates of change, these rates depend on (are functions of) the particular point under consideration.

Throughout this section we have assumed that x_2 was always greater than x_1 or, equivalently, that h was always positive. This need not be the case. As we take smaller and smaller intervals, we require only that x_1 always be in each interval. It need not be the left-hand end point. In actuality, one should show that the answer obtained for the instantaneous rate of change using x_1 as the left end point of each interval is the same answer obtained using intervals with x_1 as the right-hand end point. Only if both approaches yield the same answer can we mathematically say that an instantaneous rate of change exists. Exercise 14 is an example in which both approaches are not the same. For most functions encountered in commercial applications, however, both approaches yield the same result.

Even though the algebraic procedure given by Eq. (8) is often quicker than the tabular approach used in Examples 1 and 2, it too is formidable. The real value of Eq. (8) is that it can be used to develop other rules which are simpler and quicker for calculating instantaneous rates of change. These rules are the subject of Sections 6.5 and 6.6.

Exercises

1. Sketch the function $f(x) = x^2 - 6x + 10$ and graphically determine the slope of the tangent lines at the points (a) (1, 5), (b) (3, 1), (c) (5, 5).
2. (a) Sketch the function $f(x) = x^2 - 4x + 5$ and graphically determine the slope of the tangent line at (3, 2).
   (b) Complete Table 6.5. Do the values in the last column approach the slope determined in part (a)?

TABLE 6.5

x_1	x_2	y_1	y_2	Average rate = $\dfrac{y_2 - y_1}{x_2 - x_1}$
3	4	2		
3	3.5	2		
3	3.2	2		
3	3.1	2		

TABLE 6.6

x_1	x_2	y_1	y_2	Average rate $= \dfrac{y_2 - y_1}{x_2 - x_1}$
2	4	6		
2	3	6		
2	2.5	6		
2	2.2	6		
2	2.1	6		

3. (a) Sketch the function $f(x) = x^2 + x$ and graphically determine the slope at $x = 2$.
   (b) Complete Table 6.6 for this function. Do the values in the last column approach the slope determined in part (a)?
   (c) Use Eq. (8) to find the exact answer.

In Exercises 4 through 13, use Eq. (8) to find a general expression for the instantaneous rate of change in the given function. Use this result to find the instantaneous rate of change at $x = 1$ and $x = -4$.

4. $f(x) = 3x - 2$.
5. $f(x) = 2 - 3x$.
6. $f(x) = \frac{1}{2}x^2 - 7$.
7. $f(x) = x^2 - 2x + 10$.
8. $f(x) = x^3$.
9. $f(x) = 2x^3 - 2x^2 + 3x - 1$.
10. $f(x) = \dfrac{1}{x}$.
11. $f(x) = \dfrac{x}{x + 2}$.
12. $f(x) = \dfrac{1}{x} + x$.
13. $f(x) = \dfrac{1}{x^2}$.

14. Consider the function
$$y = f(x) = \begin{cases} 2 + 5x & (x \leq 3) \\ 23 - 2x & (x > 3), \end{cases}$$
which is sketched in Figure 6.17.

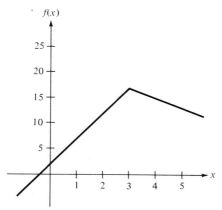

FIGURE 6.17

(a) Determine graphically whether or not this curve has a tangent at $x = 3$.
(b) Compute average rates of change for this function over the intervals having $x = 3$ as one end point and the second end point given successively by 3.2, 2.8, 3.1, 2.9, 3.05, 2.95, 3.01, 2.99, 3.005, and 2.995. Do these rates approach a fixed value? What can you conclude about the instantaneous rate of change at $x = 3$?

15. Consider the function

$$f(x) = \begin{cases} 7 & (x \leq 5) \\ 12 & (x > 5), \end{cases}$$

which is sketched in Figure 6.18. Calculate average rates of change for this function over the intervals having $x = 5$ as one end point and the second end point given successively by 3, 7, 4, 6, 4.5, 5.5, 4.7, 5.3, 4.9, 5.1, 4.95, 5.05, 4.99, and 5.01. What is the instantaneous rate of change in this function at $x = 5$?

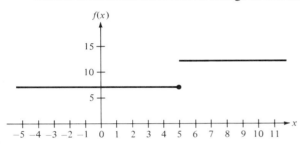

FIGURE 6.18

16. A company's sales are known to be related to advertising expenditures by the equation $S = 150{,}000 + 6000x - 50x^2$, where x denotes the monthly advertising expenditures in thousands of dollars.
   (a) Find the instantaneous rate of change in sales with respect to advertising expenditures.
   (b) Using part (a), determine whether an increase in advertising would increase sales if the present advertising budget is $45,000. Would the situation be different with a $60,000 advertising budget?

17. A firm's total sales, in millions of dollars, are given by the equation $R(x) = 3x + \frac{1}{2}x^2$, where x denotes the number of years the firm has been in operation.
   (a) Determine the firm's average growth rate in sales for its first 7 years in business.
   (b) Determine the firm's instantaneous rate of growth after its seventh year in business.
   (c) What will the firm's total sales be at the end of the tenth year if sales continue to follow the given equation?
   (d) What will the firm's total sales be at the end of the tenth year if the growth in sales after the seventh year always equals the growth achieved at the end of the seventh year?

18. Determine the values of x for the function illustrated in Figure 6.19 between or at which the instantaneous rate of change in y with respect to x is (a) positive, (b) negative, (c) zero.

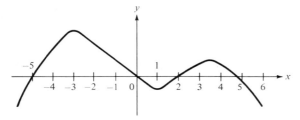

FIGURE 6.19

6.5 The Derivative

The instantaneous rate of change in a function is the derivative of that function. The following definition is an immediate consequence of Eq. (8).

Definition 6.2 The *derivative* of the function $y = f(x)$ at the point x is

$$\lim_{h \to 0} \frac{f(x+h) - f(x)}{h}.$$

Two different but equally common notations are used for the derivative of $y = f(x)$, namely, dy/dx, read "dee y dee x" and $f'(x)$, read "f prime of x." The choice of notation is strictly one of personal preference; we use both interchangeably in this text. In Example 5 in Section 6.4, we found that the instantaneous rate of change and therefore the derivative of $y = f(x) = x^2$ is $2x$. Notationally, we write either $dy/dx = 2x$ or $f'(x) = 2x$. Note, however, that dy/dx does *not* mean dy divided by dx, and $f'(x)$ does *not* mean f' times x. Both symbols are only notations for "the derivative of $y = f(x)$ with respect to the independent variable x" or, equivalently, "the instantaneous rate of change in $y = f(x)$ with respect to x."

Since the derivative is nothing more than an instantaneous rate of change, we can use the four-step procedure described on page 231 to calculate it. As we noted in Section 6.4, however, this procedure is both time-consuming and tedious. More efficiently, we use Definition 6.2 to generate simple rules for calculating derivatives, and then we use these rules to find the derivatives we need. We refer the interested reader to almost any calculus text for a derivation of the more elementary rules from the definition. The rules themselves are as follows.

Rule 1 The derivative of a constant is zero. That is, if $f(x) = c$, where c is a given fixed number, $f'(x) = 0$.

This rule is straightforward and with some thought obvious. The graph of the equation $y = f(x) = c$ is a straight line parallel to the x-axis. The tangent line to this curve at any point is the curve itself, hence its slope is zero. The slope of the tangent line is the instantaneous rate of change or the derivative.

Example 1 Determine the derivative of the function $f(x) = 10$.

Solution Using Rule 1 with $c = 10$, we obtain $f'(x) = 0$.

Example 2 Determine the derivative of the function $f(x) = 256$.

Solution Using Rule 1, $f'(x) = 0$.

Rule 2 The derivative of the function $f(x) = x^n$, where n is any real number, is $f'(x) = nx^{n-1}$.

It follows from Rule 2 that, if we are given a function x raised to a power, the derivative of this function is the power times x raised to one less power.

Example 3 Determine the derivative of $f(x) = x^2$.

Solution $f(x) = x^2$ is of the form x raised to a power. Using Rule 2 with $n = 2$, we obtain $f'(x) = 2x$. This was the function used to illustrate the derivative notation at the beginning of this section.

Example 4 Determine the derivative of $f(x) = x^{10}$.

Solution Using Rule 2, with $n = 10$, we obtain $f'(x) = 10x^9$.

Example 5 Determine the derivative of $f(x) = x^{-8}$.

Solution Using Rule 2, with $n = -8$, we obtain $f'(x) = -8x^{-9}$.

Example 6 Determine the derivative of $f(x) = x^{-1/2}$.

Solution Using Rule 2, with $n = -\frac{1}{2}$, we obtain $f'(x) = -\frac{1}{2}x^{-3/2}$.

Example 7 Determine the derivative of $f(x) = x^{3.1}$.

Solution Using Rule 2, $f'(x) = 3.1x^{2.1}$.

Having given rules for finding the derivative of a constant and of a function of the form x^n, we now combine functions. What, for example, is the derivative of the function $f(x) = 10x^8$?

Rule 3 If $f(x) = cg(x)$, where c is a fixed number and $g(x)$ is a function of x whose derivative is known, $f'(x) = cg'(x)$.

That is, the derivative of a number times a function is simply the number times the derivative of the function.

Example 8 Find $f'(x)$ if $f(x) = 20x^5$.

Solution The function $f(x) = 20x^5$ has the form of a number, 20, times another function, x^5. The derivative of x^5 is $5x^4$. (Why?) Using Rule 3, $f'(x) = 20(5x^4) = 100x^4$.

Example 9 Find $f'(x)$ if $f(x) = 6x^9$.

Solution Using Rule 3, $f'(x) = 6(9x^8) = 54x^8$.

A question that frequently arises is, Why is $g(x)$ used in Rule 3, rather than the statement, "The derivative of cx^n is c times nx^{n-1}"? The reason is that Rule 3 in its present form covers many other functions besides cx^n, as we will see later in Example 12.

Rule 4 (Addition Rule) If $f(x) = g_1(x) + g_2(x)$, where the derivatives of $g_1(x)$ and $g_2(x)$ are known, $f'(x) = g_1'(x) + g_2'(x)$.

That is, the derivative of a sum of two terms is simply the sum of the derivatives of the individual terms.

Example 10 Determine the derivative of the function $f(x) = 3x^6 + 5$.

Solution The given function is the sum of the two functions $3x^6$ and 5. Here $g_1(x) = 3x^6$, so $g_1'(x) = 18x^5$ (Rule 3), and $g_2(x) = 5$, so $g_2'(x) = 0$ (Rule 1). It now follows from Rule 4 that $f'(x) = 18x^5 + 0 = 18x^5$.

Rule 4 can be extended in the obvious fashion to include the sum of any number of terms. Furthermore, if we replace the plus signs in Rule 4 with minus signs, we obtain the *subtraction rule*: The derivative of $f(x) = g_1(x) - g_2(x)$ is $f'(x) = g_1'(x) - g_2'(x)$. The subtraction rule also can be extended to include the difference of any number of terms.

Example 11 Find $f'(x)$ for $f(x) = 2x^2 - 3x - 6$.

Solution The derivative of $f(x)$ is the difference of the derivatives of the three terms $2x^2$, $3x$, and 6. Here $g_1(x) = 2x^2$, hence $g_1'(x) = 4x$ (Rule 3), $g_2(x) = 3x$, hence $g_2'(x) = 3$ (Rule 3), and $g_3(x) = 6$, hence $g_3'(x) = 0$ (Rule 1). Therefore $f'(x) = 4x - 3 - 0 = 4x - 3$.

If all functions were polynomials, then Rules 1 through 4 would be sufficient to handle most cases. Most functions are not polynomials, however, and they cannot be differentiated using any of the previous rules. If we blindly apply Rule 2, for instance, to any function, we invariably will obtain the wrong derivative. A case in point is the exponential function $f(x) = e^x$, introduced in Section 3.5.

Rule 5 The derivative of the function $f(x) = e^x$ is $f'(x) = e^x$.

This is indeed a remarkable function. The derivative of the function equals the function itself; it is impervious to change by differentiation. Geometrically, this signifies that the slope of the tangent line to e^x at any point x numerically equals the value of e^x at that point.

Example 12 Determine the derivative of the function $f(x) = 10e^x$.

Solution Using Rules 3 and 5, we have $f'(x) = 10e^x$.

Example 13 Determine the derivative of the function $f(x) = 5x^5 + 6x^3 - 3x + 5 - 8e^x$.

Solution Differentiating $f(x)$ term by term, we obtain

$$f'(x) = 25x^4 + 18x^2 - 3 - 8e^x.$$

When the derivative of a function is needed at a particular point, it is obtained by first finding the derivative and then evaluating it at the point of interest. The notation for the derivative of $y = f(x)$ evaluated at $x = x_0$ is either

$$f'(x_0) \quad \text{or} \quad \left.\frac{dy}{dx}\right|_{x_0}.$$

Both symbols are read "the derivative of $y = f(x)$ evaluated at the point $x = x_0$."

Example 14 Find the derivative of $f(x) = 2x^2 - 3x - 6$ at both $x = 5$ and $x = -2$.

Solution Using the results from Example 11, we have $f'(x) = 4x - 3$. Evaluating this derivative at the required points, we obtain

$$f'(5) = \left.\frac{dy}{dx}\right|_5 = 4(5) - 3 = 17$$

and

$$f'(-2) = \left.\frac{dy}{dx}\right|_{-2} = 4(-2) - 3 = -11.$$

Warning: $f'(5)$ denotes the "derivative of $f(x)$" evaluated at $x = 5$ and *not* the derivative of "$f(x)$ evaluated at 5." That is, the value $x = 5$ is substituted into the expression for the derivative $f'(x)$ and not the function $f(x)$. The derivative is taken first and then evaluated; the function is *not* evaluated first and then differentiated.

Exercises

In Exercises 1 through 16, find $f'(x)$.
1. $f(x) = x^5 - 7x^3 + 4x + 2$.
2. $f(x) = x^5 - 7x^2$.
3. $f(x) = x^7 + 6x^3 - 4x^2$.
4. $f(x) = 9x^2 + 3x + 4$.

5. $f(x) = \dfrac{x^5}{5} - \dfrac{x^3}{3} - \dfrac{x^2}{2} + 10.$ 6. $f(x) = (x^2 + 3)x^5.$

7. $f(x) = (x^2 + 4)(x^5 + 3).$ 8. $f(x) = (x + 3)(x - 9).$
9. $f(x) = 10e^x.$ 10. $f(x) = x^3 + 6e^x.$
11. $f(x) = x^4 - 7x^2 + 7e^x.$ 12. $f(x) = (x + 3)(x - 9) + 10e^x.$
13. $f(x) = x^2(x + 2) + 8e^x.$ 14. $f(x) = (x^2 + 4)(x + 7) + e^x.$

15. $f(x) = \dfrac{x^2 + 12x + 20}{x + 2}.$ (*Hint:* First factor the numerator.)

16. $f(x) = \dfrac{(x^3 + 3x^2 + 21x) - 21}{x + 3}.$

In Exercises 17 through 24 find dy/dx.

17. $y = x^7 + 6x^5 + 3x + 4.$ 18. $y = x^3 - 3x + 6.$

19. $y = x^5 - \dfrac{x^4}{4} + 7e^x.$ 20. $y = \dfrac{x^4}{4} + 4e.$

21. $y = \dfrac{x^4}{4} + \dfrac{x^3}{3} - \dfrac{x^2}{2}.$ 22. $y = (x + 4)(x - 7).$

23. $y = (x^2 + 2)(x - 3).$ 24. $y = \dfrac{x^3 - 4x^2 + 5x - 20}{x - 4}.$

25. A company's sales are known to be related to advertising expenditures by the equation $s = 150{,}000 + 6000x - 50x$, where x denotes the monthly advertising expenditures in thousands of dollars.
    (a) Find the instantaneous rate of change in sales with respect to advertising expenditures.
    (b) Using part (a), determine whether an increase in advertising would increase sales if the present advertising budget is $45,000. Would the situation be different with a $60,000 advertising budget?

26. A firm's total sales, in millions of dollars, are given by the equation $R(x) = 3x + \tfrac{1}{2}x^2$, where x denotes the number of years the firm has been in operation.
    (a) Determine the firm's average rate of growth in sales for its first 7 years in business.
    (b) Determine the firm's instantaneous rate of growth after its seventh year of business.
    (c) What will the firm's total sales be at the end of the tenth year if sales continue to grow *at the same rate of growth* achieved after the seventh year?
    (d) What total sales has the firm actually reached after its tenth year of business? Compare your answer to the answer for part (c) and explain the discrepancy.

27. Figure 6.20 illustrates the distance traveled as a function of time for Mr. Williams' 2-mile trip through the Holland Tunnel. The equation of the curve illustrated is $D = -0.05t^3 + 0.25t^2 + 0.3t$, where D is the distance traveled in miles and t is measured in minutes from the start of the trip.
    (a) Determine the average rate of change in distance with respect to time for the complete 2-mile trip. Physically, what does your answer represent?

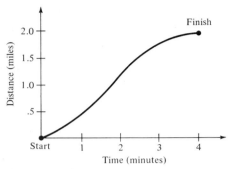

FIGURE 6.20

(b) Redraw Figure 6.20 on a separate sheet of graph paper. Draw a line on the graph whose slope represents the average speed determined in part (a).
(c) Determine the speed of Mr. Williams' car at exactly 1 minute after the start of his trip. (*Hint:* Speed is the instantaneous rate of change in distance traveled with respect to time.)
(d) Draw a line on the graph constructed for part (b), whose slope represents the instantaneous speed determined in part (c).
(e) Determine the exact speed that would be indicated on Mr. Williams' speedometer as he emerges from the tunnel ($t = 4$).

28. Determine the values of x for the function illustrated in Figure 6.21 between or at which (a) the derivative is positive, (b) the derivative is negative, (c) the derivative is zero.
29. From past experience it is known that an increase in the price of wheat leads to a decrease in the demand for wheat. Based on this information, determine which of the following equations may possibly relate the demand D for wheat to its price P.
    (a) $D = 300P^2 + 5P + 2000$
    (b) $D = 3000/P$
    (c) $D = 5P^2 - 2000$
    (d) $D = 5P^2 - 2000P$.

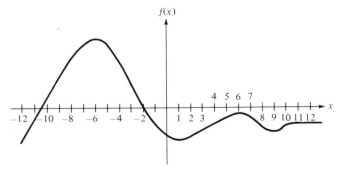

FIGURE 6.21

30. From past experience it is known that an increase in the price of silver leads to an increase in the demand for this commodity. Based on this information, determine which of the following equations may possibly relate the demand D for silver to its price P.
   (a) $D = 200/P$
   (b) $D = 300P^2 + 5P + 200$
   (c) $D = 2000 - 2P^3$
   (d) $D = 150/P^2$.

6.6 Additional Rules

The five rules of differentiation we just developed are not exhaustive. Still other important rules are the following.

Rule 6 (Product Rule) If the functions $g_1(x)$ and $g_2(x)$ both have derivatives, then the derivative of the product function $f(x) = g_1(x)g_2(x)$ is

$$f'(x) = g_1(x)g_2'(x) + g_2(x)g_1'(x). \tag{9}$$

In words, the derivative of a product is the first function times the derivative of the second function plus the second function times the derivative of the first function. The derivative of a product is *not* equal to the product of the individual derivatives.

Example 1 Find $f'(x)$ if $f(x) = x^2 e^x$.

Solution We want the derivative of the product of x^2 and e^x. Here $g_1(x) = x^2$, hence $g_1'(x) = 2x$ (Rule 2), and $g_2(x) = e^x$, hence $g_2'(x) = e^x$ (Rule 5). Substituting the appropriate terms into Eq. (9), we obtain $f'(x) = x^2 e^x + e^x(2x) = x^2 e^x + 2xe^x$.

Example 2 Determine the derivative of the function $f(x) = x^2(x^3 + 3)$.

Solution We can do this problem two ways. First, we set $g_1(x) = x^2$, so $g_1'(x) = 2x$ (Rule 2), and $g_2(x) = x^3 + 3$, so $g_2'(x) = 3x^2$ (Rule 4). Then, substituting appropriate terms into Eq. (9), we find

$$\begin{aligned} f'(x) &= x^2(3x^2) + (x^3 + 3)(2x) \\ &= 3x^4 + 2x^4 + 6x \\ &= 5x^4 + 6x. \end{aligned}$$

Alternatively, the same answer is obtained by algebraically simplifying the original function $f(x) = x^2(x^3 + 3) = x^5 + 3x^2$ and then differentiating term by term using Rule 4.

6.6 Additional Rules

Rule 7 (Division Rule) If $g_1(x)$ and $g_2(x)$ both have derivatives, with $g_2(x) \neq 0$, then the quotient function $f(x) = g_1(x)/g_2(x)$ has the derivative

$$f'(x) = \frac{g_2(x)g_1'(x) - g_1(x)g_2'(x)}{[g_2(x)]^2}. \tag{10}$$

In words, the derivative of a quotient is the denominator times the derivative of the numerator minus the numerator times the derivative of the denominator, with the entire result divided by the square of the denominator. The derivative of a quotient is *not* equal to the quotient of the individual derivatives.

Example 3 Find the derivative of $f(x) = e^x/x^5$.

Solution Set $g_1(x) = e^x$, so $g_1'(x) = e^x$ (Rule 5), and $g_2(x) = x^5$, so $g_2'(x) = 5x^4$ (Rule 2). Substituting appropriate terms into Eq. (10), we obtain

$$f'(x) = \frac{x^5(e^x) - e^x(5x^4)}{(x^5)^2} = \frac{x^5 e^x - 5x^4 e^x}{x^{10}} = \frac{xe^x - 5e^x}{x^6}.$$

Example 4 Find $f'(x)$ for $f(x) = e^x/(x^2 + 4)$.

Solution Set $g_1(x) = e^x$, so $g_1'(x) = e^x$ (Rule 5), and $g_2(x) = x^2 + 4$, so $g_2'(x) = 2x$ (Rule 4). Using Rule 7, we find

$$f'(x) = \frac{(x^2 + 4)e^x - e^x(2x)}{(x^2 + 4)^2},$$

which after simplifying becomes

$$f'(x) = \frac{(x^2 - 2x + 4)e^x}{(x^2 + 4)^2}.$$

Occasionally, we have to differentiate a function raised to a power such as $f(x) = (x^2 + 3x + 12)^3$. The following rule is applicable.

Rule 8 (Power Rule) If $f(x) = [g(x)]^n$, where n is a fixed real number and the derivative of $g(x)$ is known, then

$$f'(x) = n[g(x)]^{n-1}g'(x). \tag{11}$$

Although this rule resembles Rule 2, it is quite different. Rule 8 deals with a function to a power, while Rule 2 deals with the variable x to a power. There are similarities, however. To find the derivative of a function raised to a power, first treat the function inside the brackets, denoted here by $g(x)$, as a single term and apply Rule 2. *Then* multiply this result by the derivative of $g(x)$.

Example 5 Differentiate $f(x) = (x^2 + 3x + 12)^3$.

Solution Set $g(x) = x^2 + 3x + 12$ and take $n = 3$. Then $g'(x) = 2x + 3$, and it follows from Rule 8 that $f'(x) = 3(x^2 + 3x + 12)^2(2x + 3)$.

This answer can be obtained by a second procedure. First cube $x^2 + 3x + 12$ and then differentiate the result. Rule 8, however, is quicker.

Example 6 Differentiate $f(x) = \sqrt{x^2 + 1}$.

Solution Since $\sqrt{x^2 + 1} = (x^2 + 1)^{1/2}$, we set $g(x) = x^2 + 1$ and take $n = \frac{1}{2}$. Then $g'(x) = 2x$, and it follows from Eq. (11) that $f'(x) = \frac{1}{2}(x^2 + 1)^{(1/2)-1}(2x) = x(x^2 + 1)^{-1/2}$.

Note that the alternative procedure used in Example 5 is not applicable for this function.

Let us try to apply Rule 8 to the function $f(x) = e^{nx}$, where again n is a fixed real number. If we first rewrite $f(x)$ as $f(x) = e^{nx} = (e^x)^n$, we will have the required form for Rule 8 with $g(x) = e^x$. Now $g'(x) = e^x$ (Rule 5), and Rule 8 becomes $f'(x) = n[e^x]^{n-1}(e^x) = ne^{nx}$. We have proven Rule 9.

Rule 9 If $f(x) = e^{nx}$, where n is a fixed real number, $f'(x) = ne^{nx}$.

Example 7 Differentiate $f(x) = e^{8x}$.

Solution Take $n = 8$. Then, $f'(x) = 8e^{8x}$.

Example 8 Differentiate $f(x) = (3 + 4e^{7x})^{1.95}$.

Solution First, $f(x)$ is another function, $g(x) = 3 + 4e^{7x}$, raised to the $n = 1.95$ power. Using Rule 9 in combination with other appropriate rules, we find $g'(x) = 0 + 4(7)e^{7x} = 28e^{7x}$. It then follows from the power rule that

$$f'(x) = 1.95(3 + 4e^{7x})^{1.95-1}(28e^{7x})$$
$$= 54.6(3 + 4e^{7x})^{0.95}(e^{7x}).$$

To this point, we have considered only functions given by the one equation $y = f(x)$. Frequently, however, business problems are modeled by a set of two equations having the form $y = g_1(u)$ and $u = g_2(x)$. An example of such a set is $y = u^2 + 3u$ and $u = 6x + 2$. Since we can substitute the value for u into the equation for y, it follows that y is ultimately a function of x, and it makes sense to ask for the derivative dy/dx. Interestingly, we can find dy/dx directly without substituting first if the individual derivatives $g_1'(u)$ and $g_2'(x)$ are known. The procedure is known as the chain rule.

Rule 10 (Chain Rule) If $y = g_1(u)$ and $u = g_2(x)$, and if both the derivatives dy/du and du/dx are known, then

$$\frac{dy}{dx} = \frac{dy}{du}\frac{du}{dx}. \tag{12}$$

The term *chain rule* evolves from the ultimate derivative, dy/dx, being a "chain" of two other derivatives, namely, dy/du and du/dx.

Example 9 Determine dy/dx if $y = u^2 - 5u + 7$ and $u = 7x^2 + 10$.

Solution Using the chain rule, we first determine

$$\frac{dy}{du} = 2u - 5 \quad \text{and} \quad \frac{du}{dx} = 14x.$$

Then,

$$\frac{dy}{dx} = (2u - 5)(14x) = [2(7x^2 + 10) - 5](14x) = 196x^3 + 210x.$$

Alternatively, we could first substitute the expression for u into the equation for y, obtaining

$$y = (7x^2 + 10)^2 - 5(7x^2 + 10) + 7 = 49x^4 - 105x^2 + 57.$$

Then, differentiating this expression directly, we again determine

$$\frac{dy}{dx} = 196x^3 - 210x.$$

Example 10 A firm's monthly sales revenue S is known to be related to its advertising expenditures by the equation $S = 100 - 30E + 0.2E^2$, where E represents monthly advertising expenditures in thousands of dollars. The company allocates its monthly advertising expenditure based on the equation $E = 0.1x^2 - 0.3x + 5000$, where x represents thousands of units sold the previous month. Determine the derivative dS/dx which is the instantaneous rate of change of sales with respect to previous monthly sales.

Solution We are asked to find dS/dx, given $S = 100 - 30E + 0.2E^2$ and $E = 0.1x^2 - 0.3x + 5000$. Modifying the notation in Eq. (12) so that it is applicable to this situation, we obtain

$$\frac{dS}{dx} = \frac{dS}{dE}\frac{dE}{dx}.$$

Performing the appropriate operations, we have

$$\frac{dS}{dx} = (-30 + 0.4E)(0.2x - 0.3),$$

which upon simplification becomes

$$\frac{dS}{dx} = 0.008x^3 - 0.036x^2 + 394.036x - 591.$$

The alternative procedure suggested in Example 9 is applicable here also.

Exercises

In Exercises 1 through 12 find $f'(x)$.
1. $f(x) = (x^2 + 3x)(x^5 + x^7 + 2x)$.
2. $f(x) = e^{10x}(x^2 + 4)$.
3. $f(x) = x^{-5} + x^{-3} + x^{-2}$.
4. $f(x) = x^{5/2} + x^{1/2}$.
5. $f(x) = x^{-5/2} + x^{-1/2}$.
6. $f(x) = x^3 + x^{1/2} + x^{-5}$.
7. $f(x) = (x^7 + 6x)^5$.
8. $f(x) = (x^2 + 3x)^6$.
9. $f(x) = e^{7x}x^7$.
10. $f(x) = e^{7x}/(x^2 + 3)$.
11. $f(x) = (x^6 + 7)/(x^5 + 3x)$.
12. $f(x) = (x^2 + 3)^4/(x^7 + x^3)$.

In Exercises 13 through 20 find dy/dx.
13. $y = e^{10x}x^3$.
14. $y = (x^2 + 4x)^5$.
15. $y = x^5 + 3x^{-5}$.
16. $y = x^{-4} + x^{-3} + x^{-2}$.
17. $y = \dfrac{1}{x^5} + \dfrac{1}{x^4}$.
18. $y = \sqrt{(x^2 + 4)}$.
19. $y = x^{3/2} + 7x^{5/2}$.
20. $y = (x^{10} + 7x^2)(5x^3 + 6x)$.

In Exercises 21 through 23 find dS/dx using the chain rule.
21. $S = E^2 + 3E$, $E = x^2$.
22. $S = E^2 + 3$, $E = x^3 + 5$.
23. $S = E/(E + 2)$, $E = 3x^2 + 4$.

6.7 Higher-Order Derivatives

Just as we took the first derivative of a differentiable function, we usually can differentiate the derivative itself. This derivative of the derivative is called the *second derivative* and is commonly denoted by $f''(x)$, d^2y/dx^2, or $d^2[f(x)]/dx^2$. To obtain the second derivative of a function, we treat the first derivative as a new function and differentiate it using the rules of Sections 6.5 and 6.6.

Example 1 Determine the first derivative and second derivative of the function $y = 3x^2 + 2x + 4$.

Solution

$$\frac{dy}{dx} = 6x + 2 \quad \text{(Rule 4)}$$

and

$$\frac{d^2y}{dx^2} = 6 \quad \text{(Rule 4)}.$$

Example 2 Determine d^2y/dx^2 for the function $y = x^2e^x$.

Solution We first find dy/dx and then differentiate it to obtain d^2y/dx^2. Using the product rule (Rule 6), we have $dy/dx = x^2e^x + 2xe^x$. Differentiating dy/dx term by term (Rule 4), which requires the product rule for each of the two terms in the sum, we obtain

$$\frac{d^2y}{dx^2} = (x^2e^x + 2xe^x) + (2xe^x + 2e^x)$$
$$= x^2e^x + 4xe^x + 2e^x$$
$$= e^x(x^2 + 4x + 2).$$

Continuing in this manner, we could define third-, fourth-, fifth-, and higher-order derivatives. However, only the first two derivatives of a function have applications to business problems. We consider some of these applications in Chapter 7.

Exercises

In Exercises 1 through 6 find dy/dx and d^2y/dx^2 and evaluate the second derivative at the points $x = 1$ and $x = 2$.

1. $y = x^5 + 4x^2 + 3x$.
2. $y = x + 3$.
3. $y = x^4 + 3x^2 + 4x + 5$.
4. $y = e^x$.
5. $y = x^3e^{10x}$.
6. $y = 5x^2e^{7x}$.
7. Figure 6.22 represents the distance traveled as a function of time for Mr. Williams' 2-mile trip through the Holland Tunnel, first considered in Exercise 27 in Section 6.5. The equation of the curve illustrated is $D = -0.05t^3 + 0.25t^2 + 0.3t$, where D is the distance traveled in miles, and t is measured in minutes from the start of the trip.
   (a) Determine a general expression for the speed of Mr. Williams' car at any time during the trip.
   (b) Determine the speed that would be indicated on Mr. Williams' speedometer at $t = 1$ minute.
   (c) Determine a general expression for Mr. Williams' acceleration at any time during the trip. (*Hint:* Acceleration = $d(\text{speed})/dt = d^2D/dt^2$.)

248 6 The Derivative

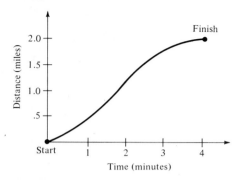

FIGURE 6.22

(d) Using your answer for part (c), determine Mr. Williams' acceleration at $t = 1$ minute and $t = 3$ minutes. What is the significance of the negative sign at $t = 3$ minutes?

Chapter 7

Applications of the Derivative

Optimization, the process of either maximizing or minimizing a commercial quantity, is the heart of most business problems. How many units of a product should be manufactured to maximize profit? How often should material be ordered to minimize total inventory cost? What is the best method for government agencies to control money supplies so as to maximize employment?

Linear programming provides a method for solving some optimization problems. But the method is restrictive; it requires that the quantity being optimized be linear and that the constraints also be linear. This is not the case in many situations. Another optimization method is based on differentiation. This method does not require linearity and is applicable to a much larger class of problems.

In Section 7.1, we develop the theory of optimization based on the derivative. Recall from Chapter 6 that differentiation is a mathematical operation performed on functions. To apply the derivative to business situations, we must first construct mathematical equations that realistically model or represent the situations of interest. We do this in Sections 7.2 through 7.4. Once the appropriate equations have been obtained, we can apply the optimization technique developed in Section 7.1 to locate the desired maxima and minima.

250 7 Applications of the Derivative

Not all applications of the derivative deal with optimization. Rates of change have direct applications themselves, and we discuss one example in Section 7.5.

7.1 Optimization through Differentiation

We seek a method for finding maxima and minima of mathematical functions that model commercial situations. In the following sections, we show how to generate these functions. For now, we assume that the appropriate function has been determined and we are interested only in finding its optimal values.

As an example, assume we know that the relationship between profit P (in dollars) and the number of items manufactured and sold x (in units) for a particular industry is modeled by the equation

$$P = -x^2 + 120x - 600. \tag{1}$$

We now wish to know the value(s) of x that will maximize P. That is, how many items should be manufactured if the objective is maximum profit?

One approach is to substitute different values of x into Eq. (1), calculate corresponding values of P, and determine the value(s) of x that maximizes P. Substituting $x = 10$ into Eq. (1), we obtain $P = -(10)^2 + 120(10) - 600 = \$500$. Substituting $x = 20$ into Eq. (1) gives $P = \$1400$. Continuing in this manner, we generate Table 7.1. It appears that $x = 60$ will produce a maximum profit of $P = \$3000$. But can we be sure? Perhaps $x = 61$ will generate a bigger profit. Or, perhaps the maximum profit occurs at $x = 137$.

The difficulty in calculating the function for only some values of the independent variable, as we did in Table 7.1, is that we are never sure that the optimum does not occur at another value.

A second and more useful procedure for finding maxima and minima is through graphing. Agreeing that the *maximum value* of a function is the largest value the function can obtain, and similarly that the *minimum value* of a function is the smallest value the function can obtain, we conclude that maxima appear as high points and minima appear as low points on the graph of a function.

Plotting the points given in Table 7.1 and realizing that $P = -x^2 + 120x - 600$ is a quadratic equation (Section 2.3), we draw Figure 7.1 as the graph of Eq. (1). It is evident from the curve that the maximum profit is about $3000 from a production run near $x = 60$ units. Although we cannot be certain that the maximum does not occur at $x = 61$, it is clear that the maximum is near $x = 60$; it certainly does not occur at $x = 137$. At this time, Eq. (1) does

TABLE 7.1

x (units)	0	10	20	40	60	70	80
P (\$)	−600	500	1400	2600	3000	2900	2600

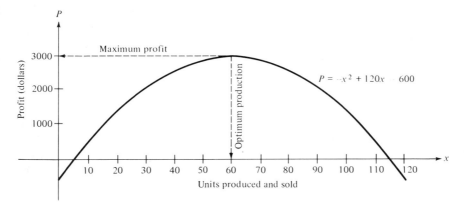

FIGURE 7.1

not have a minimum, since there is no point in Figure 7.1 that is smaller than every other point.

Example 1 Graph $y = x^2 - 10x + 16$ and determine the maximum and minimum values for y.

Solution This equation is quadratic. Arbitrarily selecting values of x, computing the corresponding values of y, and plotting these points, we obtain Figure 7.2. The minimum value of y is -9 which occurs when $x = 5$. The function does not have a maximum, since there is no point on the graph that is larger than every other point.

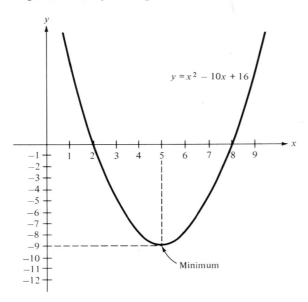

FIGURE 7.2

In both Figures 7.1 and 7.2 one of the optimal values, either the maximum or the minimum, did not exist. The reason was that the domain, the set of allowable values for the independent variable x, was infinite. The larger we allowed x to become in Figure 7.1, the smaller P became. The larger we allowed x to become in Figure 7.2, the larger y became. Both situations are unrealistic from a business point of view. In business there is always a limit beyond which it is either impossible or not feasible to go.

On an absolute scale, every product is limited in number by the amount of raw material available in the world. Realistically, production runs are also limited by the demand for the product, the time required to make the product, and the capital investment necessary to produce the product. Even under the best conditions, only a finite number (perhaps a large finite number) of each product can be produced. Similarly, every service, on an absolute scale, is limited by the time people have available to perform the service.

The point is that every business process has finite limits or boundaries. Therefore every mathematical model of a business situation must reflect these limits. Most often this is done by restricting the domain. Equation (1) is not a good model for a commercial situation, since it does not indicate limits on the values of x. A more realistic model is

$$P = -x^2 + 120x - 600 \qquad (0 \leq x \leq 150). \tag{2}$$

We now have the additional information that no less than 0 products and no more than 150 products can be produced. With these restrictions, the graph of Eq. (2) becomes Figure 7.3, where we have plotted only points associated with the domain $0 \leq x \leq 150$. It now follows that the maximum is $P = \$3000$, occurring at $x = 60$, *and* the minimum is $P = -\$5100$, occurring at $x = 150$.

Example 2 Figure 7.4 represents the relationship between the dollar value of inventory on hand for a distribution center over a 12-month time period t, where t is in months ($0 \leq t \leq 12$). Find the maximum and minimum dollar values of inventory on hand.

Solution The maximum value of inventory on hand is $7 million which occurs at both $t = 1$ and $t = 12$. The minimum value is $2 million, occurring at $t = 6$.

It follows from Example 2 that optimal values can occur at more than one place. Note that the maximum occurred at both $t = 1$ and $t = 12$.

The peaks at $t = 4$ and $t = 8$ and the valleys at $t = 2$ and $t = 9$ also are interesting. They are not maxima or minima, since there are other points on the graph that are higher ($t = 1$) and lower ($t = 6$). Nonetheless, they are high and low points for some small subset of the domain and as such are often called *relative maxima* and *relative minima*. A detailed development of their properties is given in the exercises.

Finding optimal points by graphing has two disadvantages. The first is

7.1 Optimization through Differentiation   **253**

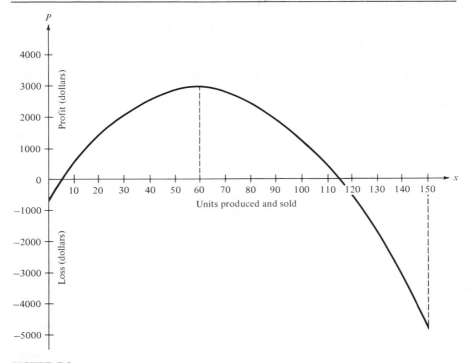

FIGURE 7.3

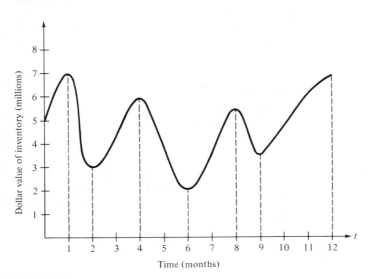

FIGURE 7.4

lack of accuracy. Can we really conclude from Figure 7.4 that the maximum occurs at $t = 6$ and not at $t = 6.1$ or 5.98? No, but since one rarely needs more accuracy than can be obtained from a graph, this disadvantage is not serious. The major difficulty is that the curve must be graphed. Accurately graphing a complicated equation is a difficult and time-consuming procedure which we would like to avoid if possible.

A third approach to optimization, often the most appealing, uses differentiation and is based on the following theorem, the proof of which can be found in most calculus texts.

Theorem 7.1 Let a function be defined on a domain $a \leq x \leq b$ and have a maximum or minimum at a point c, where $a < c < b$. If the derivative of the function exists at $x = c$, then the derivative must be zero there.

This theorem, after a little thought, is rather obvious. Note in Figures 7.2 and 7.3 that the tangents to each curve at the maximum and minimum points, if drawn, would be horizontal, and therefore have zero slopes. Since the derivative is the slope of the tangent, it follows that the derivative at such maximum or minimum points is zero.

This theorem simplifies the search for optimal points. Maxima and minima can occur only where the first derivative is zero or where the theorem is not applicable. The theorem is not applicable in two cases. First, it says nothing about optimal points at the boundaries of the domain, usually called the *end points*. In Theorem 7.1, the end points are $x = a$ and $x = b$. Second, the theorem says nothing about optimal points at places where the derivative does not exist. Accordingly, maxima and minima can occur at only one of three places:

1. Points where the first derivative is zero
2. Points where the first derivatives do not exist
3. End points

Optimal points cannot occur any place else.

To locate optimal points for any function, first find all points where the first derivative is zero, then all points where the derivative does not exist, and finally all end points. Substitute each of these points into the given function and determine which ones optimize it.

Example 3 Determine the maximum and minimum values of the function $P = -x^2 + 120x - 600$, $0 \leq x \leq 150$.

Solution Differentiating the given function, we have $dP/dx = -2x + 120$. To find values of x for which this derivative equals zero, we set the derivative equal to zero and solve for x. Accordingly,

$$-2x + 120 = 0$$
$$-2x = -120$$
$$x = 60.$$

The first derivative exists everywhere, since for each value of x the quantity $dy/dx = -2x + 120$ is defined; there are no points in category (2). The end points are $x = 0$, $x = 150$.

Gathering these points together, $x = 60$, $x = 0$, and $x = 150$, we substitute each into the given function and evaluate the corresponding values for y. This is done in Table 7.2. It is now clear that the maximum is $P = \$3000$ which occurs at $x = 60$, and the minimum is $P = -\$5100$ which occurs at $x = 150$.

TABLE 7.2

x	$P = -x^2 + 120x - 600$
60	3000††
0	-600
150	-5100†

†† Maximum.
† Minimum.

Example 4 Determine the maximum and minimum values of the function $y = x^3 - 21x^2 + 120x - 100$, $1 \leq x \leq 13$.

Solution Differentiating the given function, we find $dy/dx = 3x^2 - 42x + 120$. To locate all values of x for which this derivative is zero, we set the derivative equal to zero and solve for x. Thus

$$3x^2 - 42x + 120 = 0$$
$$x^2 - 14x + 40 = 0 \text{ (dividing by 3)}$$
$$(x - 10)(x - 4) = 0$$
$$x = 10 \quad \text{and} \quad x = 4.$$

The first derivative is zero at $x = 10$ and $x = 4$.

There are no points for which the first derivative does not exist, since $dy/dx = 3x^2 - 42x + 120$ can be evaluated at all values of x. The end points for this problem are $x = 1$ and $x = 13$.

Evaluating the given function at $x = 10$, $x = 4$, $x = 1$, and $x = 13$ successively we obtain Table 7.3. It follows that the maximum is $y = 108$ which

TABLE 7.3

x	$y = x^3 - 21x^2 + 120x - 100$
10	0†
4	108††
1	0†
12	44

† Minimum.
†† Maximum.

occurs at $x = 4$, and the minimum is $y = 0$ which occurs at two places, $x = 10$ and $x = 1$.

Example 5 Determine the maximum and minimum of $y = t^3 - 9t^2 - 120t + 1500$, $0 \leq t \leq 30$.

Solution Differentiating the given function, we find $dy/dt = 3t^2 - 18t - 120$. To locate all values of t for which this derivative is zero, we set the derivative equal to zero and solve for t. Thus

$$3t^2 - 18t - 120 = 0$$
$$t^2 - 6t - 40 = 0$$
$$(t + 4)(t - 10) = 0$$
$$t = -4 \quad \text{and} \quad t = 10.$$

The first derivative is zero at $t = -4$ and $t = 10$. Since $t = -4$ is outside our domain and not an allowable point for this problem, we disregard it.

The first derivative exists everywhere, since for each value of t the quantity $3t^2 - 18t - 120$ is defined. The end points for this problem are at $t = 0$ and $t = 30$.

Evaluating the given function at $t = 0$, $t = 10$, and $t = 30$, we obtain Table 7.4. It follows that the maximum is $y = 16,800$ occurring at $t = 30$, and the minimum is $y = 400$ occurring at $t = 10$.

TABLE 7.4

x	$y = t^3 - 9t^2 - 120t + 1500$
0	1,500
10	400†
30	16,800††

† Minimum.
†† Maximum.

Example 6 Determine the maximum and minimum values of the function $D = (P - 20)^2(P - 30)^2$, $20 \leq P \leq 35$.

Solution Differentiating this function with the product rule for differentiation, we obtain $dD/dP = 2(P - 20)(P - 30)^2 + 2(P - 20)^2(P - 30)$. Setting this quantity equal to zero and solving for P, we have

$$2(P - 20)(P - 30)^2 + 2(P - 20)^2(P - 30) = 0$$
$$2(P - 20)(P - 30)[(P - 30) + (P - 20)] = 0$$
$$2(P - 20)(P - 30)(2P - 50) = 0.$$
$$P = 20, \quad P = 30, \quad \text{and} \quad P = 25.$$

The derivative exists everywhere, and the end points are $P = 20$ and $P = 35$. It follows from Table 7.5 that the maximum is $D = 5625$ which occurs at $P = 35$, and the minimum is $P = 0$ which occurs at both $P = 20$ and $P = 30$.

TABLE 7.5

P	$D = (P - 20)^2(P - 30)^2$
20	0†
30	0†
25	625
35	5625††

† Minimum.
†† Maximum.

Exercises

In Exercises 1 through 5 find the maximum and minimum values of the given functions two ways: First, graph the functions, visually locating the optimal points and, second, use differentiation.

1. $y = 3x^2 - 24x - 5, 0 \le x \le 6$.
2. $y = -x^2 + 25x - 3.5, 4 \le x \le 16$.
3. $y = 7x^2 + 56x + 25, 0 \le x \le 5$.
4. $y = -5x^2 + 100x, 0 \le x \le 20$.
5. $y = 2x - 1, 25 \le x \le 50$.

In Exercises 6 through 15 use the first derivative to determine maximum and minimum values of the given functions.

6. $D = t^3 - 12t + 7, -3 \le t \le 3$.
7. $D = t^3 - 12t + 7, -4 \le t \le 4$.
8. $D = t^3 - 12t + 7, -5 \le t \le 5$.
9. $D = t^3 - 12t + 7, 0 \le t \le 3$.
10. $P = 2N^3 - 54N + 2800, -3 \le N \le 5$.
11. $P = 2N^3 - 54N + 2800, 0 \le N \le 6$.
12. $y = x^3 - 48x + 57, 0 \le x \le 4$.
13. $x = \frac{1}{3}t^3 - 4t^2 + 12, 0 \le t \le 9$.
14. $T = C^3 - 10C^2 + 12C + 10, 0 \le C \le 10$.
15. $y = (x - 10)(x - 20), 5 \le x \le 25$.

16. A television manufacturer has found that the profit P (in dollars) obtained from selling x television sets per week is given by the formula $P = -x^2 + 300x - 5000$. Determine how many television sets the manufacturer should produce to maximize profits. What is the maximum profit that can be realized?

17. A manufacturing company has found that the profit P (in dollars) realized in selling x items is given by

$$P = 22x - \frac{x^2}{2000} - 10{,}000.$$

   (a) How many items should the manufacturer produce to maximize profit?
   (b) Find the maximum profit.

18. The manufacturer described in Exercise 17 also has determined that the total cost of producing x items, denoted by TC (in dollars) is given by

$$TC = \frac{x^2}{2000} - 7x + 10{,}000.$$

   (a) Determine the number of items that should be produced if the manufacturer's goal is to minimize total cost rather than to maximize profit.
   (b) What is the minimum total cost? Is this answer reasonable?

19. In the course of a week, a refrigerator manufacturer can sell x refrigerators at a unit price of \$300. The total cost TC (in dollars) is given as

$$TC = 2.5x^2 - 200x + 20{,}000.$$

   (a) How many units should the firm produce each to maximize profits?
   (b) Determine the maximum profit.
   (c) Determine the total cost associated with achieving maximum profit?

20. A firm has found that the total cost TC (in dollars) of producing x items is given by the equation

$$TC = \tfrac{1}{3}x^3 - 10x^2 - 800x + 12{,}000.$$

   Determine the number of units this firm should produce to minimize its total production cost.

21. Assume that the firm described in Exercise 20 can sell all units it produces at a fixed price of \$325 per unit. Determine how many units this firm should produce to maximize its profit.

22. Find the maximum and minimum values of the function given by

$$y = \begin{cases} -x^2 + 2x + 5 & (0 \leq x < 2) \\ x + 1 & (2 < x \leq 3). \end{cases}$$

23. Find the maximum and minimum values of the function given by

$$y = \begin{cases} -x^2 + 2x + 5 & (0 \leq x \leq 2) \\ x^2 - 8x + 17 & (2 \leq x \leq 5). \end{cases}$$

24. A function $y = f(x)$ has a *relative maximum* at $x = c$ if there exists an interval (perhaps very small) centered around $x = c$ such that the maximum value of $f(x)$

over this interval occurs at $x = c$. Determine the relative maxima for the function drawn in Figure 7.4.

25. A function $y = f(x)$ has a *relative minimum* at $x = c$ if there exists an interval (perhaps very small) centered around $x = c$ such that the minimum value of $f(x)$ over this interval occurs at $x = c$. Determine the relative minima for the function drawn in Figure 7.4.
26. Can a relative maximum also be a maximum?
27. Can a relative maximum occur at an end point? Must a maximum be a relative maximum?
28. Theorem 7.1 remains true if the words "maximum" and "minimum" are replaced by "relative maximum" and "relative minimum." In addition, the following result is also valid. *If $f'(x) = 0$ and if the second derivative exists at $x = c$, then $x = c$ is a relative maximum if also $f''(c) < 0$, and $x = c$ is a relative minimum if also $f''(c) > 0$.*

    Use this second-derivative test to show that $f(x) = -x^2 + 2x + 5$ has a relative maximum at $x = 1$.
29. Use the second-derivative test given in Exercise 28 to find all relative maxima and minima for $y = x^3 - 21x^2 + 120x - 100$, $1 \le x \le 13$.
30. If both the first and second derivatives of $f(x)$ are zero at $x = c$, no conclusions can be drawn from the second-derivative test. Show that both derivatives are zero at $x = 0$ for the three functions $y = x^4$, $y = -x^4$, and $y = x^3$. Graph each function and verify that $y = x^4$ has a relative minimum at $x = 0$, $y = -x^4$ has a relative maximum at $x = 0$, and $y = x^3$ has neither at $x = 0$.

7.2 Modeling

A *model* is a representation of a particular situation. A map is a model of a geographical region, a college transcript is a model of a person's academic achievement, and an organization chart is a model of a company's management structure. Many situations, especially dynamic ones which change with time, can be modeled by mathematical equations.

Very few models are ever complete; that is, the model does not reveal every aspect of the situation it represents. A map does not detail weather conditions (unless it is a weather map, in which case it does not detail road construction), a college transcript does not indicate the ease in which grades were achieved, and an organization chart does not detail the personalities of the people filling the slots. Generally, this is of little consequence.

Each model is built for a particular purpose which usually requires representing only part of a situation. For a person planning a car trip, a road map may be an adequate model. For a person trying to locate the appropriate individual to see in a company, an organization chart is a good model. These models fit the needs of the people using them. This then is the criteria on which models are judged: Does the model adequately fit the needs of the person using the model?

The word "adequate" is important. What is considered adequate by one person may not be adequate to another. Consequently, a good model for one person is a bad model for someone else. This is a fact of life. There are no absolutes in modeling. The usefulness of a model is relative to its purpose.

Commercial situations are too complex to be modeled in their entirety. Simplifying assumptions are usually made, which neglect minor contributions. A model of price fluctuations for a given product may not include the effects of a possible strike in the industry supplying raw materials. A model of consumer spending may not include psychological factors. The choice of which factors should be neglected is often subjective and results in various models for the same situation. Neglecting a factor is reasonable if the resulting model adequately represents the given situation.

Models are important to management as an aid in decision making. Questions often arise as to which course of action out of many possible ones is the best. By applying each action to the model and observing the effects of these decisions on the model an optimal decision often can be found. In the succeeding sections, we model specific business situations with mathematical equations. We then use differentiation methods on the models to ascertain an optimal strategy for the business problem at hand.

One simplification we make throughout this chapter is the modeling of discrete variables with continuous ones. Most business quantities can assume only integer values. Automobile production is given in whole cars; profits are reported to the nearest dollar; the number of employees in an industry is a whole number. A production run of 19.73 cars, a profit of $195.73869, and 30,198.7 employees are not commercially realistic.

More formally, we say that a quantity or variable is *continuous* if, whenever it assumes two different values, it can assume all numbers between these values.* Car production is not a continuous variable; one can make 2 cars or 3 cars, but not 2.78 cars which is a number between 2 and 3. Profits are not continuous variables. A profit of $1.95 is realistic, as is a profit of $1.96, but a profit of $1.9557 is not realistic. A good example of a continuous variable is time. An order may be placed at 8:07 or at 8:08, and also at 8:07396 although we may have difficulty in measuring it to this accuracy. Variables that are not continuous are called *discrete*. Car production and profits are discrete variables.

Consider again Eq. (2):

$$P = -x^2 + 120x - 600 \qquad (0 \le x \le 150), \qquad [2]$$

which related the profit P to the number of units x produced for a particular commodity. We assumed that x could be any value between 0 and 150. Realistically, this is not the case, since production runs are discrete variables

---

* The concept of continuous variables is different from that of continuous functions. A rigorous definition of a continuous function can be found in any good calculus text.

which must be whole numbers. One cannot let $x = 49.8$. *For the model*, however, $x = 49.8$ is a perfectly good number. It can be substituted into Eq. (2) with little difficulty.

Equation (2) models a discrete variable (number of units produced) by a continuous variable x. The reason is that differentiation techniques cannot be applied to discrete variables, only to continuous ones. If we wish to optimize a commercial situation with the derivative, we first must convert all discrete independent variables to continuous ones. We consider a model, for example Eq. (2), to be valid if it adequately agrees with reality when the variable, perhaps x, assumes integer values.

Replacing discrete variables by continuous ones is a simple conceptual process. One imagines what would happen if all numbers between realistic values could be assumed. For example, what profit would be expected if 49.8 cars could be produced? A useful mathematical technique for replacing discrete variables by continuous ones is given in Appendix A.

7.3 Profit, Revenue, and Cost

A problem of great interest in production processes is to determine the size of the production run that will result in maximum profit. As we shall see, the optimal decision is not always the obvious one: Produce to capacity.

To model this problem mathematically, we first make the simplifying assumption that all goods produced can be sold. This assumption is reasonably valid in certain cases, for example, publishers of a current best seller. Next, we denote

TR = total revenue received from the sale of all items
TC = total cost incurred in manufacturing and preparing all items for sale

and

P = gross profit.

An obvious equation (model) for gross profit is

$$P = TR - TC. \qquad (3)$$

The total revenue TR and the total cost TC themselves can be decomposed into more fundamental components. We define the *unit price* of a product as the selling price of each unit of that product. It follows that

$$TR = (\text{unit price}) \times (\text{number of items sold}). \qquad (4)$$

Unit cost is defined as the cost of producing and preparing for sale one unit of a product. Thus

$$TC = (\text{unit cost}) \times (\text{number of items produced}). \tag{5}$$

Example 1 A manufacturer sells 2000 items at a unit price of $1.50. Determine the manufacturer's gross profit if the unit cost is $1.25.

Solution We first determine total revenue and total cost. From Eq. (4) we obtain

$$TR = (1.50)(2000) = \$3000.$$

Similarly, we use Eq. (5) to obtain

$$TC = (1.25)(2000) = \$2500.$$

Since $P = TR - TC$, we have

$$P = 3000 - 2500 = \$500.$$

Example 1 represents a commercial situation at its conclusion. All decisions regarding price ($1.50), cost ($1.25), and number of items sold (2000) were known, and only the final bookkeeping remained. There was no optimizing to be done. A simple optimizing problem is the following. How many items should the manufacturer sell to maximize profit if the price and cost of each item is $1.50 and $1.25, respectively.

The problem is to determine the number of items to be sold. At this point it is an unknown, so we denote it by x. It follows from Eqs. (3) through (5) that

$$TR = 1.50x$$
$$TC = 1.25x$$

and

$$P = 1.50x - 1.25x = 0.25x.$$

The gross profit is $0.25x$. Since P increases as x increases, we maximize profits by selling as many items as possible.

In practice, this is not always the case. Maximum profits do not occur at maximum sales. At some stage in an actual manufacturing operation, the added costs incurred in producing additional items become higher than the revenue received on the sale of these items.

In most manufacturing processes, the unit cost depends on the number of items produced. Each product involves two basic costs, an overhead cost and a production cost. To a large extent the overhead cost, such as insurance, lighting, security, mortgages, and equipment, does not depend on production size. It is fixed regardless of production size. Overhead costs, however, can be averaged over all items produced, so that the larger the production run, the smaller the overhead cost associated with each item.

As an example, consider the production of widgets having an overhead cost of $10,000 per month. The production cost, including raw materials and labor costs involved in the process, is 25¢ per unit. For a production run of 5000 units, the overhead cost per unit is 10,000/5000 = $2.00. The complete unit cost is the sum of the overhead and production costs, namely, 2.00 + 0.25 = $2.25 per unit. For a production run of 10,000 units, the overhead cost per unit is 10,000/10,000 = $1.00. Now the complete unit cost is 1.00 + 0.25 = $1.25 per unit. As expected, the unit cost for 10,000 items is less than the unit cost for only 5000 items.

Of course, at some point this analysis breaks down. If too many items are produced, additional plants and equipment must be obtained, which increases the total overhead cost and therefore the unit cost. Mathematically, we indicate the relationship between unit cost and production size by saying that the unit cost is a function of production size. The domain of this function is the various production sizes for which the function remains valid. In particular, letting UC denote unit cost and x the size of the production run, we can give the relationship in the widget example (an overhead cost of $10,000 and a unit production cost of $0.25) as

$$UC = \frac{10,000}{x} + 0.25. \tag{6}$$

If we also know that the capacity of the plant is 100,000 widgets per month, we add to this equation the domain

$$0 \leq x \leq 100,000. \tag{7}$$

Similarly, it should not be surprising that price and demand are related. Usually, high prices are associated with low demand and low prices are associated with high demand. That is, unit price is also a function of demand. But we are assuming that the demand is always sufficient to absorb any production run. Therefore, the production run becomes the limiting factor. Consequently, unit price is a function of production size x.

In the general problem, we wish to find the value of x, production size, that maximizes profit. Although both unit cost and unit price are functions of x, these functions are known, so Eqs. (3) through (5) can be used to generate a relationship between profit P and production size x. We find the value of x that maximizes P by using the optimization method developed in Section 7.1.

As a concrete example, we reconsider the widget problem involving a unit cost given by Eqs. (6) and (7). We also assume that the unit price UP is given as

$$UP = 2.25 - (2.5 \times 10^{-5})x.$$

Using Eq. (4), we determine

$$TR = (UP)(x) = [2.25 - (2.5 \times 10^{-5})x]x = 2.25x - (2.5 \times 10^{-5})x^2,$$

and using Eqs. (5) and (6) we find

$$TC = (UC)(x) = \left(\frac{10,000}{x} + 0.25\right)x = 10,000 + 0.25x.$$

The gross profit is given by Eq. (3) as

$$P = TR - TC = [2.25x - (2.5 \times 10^{-5})x^2] - (10,000 + 0.25x)$$
$$= -(2.5 \times 10^{-5})x^2 + 2.00x - 10,000.$$

The appropriate domain for this problem is given by Eq. (7) as $0 \leq x \leq 100{,}000$. We seek the value of x that maximizes

$$P = -(2.5 \times 10^{-5})x^2 + 2.00x - 10,000 \qquad (0 \leq x \leq 100{,}000).$$

Using the differentiation procedure detailed in Section 7.1, we first find

$$\frac{dP}{dx} = -(5 \times 10^{-5})x + 2.00.$$

Setting this derivative equal to zero, we calculate

$$-(5 \times 10^{-5})x + 2.00 = 0$$
$$(5 \times 10^{-5})x = 2.00$$
$$x = \frac{2.00}{5 \times 10^{-5}}$$
$$x = 40{,}000.$$

Other points where a maximum could occur are the end points $x = 0$ and $x = 100{,}000$ and points where the first derivative does not exist, of which there are none. It follows from Table 7.6 that the largest monthly profit is $P = \$30{,}000$ at a production run of 40,000 units, far from capacity.

Example 2 A publisher sells a certain book for \$15 per copy. The unit cost for this book is related to the number published per week x by the equation

$$UC = \frac{x}{5000} + 8 + \frac{12{,}000}{x}.$$

TABLE 7.6

x	$P = -(2.5 \times 10^{-5})x^2 + 2.00x - 10,000$
0	$-10,000$
40,000	$30,000$
100,000	$-60,000$

Determine the production run that will generate maximum profit. Assume that all books published can be sold, the capacity of the publishing house is 20,000 copies per week, and the publisher has contractual obligations to a distributor for 5000 copies per week.

Solution Here the unit cost depends on the production run, but the unit price does not; it is $15 per copy regardless of size. Using Eqs. (3) through (5), we calculate

$$TR = (15)x = 15x$$

$$TC = (UC)x = \left(\frac{x}{5000} + 8 + \frac{12,000}{x}\right)x = \frac{x^2}{5000} + 8x + 12,000$$

and

$$P = TR - TC = 15x - \left(\frac{x^2}{5000} + 8x + 12,000\right) = -\frac{x^2}{5000} + 7x - 12,000.$$

To determine a suitable domain, we note that production capacity limits the run to 20,000 copies, so $x \leq 20,000$. Contractual obligations require a run of at least 5000 copies, hence $x \geq 5000$. Together, we have $5000 \leq x \leq 20,000$. The problem is to find values of x that will maximize

$$P = -\frac{x^2}{5000} + 7x - 12,000 \quad (5000 \leq x \leq 20,000).$$

Setting the first derivative

$$\frac{dP}{dx} = -\frac{2x}{5000} + 7$$

equal to zero, we calculate

$$-\frac{2x}{5000} + 7 = 0$$

$$\frac{2x}{5000} = 7$$

$$2x = 35,000$$

$$x = 17,500.$$

Other points where the maximum may exist include the end points $x = 5000$ and $x = 20,000$ and points where dP/dx does not exist, of which there are none. It then follows from Table 7.7 that the maximum profit of $P = \$49,250$ is realized with a production run of $x = 17,500$ copies per week.

TABLE 7.7

x	$P = -\dfrac{x^2}{5000} + 7x - 12,000$
5,000	18,000
17,500	49,250
20,000	48,000

Exercises

1. A television manufacturer can sell all units produced at $200 per unit. The total cost TC (in dollars) in producing x units per week is given by

   $TC = 5000 + 20x + \frac{1}{2}x^2$.

   (a) Determine an expression (model) for total profit as a function of x.
   (b) Determine an appropriate domain for x if the weekly production capacity is limited to 3000 units.
   (c) Determine the maximum weekly profit.
2. Redo Exercise 1 if the weekly production capacity is only 150 units.
3. A company can sell all units of a particular product. The unit cost (in dollars) of producing x units is given by

   $UC = \dfrac{x}{4000} - 5 + \dfrac{50,000}{x}$.

   The unit price is fixed at $10 per item.
   (a) Determine the total revenue and total cost.
   (b) Determine an expression for the total profit as a function of x.
   (c) Determine an appropriate domain for x if the production capacity is limited to 50,000 units.
   (d) Find the maximum profit.
4. Redo Exercise 3 if in addition the manufacturer has contractual obligations for 40,000 units per week.
5. Assume that the manufacturer described in Exercise 3 is interested in minimizing costs rather than maximizing profits. How many units should be produced to achieve this objective?
6. A refrigerator manufacturer can sell all the refrigerators it can produce. The total cost for producing x refrigerators per week is given by the equation $TC = 300x + 2000$. The unit price also is related to the number produced by the formula $UP = 500 - 2x$.
   (a) Determine a formula for the total revenue.
   (b) Determine a formula for the profit.

(c) Determine an appropriate domain if the production capacity is 100 units per week.
(d) How many units should be produced to maximize profits?
7. Redo Exercise 6 if the production capacity is only 40 refrigerators per week.
8. A clothing manufacturer can sell all the suits produced each month up to a limit of 500. The unit cost and unit price are both related to the number of suits x produced by the formulas

$$UC = 50 + \frac{5000}{x} \quad \text{and} \quad UP = 200 - \tfrac{1}{4}x.$$

Determine the maximum profit the company can realize each month.

9. *Marginal revenue*, denoted by MR, is defined as the rate of change in total revenue with respect to units sold. That is, $MR = d(TR)/dx$, where x denotes units sold. Find the marginal revenue if
   (a) $TR = 15x$
   (b) $TR = 200x - \tfrac{1}{4}x^2$
   (c) $TR = 2.25x - (2.5 \times 10^{-5})x^2.$

10. Evaluate the marginal revenue found in Exercise 9c for both $x = 30{,}000$ and $x = 50{,}000$.

11. *Marginal cost*, denoted by MC, is defined as the rate of change in total cost with respect to units produced. That is, $MC = d(TC)/dx$. Find the marginal cost if
    (a) $TC = \dfrac{x^2}{5000} + 8x + 12{,}000$
    (b) $TC = 50x + 5000$
    (c) $TC = 10{,}000 + 0.25x.$

12. Evaluate the marginal cost found in Exercise 11c for both $x = 30{,}000$ and $x = 50{,}000$.

13. Intuitively, profit should increase with sales as long as the change in revenue with respect to sales is larger than the change in cost with respect to sales. As soon as MR becomes less than MC, it ceases to be profitable to produce more goods. Use Exercises 10 and 12 to verify this for the widget problem discussed in the text.

14. To verify the contention in Exercise 13 mathematically, differentiate Eq. (3) and show that the maximum profit is realized either where $MR = MC$ or where dP/dx does not exist or at an end point.

15. Use the results from Exercise 14 in Example 2.

7.4 Inventory Control*

Optimizing techniques are valuable in inventory control problems for minimizing storage costs while simultaneously ensuring that enough items are on hand to meet current demand. In this section, we present a simplified inventory model and then determine the optimum number of items to stock. Inherently, we

---
* This section contains more advanced optional material.

wish to avoid both overstockage with its resulting increase in storage costs, and understockage with its resulting loss in sales.

As an aid to modeling such situations, we make the following simplifying asumptions:

A. The demand for items is known. For example, one can sell exactly 2500 items per year.
B. The price of each item is fixed. For example, each case sells for $15.
C. The demand for the product is uniform. That is, as many items are sold during the first day as are sold during the 200th day, and as many items are sold during the fourth week as are sold during the seventeenth week.
D. The total cost associated with inventory consists of reordering costs and warehousing (storage) costs.
E. Inventory is reordered at equal time intervals and in equal lots. For example, 700 cases every 2 weeks.
F. We are interested in minimizing inventory costs over a complete year.

A uniform demand for a given product means that the rate of sales or the rate of depletion is a constant. Since rates of change are derivatives, we can rephrase this statement as, "The derivative of inventory with respect to time is a constant." We will see in Chapter 8 that, whenever the derivative is a constant, the function itself represents a straight line.* Thus the graph of inventory as a function of time must be basically a straight line.

It is not a complete straight line, however. Each time a new order is received, there is an immediate jump in the inventory on hand. Ideally, the inventory should decline to zero just as each new order arrives. Once this new order is stored, it will be depleted uniformly (graphically as a straight line) until the next order arrives. Thus, assumptions (C) and (E) imply a depletion of inventory as illustrated in Figure 7.5.

Note that the assumptions in some cases are rather restrictive. Rarely in practice are demand and supply quantities known with absolute precision. Nevertheless, these assumptions allow us to construct a mathematical model which can be used to derive specific rules governing the timing and size of reordering items. Our results will be accurate only to the extent that our assumptions closely model real life. Refinements to the model (which are beyond the scope of this text) can be made where necessary.

Clearly, if there were no reordering costs associated with accounting, processing, and shipping, we could order each item individually, timing its arrival to coincide with its sale, and thereby eliminate all storage costs. However,

---

* The converse is already known. If the graph of inventory as a function of time is a straight line, its equation is linear (see Chapter 2) and is given by $I = mt + b$, where I and t denote inventory and time, respectively. Using the rules of differentiation, we see that $dI/dt = m$, a constant.

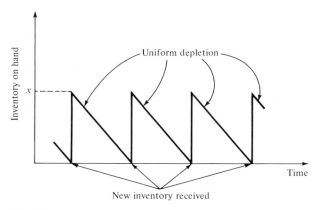

FIGURE 7.5

since reorder costs generally decline with increasing order size, it frequently is cost effective to order larger quantities. Unfortunately, storage costs mount with increasing order size, militating against exceedingly large orders.

To begin formulating a model, we let

TC = total inventory cost
TRC = total reordering cost

and

TSC = total storage cost.

It then follows from assumption (D) that

$$TC = TRC + TSC. \tag{8}$$

The total reordering cost TRC and the total storage cost TSC themselves can be decomposed into more fundamental components. We define the *individual reorder cost IRC* as the cost of reordering one complete shipment. Individual reorder costs consist of fixed costs covering such expenses as processing and accounting charges, and variable costs covering shipping, delivery, and interest charges. The variable charges are dependent on the size of the order.

Letting b denote the fixed costs per order, M the variable costs, and x the size of each order, we have

$$IRC = b + Mx. \tag{9*}$$

* An additional assumption has been made here which is expanded on in Exercise 13.

Although the size of an order x is unknown, it follows from assumption (E) that it is the same for each order. The total reorder cost TRC is equal to the individual reorder cost per order IRC times the number of orders placed during a year. We can determine the number of orders placed during a year by using assumptions (A) and (E). Denote the total number of items needed during a year by Q. Then Q/x is the number of orders. It follows from this result and Eq. (9) that

$$TRC = \text{(individual reorder cost)} \times \text{(number of orders placed)}$$
$$= (b + Mx)\left(\frac{Q}{x}\right)$$

or

$$TRC = \frac{bQ}{x} + QM. \tag{10}$$

Example 1 From past experience, a plastics distributor knows 25,000 drums of fiberglass resin will be sold in a year. Find an expression for the total yearly reorder cost if fixed costs are $10 per order and variable costs average $2 per drum.

Solution Here $Q = 25{,}000$, $M = \$2$, and $b = \$10$. The size of each order is not known, so it remains x. Substituting these quantities into Eq. (10), we obtain

$$TRC = \frac{250{,}000}{x} + 50{,}000.$$

Total storage cost, denoted TSC, is the cost of storing one unit of inventory for a year times the average yearly inventory stored. Careful consideration of Figure 7.5 reveals that the average inventory on hand is simply one-half the number of units received with each order. (Why?) That is, average inventory is $x/2$. Letting k denote the cost of storing one unit of inventory for a period of 1 year, we find

$$TSC = k\left(\frac{x}{2}\right). \tag{11}$$

Substituting Eqs. (10) and (11) into Eq. (8), we obtain a formula for the total inventory cost in terms of the size of each individual order:

$$TC = QM + \frac{bQ}{x} + \frac{kx}{2}. \tag{12}$$

To determine an appropriate domain for x, we first realize that the smallest conceivable order is $x = 1$, corresponding to single items being ordered as they are needed. The largest possible order is $x = Q$, which corresponds to all items needed for the entire year being ordered in one lot. Accordingly,

$$1 \leq x \leq Q \tag{13}$$

To calculate the optimum value of x that minimizes the total cost TC, we apply the optimization procedure in Section 7.1 directly to Eqs. (12) and (13). In particular,

$$\frac{d(TC)}{dx} = -\frac{bQ}{x^2} + \frac{k}{2}.$$

Setting this derivative equal to zero, we calculate

$$-\frac{bQ}{x^2} + \frac{k}{2} = 0$$

$$\frac{k}{2} = \frac{bQ}{x^2}$$

$$kx^2 = 2bQ$$

$$x^2 = \frac{2bQ}{k}$$

$$x = \pm\sqrt{\frac{2bQ}{k}}.$$

Since $x = -\sqrt{2bQ/k}$ is negative and therefore not in the domain defined by Eq. (13), we disregard it. The first derivative $d(TC)/dx$ does *not* exist at $x = 0$, but since this point also is not in the domain we can disregard it too. A minimum value can occur only at $x = \sqrt{2bQ/k}$ or at the end points $x = 1$ and $x = Q$.

Example 2 A distributor estimates requirements for television sets to be 1000 units over the next year. Fixed costs associated with each order to the manufacturer average $10 per order; variable costs average $2 per unit. Storage costs per unit per year are $8. Determine the optimum order size and how often each order should be placed to minimize inventory costs.

Solution Here $Q = 1000$, $b = 10$, $M = 2$, and $k = 8$, so Eq. (12) becomes

$$TC = 2000 + \frac{10{,}000}{x} + 4x.$$

The minimum occurs either at $x = \sqrt{2bQ/k} = \sqrt{2(10)(1000)/8} = \sqrt{2500} = 50$ or at the end points $x = 1$ and $x = 1000$. Substituting these values of x into the equation for TC, we generate Table 7.8 from which it follows that the optimal order size is $x = 50$.

TABLE 7.8

x	$TC = 2000 + \dfrac{10{,}000}{x} + 4x$
1	12,004
50	2,400
1000	6,010

Each order should be for 50 units. One thousand units are required over the year; hence the distributor must place $1000/50 = 20$ orders. Since orders must be placed at equal time intervals [assumption (E)], each order must be placed every one-twentieth of a year or, based on a 360-day year, every 18 days.

Example 3 A buyer estimates that a store will sell 200 units of a particular product during the year. Fixed costs associated with each order to the manufacturer average \$50 per order, while variable costs average 10¢ per unit. Storage costs per year per unit average $12\frac{1}{2}$¢. Determine the optimum size required to minimize total inventory costs.

Solution Here $Q = 200$, $b = 50$, $M = 0.10$, and $k = 0.125$, where all monetary figures have been expressed as dollars. With these values, Eq. (12) becomes

$$TC = 20 + \frac{10{,}000}{x} + 0.0625x.$$

The minimum occurs either at

$$x = \sqrt{\frac{2bQ}{k}} = \sqrt{\frac{2(50)(200)}{0.125}} = \sqrt{160{,}000} = 400,$$

or at the end points $x = 1$ and $x = 200$. We disregard the point $x = 400$, since it is outside the acceptable domain for this problem. The minimum can occur only at the end points. At $x = 1$, $TC = \$10{,}020.0625$, whereas at $x = 200$, $TC = \$82.50$. Minimum inventory cost is achieved by ordering all 200 units in one lot.

In general, the minimum storage cost is obtained when $x = \sqrt{2bQ/k}$, and this value is referred to as the *optimum order size* or *economic order quantity*. As Example 3 illustrates, however, the minimum is not always achieved at this point. We explore this fact in greater detail in Exercise 11.

Exercises

1. The cost of placing an order is $10, and the cost of storing one item for one year is $3. Determine the optimum order size if 540 items are required during a year.
2. The cost of placing an order is $1, and the cost of storing one item for one year is $25. Determine the economic order quantity if 31,250 items are required in a year.
3. The cost of placing an order is $4.80, and the storage cost per unit for one month is $2. Find the optimum order size if 500 units are needed over the next 6 months.
4. Suppose the variable costs associated with the item described in Exercise 3 are $1 per item but other features remain the same.
    (a) Determine the total reorder costs over the course of a year.
    (b) Find the total storage costs over the course of a year.
    (c) Find the total costs incurred over the course of a year.
5. The cost of placing an order is $9.60. One thousand units are required during a year, and the cost of carrying one item for 1 year is $1.92.
    (a) Determine the economic order quantity.
    (b) Find the number of orders that will be placed during a year.
6. For the situation described in Exercise 5 and the results obtained in parts (a) and (b), construct a figure similar to Figure 7.5.
7. Assume that Figure 7.6 adequately describes the inventory situation of a medium-sized appliance dealer.
    (a) Determine the economic order quantity for the dealer.
    (b) Find the total number of orders placed during a year.
    (c) Determine the total number of appliances sold during a year.
8. A concept closely related to the economic order quantity is the reorder point. If inventory on hand falls below a specified level, called the reorder point, a reorder is placed (the size of the reorder is of course the economic order quantity). Referring to Figure 7.6 and assuming it takes 1 month between the time an order is placed and receipt of the order, determine the reorder point.
9. Determine the reorder point for Exercise 5, assuming the time between placement and receipt of an order is 2 days. (*Hint:* Construct a graph similar to Figure 7.6.)
10. Assume the appliance dealer described in Exercise 7 purchases all his appliances from one manufacturer. Of what value would Figure 7.6 be to the manufacturer in calculating production schedules and raw material requirements?

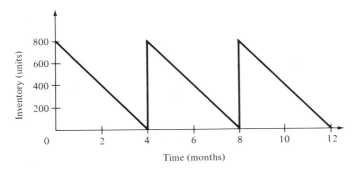

FIGURE 7.6

11. Using the second-derivative test described in Exercise 28 in Section 7.1, show that the economic order quantity $x = \sqrt{2bQ/k}$ is a relative minimum for Eq. (12). From this, conclude that whenever the economic order quantity lies in the domain $1 \leq x \leq Q$ it is also the minimum.
12. The minimum value for TC occurs at $x = 1$, $x = Q$, or $x = \sqrt{2bQ/2}$. None of these sizes involve the quantity M. Explain why this is reasonable.
13. Equation (9) presupposes a linear relationship between the order size x and the individual reorder cost IRC. Derive an expression for the total storage cost TC if the relationship between x and IRC is quadratic; that is, find a formula for TC if $IRC = b + Mx^2$. What is the economic order quantity in this case?

7.5 Econometrics*

An important but simple application of the derivative as a rate of change occurs in economics. By applying differentiation techniques to a simplified model of a national economy, we can gain interesting insights into the effect of an increase in overall business investment on the total economy. These insights are the beginnings of Keynesian theory.

Our first assumption is that the economy being modeled consists of only consumers (i.e., individuals) and businesses, ignoring the effects of government and foreign sectors. Such a model approximates conditions in the United States during the previous century. In the exercises, we refine this model to include the government.

Our second assumption is that businesses do not save money. All money received by business is distributed to banks to repay past loans or to stockholders in the form of dividends. Banks, being themselves businesses, return their receivables to individuals as interest payments. Therefore all money received by businesses is eventually distributed to individuals as dividends or interest. Our second assumption can be restated as, "All money spent by consumers and businesses is received only by consumers."

Each dollar received by consumers is either spent or saved. Since, by our second assumption, banks cannot save money, they lend all available funds to businesses to use for commercial expenditures. Businesses borrow money from banks for investment. As these investments produce money, the profits are used to repay loans from the banks and as dividends. Consumers receive money from other consumers, or from banks in the form of interest payments, or from businesses in the form of dividends. This money is either respent directly or saved. Saved money is lent to business for investment, and the cycle begins again.

In summary, we are making the following simplifying assumptions:

1. The economy consists of only consumers and businesses.
2. All money is received only by consumers.

* This section contains more advanced optional material.

3. Businesses borrow from consumer savings to obtain funds for business expenditures.

We are interested in the question, What are the effects of a change in business investment on the total economy?

Before attempting an answer, we must determine how we will measure the total economy. A reasonable measure is the gross national product which is simply the total amount of money spent within the economy. We denote this quantity by T. If we let B denote the amount of money spent by businesses, the question of interest can be restated: What is the change in T with respect to a change in B? Or, mathematically, what is dT/dB?

To calculate this derivative, we first must express T as a function of B. That is, we must model our economy in such a way that we can ultimately obtain one equation for T in terms of B. We have already

T = total expenditures within the economy

and

B = total business expenditures.

In addition, we now define

C = total consumer expenditures.

It then follows directly from assumption (1) that

$$T = C + B. \tag{14}$$

Equation (14) is a model of the economy, but it is not sufficient for our purpose. We need an equation for T strictly in terms of B (so we can differentiate), whereas Eq. (14) relates T to both B and C. More information is needed.

From all available data, it appears that consumer spending C can be decomposed into two parts, necessities and luxuries. Necessities are the absolute basic necessities, and luxuries are everything else. For example, a family may have to spend $30 per week for food to survive but, if more funds were available, they might spend $100 per week for food. This additional $70 for food would be classified as a luxury. Expenditures for basic necessities, which we denote as C_0, are reasonably fixed. Expenditures for luxuries are a function of the amount of money available; the more money available, the more money spent on luxuries.

Since all money spent by consumers and businesses T is received by consumers [assumption (2)], we see that expenditure for luxuries is a function of T. Furthermore, if we assume that this portion of consumer spending is proportional to the amount of money available, we have

$$C = mT + C_0, \tag{15}*$$

where m denotes the fraction of total income spent on luxuries.

To simplify the problem even further, we assume we are modeling an affluent society in which expenditures for basic necessities are much less than expenditures for luxuries. In such a case, we can replace Eq. (15) with

$$C = mT, \tag{16}$$

with little error.

Substituting Eq. (16) into Eq. (14) and rearranging, we find $T = mT + B$, or

$$T = \left(\frac{1}{1-m}\right)B. \tag{17}$$

Equation (17) is the model we were seeking. Differentiating it with respect to B, we obtain the desired rate of change as

$$\frac{dT}{dB} = \frac{1}{1-m}. \tag{18}$$

Equation (17) is the equation of a straight line, and Eq. (18) is its slope. It follows directly from our knowledge of slopes (Section 2.2) that an increase in business expenditures B by an amount Q forces a corresponding increase in T by $1/(1-m)$ times Q.

Example 1 Determine the effect of a $10,000 increase in business investment on the total national income if individuals as a group spend 75% of the national income.

Solution Since all monies spent within the economy are received by consumers [assumption (2)], total national income is the same as total expenditures T. A $10,000 increase in business investment is the same as an increase in business expenditures B of $10,000. From Eq. (18), the change in T with respect to B is $1/(1-m)$. Here $m = 0.75$, hence

$$\frac{1}{1-m} = \frac{1}{1-0.75} = \frac{1}{0.25} = 4.$$

---

* This was one of the key assumptions made by Lord Keynes in his general theory of employment, interest, and money.

That is, T must increase 4 units with every 1-unit change in B. A $10,000 change in B results in a change of $40,000 in T.

At first glance, this result is startling. Example 1 indicates that the total income for society as a whole can be increased by $4 with only a $1 increase in business investment, assuming $m = 0.75$. A little thought reveals why this is so.

After the initial expenditure of $1, the recipient, a consumer, receives his dollar as income. He or she spends 75¢ (75% of $1) and saves the rest. A second consumer receives the 75¢ spent by the first consumer. He or she in turn spends 56¢ (75% of 75¢) and saves the rest. A third consumer receives the 56¢ spent by the second consumer and continues the cycle. As this income is received and respent, the net effect will be an increase of $4 in the total income of society.

Of course these results are valid only to the extent that our model adequately represents a given society. It is unlikely that any economy including the United States economy during the previous century can be adequately modeled by Eqs. (14) and (16). The model does not include the time period required to respend increased income, the economic state of the people receiving this income, or the ability of society to produce the extra goods demanded by consumers having additional money. Unemployment and inflation are not parts of our model. Population growth, food supplies, and available resources have also been neglected. Nevertheless, the construction presented here is typical of that applied to other more sophisticated models.

Exercises

1. Determine the increase in the total national income of a society modeled by Eqs. (14) and (16) if $m = 0.6$ and business investment is increased by $25 million.
2. Determine the decrease in the total national income of a society modeled by Eqs. (14) and (16) if $m = 0.7$ and business investment is decreased by $10 million.
3. Redo Exercise 1 for a society modeled by Eqs. (14) and (15).
4. The constant m used in Eq. (14) is referred to as the *marginal propensity to consume*. The term $1/(1 - m)$ derived in Eq. (18) is known as the *multiplier*. Determine the value of the multiplier if the marginal propensity to consume is 0.8. Why do you think the term "multiplier" is used?
5. Determine the net effect on the income of a society after a $1000 increase in business expenditures has been cycled through five people if $m = 0.6$.
6. Consider a three-sector economy consisting of individual consumers, businesses, and governments each making expenditures denoted as C, B, and G, respectively. Assume that the following three equations adequately describe the relationship between these quantities:

$T = C + B + G$
$C = mT + C_0$

and

$B = B_0,$

where B_0 denotes a constant business expenditure and the other symbols are as previously defined in the text.
(a) Describe, in words, what each of these equations means physically.
(b) Determine the rate of change in total expenditures with respect to a change in government expenditures.
(c) What is the expected increase in total expenditures associated with an increase of $5 million in government expenditures if $m = 0.75$?

7. Redo Exercise 6 if the economy is modeled by the equations

$$T = C + B + G$$
$$C = mT + C_0$$

and

$$B = kT + B_0,$$

where k represents that fraction of total income spent by businesses for investment over and above some fixed amount B_0 which is always required. Take $k = 0.2$.

Chapter 8

The Integral

The process of finding tangents to curves leads, by a limiting process, to the definition of the derivative. Similarly, the process of finding areas under a curve leads, by another limiting process, to the definition of the integral. Although the problems of finding derivatives and integrals seem unrelated, they really form two sides of the same coin and together constitute the discipline of calculus.

From a business perspective, derivatives, when viewed as rates of change, play a much more directly important role than integrals. Applications do exist, however, requiring methods of integration in their solution. More important, integrals, when viewed as areas under a curve, are essential in defining probabilities and solving advanced statistical problems.

8.1 Areas

The general problem of finding an area under a curve is illustrated in Figure 8.1. We are given a function $y = f(x)$, and we want the area under its graph and above the horizontal x-axis between two fixed values of x; here $x = a$ and $x = b$. The function is known and presumed nonnegative; that is, its graph lies on or above the x-axis.

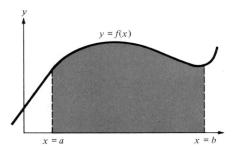

FIGURE 8.1

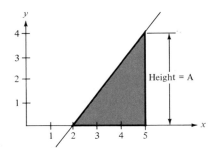

FIGURE 8.2

As a specific example, let us find the area under the curve given by $y = \frac{4}{3}x - \frac{8}{3}$ between $x = 2$ and $x = 5$. The curve itself is a straight line (why?), and the area of interest appears shaded in Figure 8.2. This area is contained within a right triangle having height 4 and base 3. Since the area enclosed by any triangle is one-half the base times the height, this particular area is $A = \frac{1}{2}(3)(4) = 6$.

As a second example, let us find the area under the curve given by $y = 0.2x + 2.8$ from $x = 1$ to $x = 6$. We have plotted the curve in Figure 8.3. By dividing the area into two parts as indicated in Figure 8.4, the total area is calculated easily as the sum of the area of a rectangle and a right triangle. Thus total area $= A_1 + A_2 = (5)(3) + \frac{1}{2}(5)(1) = 17.5$.

In the last example, we partitioned the desired area under the curve into two subregions, "stacking" one area above the other and then summing the two individual areas. Often it is more convenient to partition the area into horizontal sections.

Example 1 Find the area under the curve illustrated in Figure 8.5 from $x = 1$ to $x = 7$.

Solution The area under the curve is not a simple geometric region. Nonetheless, by partitioning this region as illustrated in Figure 8.6, we find that the desired area is the sum of three areas enclosed by two right triangles and one

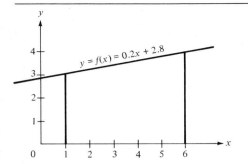

FIGURE 8.3

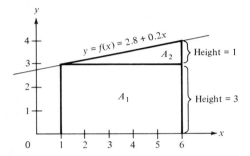

FIGURE 8.4

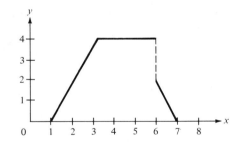

FIGURE 8.5

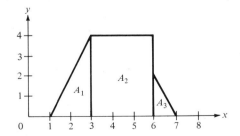

FIGURE 8.6

rectangle. Therefore

Total area $= A_1 + A_2 + A_3 = \frac{1}{2}(2)(4) + (3)(4) + \frac{1}{2}(1)(2) = 17$.

Horizontal partitioning into subareas followed by a summation of the individual parts is a useful procedure for *approximating* the areas under many curves. In particular, consider the general curve illustrated in Figure 8.7. It is not obvious how to divide the area under this curve into subareas which can be calculated easily. Indeed, it probably is not possible to do so. Instead we arbitrarily divide the area into n rectangles of equal width along the x-axis as in Figure 8.8. The height of each rectangle is the value of $y = f(x)$ at the midpoint of the base. The value of n, which represents the number of subrectangles we are using, is ours to choose. As we shall see, the larger we choose n (i.e., the more rectangles we use), the better our final approximation of the total area will be.

To simplify notation, we set $\Delta x = (b - a)/n$ and m_i equal to the midpoint

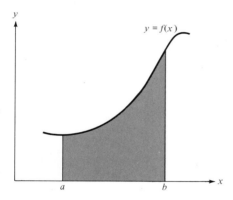

FIGURE 8.7

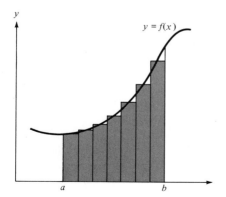

FIGURE 8.8

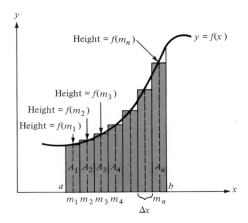

FIGURE 8.9

of the ith base. Accordingly, m_1 is the midpoint of the first rectangular base, m_2 is the midpoint of the second rectangular base, and so on. The width of each rectangle in Figure 8.8 is Δx. The height of the first rectangle is $y = f(m_1)$, the height of the second rectangle is $f(m_2)$, and the height of the nth rectangle is $f(m_n)$. These quantities are detailed in Figure 8.9.

The area of the first rectangle A_1 is $f(m_1) \Delta x$. The area of the second rectangle A_2 is $f(m_2) \Delta x$. The area of the nth and last rectangle A_n is $f(m_n) \Delta x$. An approximation to the complete area under the curve is the sum of these subareas, or

$$A \approx f(m_1) \Delta x + f(m_2) \Delta x + f(m_3) \Delta x + \cdots + f(m_n) \Delta x. \tag{1}$$

where the symbol $\approx$ denotes "approximately equals." Using the sigma notation introduced in Chapter 1, we can rewrite Eq. (1) as

$$A \approx \sum_{i=1}^{n} f(m_i) \Delta x. \tag{2}$$

Example 2 Approximate the area under the curve $y = x^2 + 5$ from $x = 0$ to $x = 6$ using three partitions.

Solution Here $a = 0$, $b = 6$, and $n = 3$ (since we are using three partitions), so $\Delta x = (6 - 0)/3 = 2$. The curve and the three partitions of equal width are drawn in Figure 8.10. With $f(x) = x^2 + 5$, we calculate the heights of the three rectangles as

$f(m_1) = f(1) = (1)^2 + 5 = 6$
$f(m_2) = f(3) = (3)^2 + 5 = 14$

and

$f(m_3) = f(5) = (5)^2 + 5 = 30.$

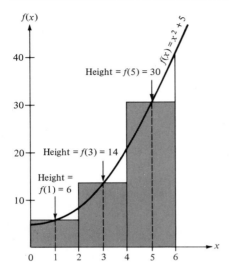

FIGURE 8.10

Then

$A_1 = f(1)\,\Delta x = (6)(2) = 12$
$A_2 = f(3)\,\Delta x = (14)(2) = 28$
$A_3 = f(5)\,\Delta x = (30)(2) = 60,$

and the total area is approximated by

$A \approx A_1 + A_2 + A_3 = 12 + 28 + 60 = 100.$

We will see in Section 8.3 that the actual area under the curve in Example 2 is 102. Therefore we have approximated this area reasonably well with only three partitions. Would we do even better with more partitions?

Example 3 Redo Example 2 using six partitions.

Solution Again $a = 0$ and $b = 6$, but now $n = 6$ (since we are using six partitions), so $\Delta x = (6 - 0)/6 = 1$. The curve and the six partitions of equal width are drawn in Figure 8.11. We calculate the six subareas as

$A_1 = f(\tfrac{1}{2})\,\Delta x = (\tfrac{21}{4})(1) = \tfrac{21}{4}$
$A_2 = f(\tfrac{3}{2})\,\Delta x = (\tfrac{29}{4})(1) = \tfrac{29}{4}$
$A_3 = f(\tfrac{5}{2})\,\Delta x = (\tfrac{45}{4})(1) = \tfrac{45}{4}$
$A_4 = f(\tfrac{7}{2})\,\Delta x = (\tfrac{69}{4})(1) = \tfrac{69}{4}$
$A_5 = f(\tfrac{9}{2})\,\Delta x = (\tfrac{101}{4})(1) = \tfrac{101}{4}$

and

$A_6 = f(\tfrac{11}{2})\,\Delta x = (\tfrac{141}{4})(1) = \tfrac{141}{4}.$

8.1 Areas **285**

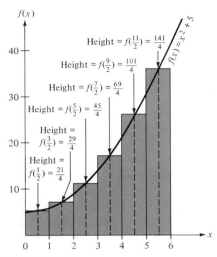

FIGURE 8.11

Thus

$$A \approx \tfrac{21}{4} + \tfrac{29}{4} + \tfrac{45}{4} + \tfrac{69}{4} + \tfrac{101}{4} + \tfrac{141}{4} = \tfrac{406}{4} = 101.5,$$

which is a better approximation than the result from Example 2.

We can save a great deal of time in computing approximate areas if we note that the graph of the function is not needed. Since the area of each rectangle is $f(m_i) \, \Delta x$, all that is necessary are the numbers Δx and $f(m_i)$, which is simply the function $f(x)$ evaluated at the midpoint of the ith base.

Example 4 Approximate the area under the curve $y = x^2 + 3x + 2$ from $x = 0$ to $x = 8$ using four partitions.

Solution Here $a = 0$, $b = 8$, and $n = 4$, so $\Delta x = (8 - 0)/4 = 2$. Each rectangle has a base of width 2. The midpoints of the four rectangles, illustrated in Figure 8.12, are located at $x = 1$, $x = 3$, $x = 5$, and $x = 7$. It follows that

$$f(m_1) = f(1) = (1)^2 + 3(1) + 2 = 6$$
$$f(m_2) = f(3) = (3)^2 + 3(3) + 2 = 20$$
$$f(m_3) = f(5) = (5)^2 + 3(5) + 2 = 42$$

and

$$f(m_4) = f(7) = (7)^2 + 3(7) + 2 = 72.$$

286   8 The Integral

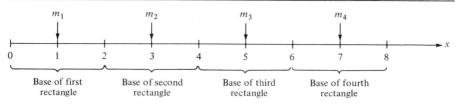

FIGURE 8.12

Then

$$A_1 = f(m_1)\,\Delta x = 6(2) = 12$$
$$A_2 = f(m_2)\,\Delta x = 20(2) = 40$$
$$A_3 = f(m_3)\,\Delta x = (42)(2) = 84$$
$$A_4 = f(m_4)\,\Delta x = 72(2) = 144$$

and

$$A \approx 12 + 40 + 84 + 144 = 280.$$

We show in Section 8.3 that the true area for this problem is $282\frac{2}{3}$. Thus we have generated a reasonably good approximation with just a few partitions. Again, would we do even better with more partitions?

Example 5   Redo Example 4 using eight partitions.

Solution   As in the previous example, we divide interval $[0, 8]$ into eight partitions of equal width, $\Delta x = (8 - 0)/8 = 1$, and then locate the midpoints of each partition. This is illustrated in Figure 8.13. Since $f(x) = x^2 + 3x + 2$, we compute

$$A_1 = f(m_1)\,\Delta x = f(\tfrac{1}{2})(1) = \tfrac{15}{4}$$
$$A_2 = f(m_2)\,\Delta x = f(\tfrac{3}{2})(1) = \tfrac{35}{4}$$
$$A_3 = f(m_3)\,\Delta x = f(\tfrac{5}{2})(1) = \tfrac{63}{4}$$
$$A_4 = f(m_4)\,\Delta x = f(\tfrac{7}{2})(1) = \tfrac{99}{4}$$
$$A_5 = f(m_5)\,\Delta x = f(\tfrac{9}{2})(1) = \tfrac{143}{4}$$
$$A_6 = f(m_6)\,\Delta x = f(\tfrac{11}{2})(1) = \tfrac{195}{4}$$
$$A_7 = f(m_7)\,\Delta x = f(\tfrac{13}{2})(1) = \tfrac{255}{4}$$

and

$$A_8 = f(m_8)\,\Delta x = f(\tfrac{15}{2})(1) = \tfrac{323}{4}.$$

8.1 Areas **287**

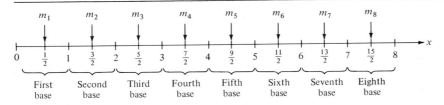

FIGURE 8.13

Then

$$A \approx \tfrac{15}{4} + \tfrac{35}{4} + \tfrac{63}{4} + \tfrac{99}{4} + \tfrac{143}{4} + \tfrac{195}{4} + \tfrac{255}{4} + \tfrac{323}{4}$$
$$= \tfrac{1128}{4} = 282,$$

which represents a better approximation than that obtained with four partitions.

From the past four examples, it appears that the approximations improve as we increase the number of partitions. One might guess that as the number of partitions becomes arbitrarily large, that is, as $n \to \infty$ (read "as n goes to infinity"), the corresponding approximations approach the true area and actually equal the true area in the limit. Mathematically,

$$\text{Area} = \lim_{n \to \infty} \sum_{i=1}^{n} f(m_i)\,\Delta x. \tag{3}$$

For all functions used in business situations, this is indeed the case; that is, as we take more partitions ($n \to \infty$), we obtain better approximations, and the true area is given by Eq. (3).

Unfortunately, Eq. (3) is unmanageable as a computational procedure, and we must find other methods for calculating exact areas. We do this in the next two sections. Nevertheless, Eq. (2) provides a good computational procedure for *approximating* a given area (see Examples 2 through 5), and it is often used for just such purposes.

Exercises

In Exercises 1 through 9 find the exact area under the given curves over the prescribed interval. Draw a graph for each problem.
1. $y = 5x$, from $x = 0$ to $x = 10$.
2. $y = x - 4$, from $x = 4$ to $x = 8$.

3. $y = x + 4$, from $x = -4$ to $x = 4$.
4. $y = x + 4$, from $x = 0$ to $x = 4$.
5. $y = 3x + 6$, from $x = 0$ to $x = 6$.
6. $y = 10$, from $x = -5$ to $x = 15$.
7. $y = \begin{cases} 2x & x < 1 \\ 4 - 2x & x \geq 1 \end{cases}$ (from $x = 0$ to $x = 2$).
8. $y = \begin{cases} 2x + 2 & x < 2 \\ -\tfrac{1}{2}x + 7 & x \geq 2 \end{cases}$

    (a) From $x = -1$ to $x = 14$ (b) From $x = 1$ to $x = 10$
    (c) From $x = 0$ to $x = 14$ (d) From $x = 0$ to $x = 10$.

9. $y = \begin{cases} x & (0 \leq x < 2) \\ 2 & (2 \leq x < 4) \\ -\tfrac{1}{2}x + 5 & (x \geq 4) \end{cases}$

    (a) From $x = 0$ to $x = 10$ (b) From $x = 0$ to $x = 8$.

In Exercises 10 through 20 find an approximation for the area under the given curve over the prescribed interval, using the number of partitions specified.

10. $y = x^2 + 3x + 3$, from $x = -5$ to $x = 7$, with $n = 2$.
11. $y = x^2 + 3x + 3$, from $x = -5$ to $x = 7$, with $n = 4$.
12. $y = x^2 + 3x + 3$, from $x = -5$ to $x = 7$, with $n = 6$.
13. $y = 2x^2 - 5$, from $x = 3$ to $x = 6$, with $n = 3$.
14. $y = 2x^2 - 5$, from $x = 3$ to $x = 6$, with $n = 6$.
15. $y = e^x$, from $x = 0$ to $x = 4$, with $n = 2$. (*Hint:* Use Appendix C.2.)
16. $y = e^x$, from $x = 0$ to $x = 4$, with $n = 4$.
17. $y = e^x$, from $x = -0.5$ to $x = 7.5$, with $n = 8$.
18. $y = x^3 - x + 4$, from $x = 2$ to $x = 8$, with $n = 4$.
19. $y = 2x^4 + 1$, from $x = -1$ to $x = 5$, with $n = 3$.
20. $y = xe^x$, from $x = 0$ to $x = 4$, with $n = 2$. (*Hint:* Use Appendix C.2.)

8.2 Antiderivatives

One quick method for calculating areas under curves rests on reversing the differentiation process described in Chapter 6. To find the area under a given curve defined by the equation $y = f(x)$, we first seek a new function $F(x)$ whose derivative is $f(x)$. Once we have determined $F(x)$, evaluating the area under the curve $y = f(x)$ is an easy matter.

Definition 8.1 A function $F(x)$ is an *antiderivative* of another function $f(x)$ if $F'(x) = f(x)$.

Definition 8.1 is rather straightforward. It simply states that, if $f(x)$ is the derivative of $F(x)$, then $F(x)$ is called the antiderivative of $f(x)$.

Example 1 Verify that $F(x) = 5x^3 + 3x^2 + 10x + 6$ is an antiderivative of $f(x) = 15x^2 + 6x + 10$.

Solution $F(x)$ is an antiderivative of $f(x)$ if and only if $F'(x) = f(x)$. Differentiating $F(x)$, we find

$$F'(x) = \frac{d}{dx}(5x^3 + 3x^2 + 10x + 6) = 15x^2 + 6x + 10,$$

which is $f(x)$. Therefore $F(x)$ is an antiderivative of $f(x)$.

Example 2 Verify that $F(x) = 5x^3 + 3x^2 + 10x + 17$ is also an antiderivative of the function $f(x)$ given in Example 1.

Solution Again $F(x)$ is an antiderivative of $f(x)$ if and only if $F'(x) = f(x)$. Differentiating this $F(x)$, we obtain

$$F'(x) = \frac{d}{dx}(5x^3 + 3x^2 + 10x + 17) = 15x^2 + 6x + 10,$$

which is $f(x)$. Therefore this $F(x)$ is also an antiderivative of $f(x)$.

We see from Examples 1 and 2 that a given function $f(x)$ can have more than one antiderivative. In fact, since the derivative of a constant is zero, any constant added to one antiderivative is another antiderivative. Accordingly, once one antiderivative has been found, we can construct infinitely many other antiderivatives by adding an arbitrary constant c to it. If $F(x)$ is one antiderivative of $f(x)$, then $F(x) + c$ is called the *complete antiderivative* or, more commonly, the *indefinite integral* of $f(x)$. The terms "complete antiderivative" and "indefinite integral" are synonomous. If follows from the previous examples that the complete antiderivative or indefinite integral of $f(x) = 15x^2 + 6x + 10$ is $F(x) = 5x^3 + 3x^2 + 10x + c$, where c denotes an arbitrary constant. Every choice of c (in particular, $c = 6$ in Example 1 and $c = 17$ in Example 2) results in a bona fide antiderivative.

The usual notation for the indefinite integral of $f(x)$ is

$$\int f(x)\, dx \tag{4}$$

The symbol $\int$ is called an *integral sign*. The dx term is only notation; it does have mathematical significance, but not to any work in the business world. The statement, "Evaluate $\int f(x)\, dx$," means "Find the complete antiderivative of $f(x)$."

The process of finding an indefinite integral is simplified by a set of rules analogous to the rules of differentiation. This is not surprising, since indefinite integration is the reverse process of differentiation. In fact, every differentiation formula has a counterpart in integration. In the remainder of this section, we simply list and illustrate some of the more common rules. We defer until Section 8.3 the link between indefinite integration and computing areas under curves.

Rule 1 For any real number n, $n \neq -1$,*

$$\int x^n \, dx = \frac{x^{n+1}}{n+1} + c \quad (c = \text{an arbitrary constant}).$$

Example 3 Evaluate $\int x^2 \, dx$.

Solution We seek the complete antiderivative of $f(x) = x^2$. Using Rule 1 with $n = 2$, we have

$$\int x^2 \, dx = \frac{x^{2+1}}{2+1} + c = \tfrac{1}{3}x^3 + c,$$

where c is an arbitrary constant.

To verify that $F(x) = \tfrac{1}{3}x^3 + c$ is indeed the complete antiderivative of $f(x) = x^2$, we differentiate. Since

$$F'(x) = \frac{d}{dx}(\tfrac{1}{3}x^3 + c) = x^2,$$

which is $f(x)$, $F(x)$ is the required function.

Example 4 Evaluate $\int \sqrt{x} \, dx$.

Solution We seek the complete antiderivative or indefinite integral of $f(x) = \sqrt{x}$. Noting that $\sqrt{x} = x^{1/2}$, we can rewrite $\int \sqrt{x} \, dx = \int x^{1/2} \, dx$. Then, using Rule 1 with $n = \tfrac{1}{2}$, we obtain

$$\int x^{1/2} \, dx = \frac{x^{(1/2)+1}}{\tfrac{1}{2}+1} + c = \frac{x^{3/2}}{\tfrac{3}{2}} + c = \tfrac{2}{3}x^{3/2} + c.$$

Example 5 Evaluate $\int 1 \, dx$.

Solution Since $x^0 = 1$, we have

$$\int 1 \, dx = \int x^0 \, dx = \frac{x^{0+1}}{0+1} + c = x + c.$$

---

* The special case $n = -1$ is covered by the formula $\int (1/x) \, dx = \ln x + c$, where $\ln x$ is the natural logarithm of x. This logarithm is different from the common logarithm discussed in Chapter 1.

8.2 Antiderivatives

Rule 2 If k is a known constant,

$$\int kf(x)\, dx = k \int f(x)\, dx.$$

Example 6 Evaluate $\int 6x^2\, dx$.

Solution Using Rule 2 with $k = 6$, we have

$$\int 6x^2\, dx = 6 \int x^2\, dx.$$

From Rule 1, $\int x^2\, dx = \tfrac{1}{3}x^3 + c$, hence

$$\int 6x^2\, dx = 6(\tfrac{1}{3}x^3 + c) = 2x^3 + 6c.$$

It is common to replace $6c$ simply by c, rationalizing that, since c denotes an arbitrary constant, 6 times c also represents an arbitrary constant. Therefore the two can be used interchangeably. Then

$$\int 6x^2\, dx = 2x^3 + c.$$

Rule 3 $\quad \int [f_1(x) \pm f_2(x)]\, dx = \int f_1(x)\, dx \pm \int f_2(x)\, dx.$

Example 7 Evaluate $\int (x^2 + x^5)\, dx$.

Solution Using Rule 3, we have

$$\begin{aligned}\int (x^2 + x^5)\, dx &= \int x^2\, dx + \int x^5\, dx \\ &= (\tfrac{1}{3}x^3 + c_1) + (\tfrac{1}{6}x^6 + c_2) \quad \text{(Rule 1 twice)} \\ &= \tfrac{1}{3}x^3 + \tfrac{1}{6}x^6 + (c_1 + c_2).\end{aligned}$$

Since c_1 and c_2 are both arbitrary constants, so too is $c_1 + c_2$. The sum $c_1 + c_2$ can be replaced by the one arbitrary constant c. Accordingly,

$$\int (x^2 + x^5)\, dx = \tfrac{1}{3}x^3 + \tfrac{1}{6}x^6 + c.$$

It is frequently convenient to combine Rules 2 and 3 into the following single formula:

$$\int [k_1 f_1(x) + k_2 f_2(x)]\, dx = k_1 \int f_1(x)\, dx + k_2 \int f_2(x)\, dx. \tag{5}$$

Example 8 Evaluate $\int (5x^2 - 3x)\, dx$.

Solution Using Eq. (5) with $k_1 = 5$ and $k_2 = -3$, we have

$$\int (5x^2 - 3x)\, dx = 5\int x^2\, dx - 3\int x\, dx$$
$$= 5(\tfrac{1}{3}x^3 + c_1) - 3(\tfrac{1}{2}x^2 + c_2) \qquad \text{(Rule 1 twice)}$$
$$= \tfrac{5}{3}x^3 - \tfrac{3}{2}x^2 + (5c_1 - 3c_2).$$

Since c_1 and c_2 are both arbitrary constants, so too is $5c_1 - 3c_2$. The entire quantity can be replaced by the one arbitrary constant c. Accordingly,

$$\int (5x^2 - 3x)\, dx = \tfrac{5}{3}x^3 - \tfrac{3}{2}x^2 + c.$$

Equation (5) can be generalized in a straightforward manner to the sums (and differences) of more than two functions. The resultant integrations usually lead to a combination of arbitrary constants like $5c_1 - 3c_2$ in Example 8. These combinations can always be replaced by the one arbitrary constant c.

Example 9 Evaluate $\int (2x^4 - 3x^2 + 1)\, dx$.

Solution

$$\int (2x^4 - 3x^2 + 1)\, dx = 2\int x^4\, dx - 3\int x^2\, dx + \int 1\, dx$$
$$= 2\left(\frac{x^5}{5} + c_1\right) - 3\left(\frac{x^3}{3} + c_2\right) + (x + c_3)$$
$$= \tfrac{2}{5}x^5 - x^3 + x + (2c_1 - 3c_2 + c_3)$$
$$= \tfrac{2}{5}x^5 - x^3 + x + c.$$

Example 10 Evaluate $\int (x^{9.9} - 24x^{5.7} + 35x^{-1.6} + 19x^{-2.1})\, dx$.

Solution

$$\int (x^{9.9} - 24x^{5.7} + 35x^{-1.6} + 19x^{-2.1})\, dx$$
$$= \int x^{9.9}\, dx - 24\int x^{5.7}\, dx + 35\int x^{-1.6}\, dx + 19\int x^{-2.1}\, dx$$
$$= \left(\frac{x^{10.9}}{10.9} + c_1\right) - 24\left(\frac{x^{6.7}}{6.7} + c_2\right) + 35\left(\frac{x^{-0.6}}{-0.6} + c_3\right)$$
$$+ 19\left(\frac{x^{-1.1}}{-1.1} + c_4\right)$$

$$= \left(\frac{1}{10.9}\right)x^{10.9} - \left(\frac{24}{6.7}\right)x^{6.7} - \left(\frac{35}{0.6}\right)x^{-0.6} - \left(\frac{19}{1.1}\right)x^{1.1}$$
$$+ (c_1 - 24c_2 + 35c_3 + 19c_4)$$
$$= 0.09x^{10.9} - 3.58x^{6.7} - 58.33x^{-0.6} - 17.27x^{1.1} + c$$

Rule 4 For any real number m, $m \neq 0$,

$$\int e^{mx} \, dx = \frac{e^{mx}}{m} + c. \tag{6}$$

To evaluate $\int e^{7x} \, dx$, we use Rule 4 with $m = 7$. Then,

$$\int e^{7x} \, dx = \tfrac{1}{7}e^{7x} + c.$$

Similarly, with $m = -\tfrac{1}{2}$ we have

$$\int e^{-(1/2)x} \, dx = \frac{e^{-(1/2)x}}{-\tfrac{1}{2}} + c = -2e^{-(1/2)x} + c.$$

The special case of $m = 0$ in Rule 4 is handled by the formula

$$\int e^{0x} \, dx = \int e^0 \, dx = \int 1 \, dx = x + c.$$

We illustrate the combined use of the previous four rules with the following example.

Example 11 Evaluate $\int (x^2 - 3e^{2x} + 4) \, dx$.

Solution

$$\int (x^2 - 3e^{2x} + 4) \, dx = \int x^2 \, dx - 3\int e^{2x} \, dx + 4\int 1 \, dx$$
$$= (\tfrac{1}{3}x^3 + c_1) - 3(\tfrac{1}{2}e^{2x} + c_2) + 4(x + c_3)$$
$$= \tfrac{1}{3}x^3 - \tfrac{3}{2}e^{2x} + 4x + (c_1 - 3c_2 + 4c_3)$$
$$= \tfrac{1}{3}x^3 - \tfrac{3}{2}e^{2x} + 4x + c.$$

Exercises

Evaluate the following integrals.

1. $\int x^2 \, dx$
2. $\int x^3 \, dx$
3. $\int x \, dx$
4. $\int 7 \, dx$
5. $\int (x^3 + x^2) \, dx$
6. $\int (x^3 + x^2 + x + 7) \, dx$
7. $\int (6x^2 + 3x) \, dx$
8. $\int (3x^2 + 2x + 5) \, dx$
9. $\int e^x \, dx$
10. $\int e^{5x} \, dx$
11. $\int (6x^2 + e^{5x}) \, dx$
12. $\int (e^x + e^{5x}) \, dx$
13. $\int (2x - 3e^x) \, dx$
14. $\int (x^{-3} + x^{-2}) \, dx$
15. $\int e^{-x} \, dx$
16. $\int (3 - e^{5x}) \, dx$
17. $\int \left(\dfrac{e^x + e^{-x}}{3} \right) dx$
18. $\int (5 + 1/x^2) \, dx$
19. $\int (4x^{1/2} + 3x^{1/3}) \, dx$
20. $\int (x^{1/2} + x^{-1/2}) \, dx$

8.3 The Definite Integral

We saw in Section 8.1 that the area under a positive curve is given mathematically by the formidable expression

$$\text{Area} = \lim_{n \to \infty} \sum_{i=1}^{n} f(m_i) \, \Delta x. \qquad [3]$$

The difficulty is in evaluating the right side of Eq. (3). Nonetheless, since it gives the *exact* area under a positive curve, it is an important mathematical quantity having its own name.

Definition 8.2 The *definite integral* of $f(x)$ between $x = a$ and $x = b$, denoted by $\int_a^b f(x) \, dx$, is

$$\int_a^b f(x) \, dx = \lim_{n \to \infty} \sum_{i=1}^{n} f(m_i) \, \Delta x. \qquad (7)$$

Warning: Although the notations for the indefinite integral and the definite integral are similar, the two quantities are completely different. The indefinite integral $\int f(x)\,dx$ is a function $F(x)$ whose derivative is $f(x)$; the definite integral $\int_a^b f(x)\,dx$ is a number defined by Eq. (7). Furthermore, we have methods for obtaining $\int f(x)\,dx$; we have no methods yet for calculating Eq. (7). There is, however, a powerful theorem which allows one to compute the definite integral of $f(x)$ from a knowledge of the indefinite integral of $f(x)$. It is precisely this result that justifies the similar notations. The relevant theorem is the following.

Theorem 8.1 (The Fundamental Theorem of Integral Calculus) If $F(x)$ is an antiderivative of $f(x)$ and if $F(x)$ is continuous (its graph has no holes),

$$\int_a^b f(x)\,dx = F(b) - F(a). \tag{8}$$

That is, to evaluate Eq. (7), first find one antiderivative of $f(x)$, call it $F(x)$. Any antiderivative will do; it need not be the complete antiderivative. Evaluate $F(x)$ at $x = b$ and $x = a$ and then subtract $F(a)$ from $F(b)$.

Example 1 Evaluate $\int_0^2 x^2\,dx$.

Solution From Example 3 in Section 8.2 one antiderivative of $f(x) = x^2$ is $F(x) = \frac{1}{3}x^3$ (take $c = 0$). Here $b = 2$ and $a = 0$, so $F(b) = F(2) = \frac{1}{3}(2)^3 = \frac{8}{3}$ and $F(a) = F(0) = \frac{1}{3}(0)^2 = 0$. Therefore

$$\int_0^2 x^2\,dx = \tfrac{8}{3} - 0 = \tfrac{8}{3}.$$

Example 2 Evaluate $\int_1^7 e^{5x}\,dx$.

Solution One antiderivative of $f(x) = e^{5x}$ is $F(x) = \frac{1}{5}e^{5x}$. Here $b = 7$ and $a = 1$, hence $F(b) = F(7) = \frac{1}{5}e^{35}$ and $F(a) = F(1) = \frac{1}{5}e^5$. Then,

$$\int_1^7 e^{5x} = \tfrac{1}{5}e^{35} - \tfrac{1}{5}e^5.$$

Every antiderivative of $f(x)$ when substituted into the right side of Eq. (7) yields the same answer for the definite integral of $f(x)$. To see why, let us redo Example 1 with the complete antiderivative of $f(x)$, namely, $F(x) = \frac{1}{3}x^3 + c$. Now $F(b) = F(2) = \frac{1}{3}(2)^3 + c = \frac{8}{3} + c$ and $F(a) = F(0) = \frac{1}{3}(0)^3 + c = c$. Then,

$$\int_0^2 f(x)\,dx = (\tfrac{8}{3} + c) - c = \tfrac{8}{3},$$

as before. Observe that the arbitrary constant c appeared in both $F(b)$ and $F(a)$. Once the difference $F(b) - F(a)$ was computed, the two c's subtracted to zero.

8 The Integral

This is equally true in general: The arbitrary constant c appears in both $F(b)$ and $F(a)$ and subtracts to zero when $F(b) - F(a)$ is calculated. Therefore, if we evaluate Eq. (7) with $c = 0$, we obtain the correct answer with a minimum number of calculations.

Example 3 Evaluate $\int_{-1}^{3} (5x^2 - 3x) \, dx$.

Solution From Example 8 in Section 8.2 we have $F(x) = \frac{5}{3}x^3 - \frac{3}{2}x^2$ (with $c = 5c_1 - 3c_2 = 0$). Then,

$$F(3) = \frac{5}{3}(3)^3 - \frac{3}{2}(3)^2 = \frac{135}{3} - \frac{27}{2} = \frac{189}{6}$$
$$F(-1) = \frac{5}{3}(-1)^3 - \frac{3}{2}(-1)^2 = -\frac{5}{3} - \frac{3}{2} = -\frac{19}{6}$$

and

$$\int_{-1}^{3} (5x^2 - 3x) \, dx = \frac{189}{6} - \left(-\frac{19}{6}\right) = \frac{189}{9} + \frac{19}{6} = \frac{208}{6} = 34\frac{2}{3}.$$

It is now a direct consequence of both Eqs. (3) and (7) that

$$\text{Area} = \int_a^b f(x) \, dx, \qquad (9)$$

when $f(x)$ represents a positive curve. Therefore we can use definite integration for computing *exact* areas.

Example 4 Find the area under the curve $y = x^2 + 5$ from $x = 0$ to $x = 6$.

Solution The curve has been drawn in Figure 8.14 and it is positive. One antiderivative of $x^2 + 5$ is $F(x) = \frac{1}{3}x^3 + 5x$. Using Eq. (8), we compute

$$\text{Area} = \int_0^6 (x^2 + 5) \, dx = F(6) - F(0) = \left(\frac{216}{3} + 30\right) - 0 = 102.$$

Compare this answer with the result from Example 2 in Section 8.1.

Example 5 Find the area under the curve $y = x^2 + 3x + 2$ from $x = 0$ to $x = 8$.

Solution One antiderivative of $x^2 + 3x + 2$ is $F(x) = \frac{1}{3}x^3 + \frac{3}{2}x^2 + 2x$. It follows from Eq. (8) that

$$\text{Area} = \int_0^8 (x^2 + 3x + 2) \, dx = F(8) - F(0)$$
$$= \left[\frac{1}{3}(8)^3 + \frac{3}{2}(8)^2 + 2(8)\right] - \left[\frac{1}{3}(0)^3 + \frac{3}{2}(0)^2 + 2(0)\right] = 282\frac{2}{3}.$$

Compare this answer with the results from Example 4 in Section 8.1.

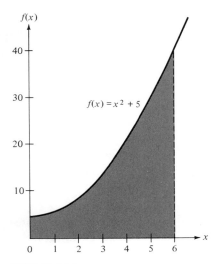

FIGURE 8.14

If the curve under consideration is not positive, Eq. (9) is not valid and the definite integral does not give the area. In such cases the definite integral is the area above the x-axis *minus* the area below the x-axis. In particular, consider $\int_{-1}^{1} x \, dx$. Here $F(x) = \frac{1}{2}x^2$ and

$$\int_{-1}^{1} x \, dx = \frac{(1)^2}{2} - \frac{(-1)^2}{2}$$

$$= \tfrac{1}{2} - \tfrac{1}{2} = 0.$$

Geometrically, this answer signifies that there is as much area above the x-axis as there is below the x-axis. See Figure 8.15.

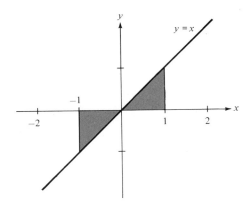

FIGURE 8.15

Up to this point, we have integrated only functions of the variable x. However, all the rules of integration are equally valid for functions of t, functions of z, or indeed functions of any variable. In particular,

$$\int_0^1 t^2 \, dt = \frac{t^3}{3}\bigg|_0^1 = \frac{(1)^3}{3} - \frac{(0)^3}{3} = \frac{1}{3}$$

and

$$\int_{-3}^2 e^{2z} \, dz = \frac{e^{2z}}{2}\bigg|_{-3}^1 = \frac{e^4}{2} - \frac{e^{-6}}{2}$$

where we have used the notation $F(x)|_a^b$ to denote $F(b) - F(a)$.

Finally, as a consequence of Theorem 8.1, it follows that the rules of integration developed in Section 8.2 are equally applicable for definite integration. For example,

$$\int_{-1}^1 2x \, dx = 2 \int_{-1}^1 x \, dx \quad \text{(Rule 2)}$$

and

$$\int_0^6 (x^2 + 1) \, dx = \int_0^6 x^2 \, dx + \int_0^6 1 \, dx \quad \text{(Rule 3)}.$$

Exercises

Evaluate the following integrals.

1. $\int_2^{10} x^2 \, dx$
2. $\int_2^{10} x^3 \, dx$
3. $\int_1^8 x \, dx$
4. $\int_3^9 7 \, dx$
5. $\int_0^3 (x^3 + x^2) \, dx$
6. $\int_0^3 (x^3 + x^2 + x + 7) \, dx$
7. $\int_{-2}^2 (6x^2 + 3x) \, dx$
8. $\int_{-4}^5 (3x^2 + 2x + 5) \, dx$
9. $\int_0^6 e^x \, dx$
10. $\int_0^3 e^{5x} \, dx$
11. $\int_{-1}^4 (6t^2 + e^{5t}) \, dt$
12. $\int_3^6 (e^t + e^{5t}) \, dt$
13. $\int_0^5 (2t - 3e^t) \, dt$
14. $\int_1^2 (z^{-3} + z^{-2}) \, dz$
15. $\int_{-3}^3 e^{-x} \, dx$
16. $\int_0^3 (3 - e^{2z}) \, dz$

17. $\int_0^{10} \left(\frac{e^t + e^{-t}}{3}\right) dt$ 18. $\int_1^2 (5 + 1/x^2) \, dx$

19. $\int_0^{16} (4z^{1/2} + 3z^{1/3}) \, dz$ 20. $\int_0^4 (t^{1/2} + t^{-1/2}) \, dt$

21. Why is the following formula valid?

$$\int_a^b \frac{du(x)}{dx} dx = u(b) - u(a).$$

In Exercises 22 through 28, find the exact area under the given curve over the prescribed interval.

22. $y = x^2 + 3x + 3$, from $x = -5$ to $x = 7$. Compare your answer with those for Exercises 10 through 12 in Section 8.1.
23. $y = 2x^2 - 5$, from $x = 3$ to $x = 6$. Compare your answer with those for Exercises 13 and 14 in Section 8.1.
24. $y = e^x$, from $x = 0$ to $x = 4$. Compare your answer with those for Exercises 15 and 16 in Section 8.1.
25. $y = x^3 - x + 4$, from $x = 0$ to $x = 2$.
26. $y = x^3 - x + 4$, from $x = 0$ to $x = 1$. Is this answer half the area obtained in Exercise 25? Did you expect it to be?
27. $y = e^{-x}$, from $x = 0$ to $x = 1$.
28. $y = x + 5e^{2x}$, from $x = 1$ to $x = 4$.

8.4 Applications of the Integral

In this section, we present three business problems which require integration in their solutions. The first two emphasize the relationship between integration and differentiation, while the third is a practical application of the integral to inventory holding cost analysis. Other important applications occur in statistical analysis and quality control, but these topics also require an understanding of probability at a level beyond the scope of this book.

Total Revenue from Marginal Revenue

In Section 7.3, we found that the derivative of total revenue with respect to sales (in units) is marginal revenue. Therefore, once we know the total revenue as a function of sales, we can find the marginal revenue by differentiation. Can the process be reversed? That is, can we find the total revenue from a knowledge of the marginal revenue?

Consider a situation in which the marginal revenue function is known to be $MR = \$10$ per unit. The function is plotted in Figure 8.16. Here the revenue earned for each unit sold is $10. If one unit is sold, total revenue is $10; if two units are sold, total revenue is $20; and so on. If d units are sold, the total revenue is d times $10, which graphically is given by the area under the marginal

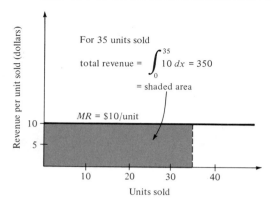

FIGURE 8.16

revenue curve from $x = 0$ to $x = d$. The shaded area in Figure 8.16 represents the total revenue earned from sales of 35 units, which is $350.

The same analysis holds for more complicated marginal revenue functions such as the one illustrated in Figure 8.17. The area under the marginal revenue curve represents the total revenue. Since area under a curve is obtained mathematically by integration, we have the formula

$$\text{Total revenue for } d \text{ units sold} = \int_0^d (\text{marginal revenue}) \, dx. \tag{10}$$

Example 1 Determine the total revenue for the marginal revenue illustrated in Figure 8.17, assuming total sales of 1000 units.

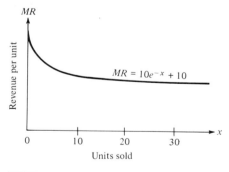

FIGURE 8.17

Solution

$$\text{Total revenue} = \int_0^{1000} (10e^{-x} + 10) \, dx$$

$$= 10 \int_0^{1000} e^{-x} \, dx + 10 \int_0^{1000} 1 \, dx$$

$$= 10(-e^{-x}|_0^{1000}) + 10(x|_0^{1000})$$
$$= 10(-e^{-1000} + 1) + 10(1000 - 0)$$
$$= \$10{,}010.$$

Example 2 The marginal revenue (in thousands of dollars) of XYZ Corporation is $MR = 1300 - 0.2e^{10x}$, where x denotes the number of units (in thousands) produced and sold. Determine the total revenue for 1000 units sold.

Solution One thousand units sold corresponds to $x = 1$. Therefore

$$\text{Total revenue} = \int_0^1 (1300 - 0.2e^{10x})\,dx$$
$$= 1300\int_0^1 1\,dx - 0.2\int_0^1 e^{10x}\,dx$$
$$= 1300(x|_0^1) - 0.2(\tfrac{1}{10}e^{10x}|_0^1)$$
$$= 1300(1 - 0) - 0.2(\tfrac{1}{10}e^{10} - \tfrac{1}{10})$$
$$= 859.47 = \$859{,}470.$$

Total Cost from Marginal Cost

In Section 7.3, we found that the marginal cost was the derivative of total production costs with respect to output in units. If the total production cost is known as a function of output, the marginal cost can be obtained by differentiation. Can we reverse the process? That is, if we know the marginal cost, can we use it to calculate the total production cost?

This situation is completely analogous to the one involving marginal and total revenues. Repeating that analysis we conclude

$$\text{Total production cost} = \int_0^d (\text{marginal cost})\,dx, \qquad (11)$$

where d is the number of units being produced.

Example 3 Determine the cost of producing 5000 textbooks if the marginal cost (in dollars per unit) is $C(x) = (x/5000) + 1.50$.

Solution

$$\text{Total cost} = \int_0^{5000} \left(\frac{x}{5000} + 1.50\right) dx$$
$$= \frac{1}{5000}\int_0^{5000} x\,dx + 1.5\int_0^{5000} 1\,dx$$
$$= 2500 + 7500$$
$$= \$10{,}000.$$

Inventory Carrying Costs

For many corporations, the cost of storing or holding inventory is a large portion of the overall cost of doing business. Rent payments for warehouse facilities, utilities, insurance, and security are only some of the factors involved. These costs are especially relevant to manufacturers who must store large amounts of their product prior to shipment and to distributors who do not manufacture products (and do not incur production costs) but who buy and store large quantities of goods for distribution to smaller retailers.

The key variables in determining holding costs are (1) the unit holding cost, which was defined in Chapter 7 as the cost of holding one unit of inventory for a specified time period, and (2) the actual time an item is held. In most situations, both these factors are known. The problem is to use this information to determine the total cost of holding a quantity of items.

As a specific example, consider a distributor of office supplies who receives 750 cases of pencils from a manufacturer every 30 days. Once the shipment has been received, the distributor begins to sell the pencils to retailers, continually reducing inventory until a new shipment of 750 cases arrives. From past experience, the inventory I is related to the number of days x since the last shipment by the equation $I = 750 - 25x$. This function is plotted in Figure 8.18. Note that, at $x = 0$, $I = 750 - 25(0) = 750$, indicating that the full delivery is in stock. At $x = 30$, however, $I = 750 - 25(30) = 0$, indicating that the entire stock has been depleted. The problem is to determine the cost of keeping inventory over a 30-day period, the carrying cost, given that the cost of storing one case of pencils for 1 day is K_c.

As a first approximation, we assume that all sales are made at the end of the day. Using the equation $I = 750 - 25x$, we find that 725 cases were being held after 1 day, 700 cases after 2 days, 675 cases after 3 days, and so on. There-

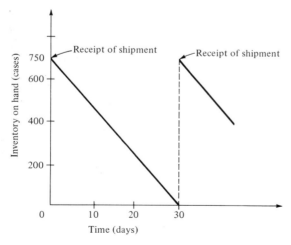

FIGURE 8.18

8.4 Applications of the Integral

TABLE 8.1

Day	Cases Held at the End of the Day	Cases Held during the Day	Daily Holding Cost
1	725	750	K_c times 750
2	700	725	K_c times 725
3	675	700	K_c times 700
⋮	⋮	⋮	⋮
30	0	25	K_c times 25

fore 25 cases were distributed each day. In reality, these cases are distributed at various time throughout the day but, under our simplifying assumption that the distributions are made at the end of the day, we can tabulate Table 8.1. Summing the daily costs, we obtain

$$\begin{aligned}\text{Total carrying cost} &= (K_c)(750) + (K_c)(725) + (K_c)(700) + \cdots + (K_c)(25) \\ &= (K_c)(750 + 725 + 700 + \cdots + 25) \\ &= (K_c)(11{,}625).\end{aligned}$$

Observe that the sum $750 + 725 + 700 + \cdots + 25$ represents an approximation to the area under the curve $I = 750 - 25x$ as illustrated by the shaded region in Figure 8.19. If instead of assuming that all distributions are made at the end of the day, we assume that half the distributions are made in the middle of the day and the other half at the end of the day, we would

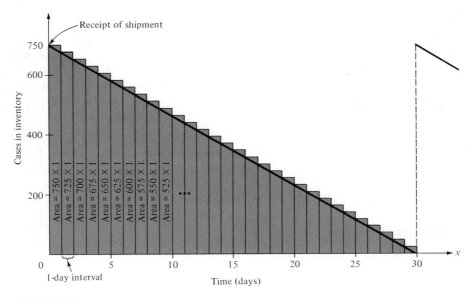

FIGURE 8.19

obtain a better approximation of the actual carrying cost. We would also obtain a better approximation to the area under the curve $I = 750 - 25x$. In fact, the actual carrying cost is K_c times the actual area under the curve. Since this area can be obtained exactly through integration, we have

$$\text{Total carrying cost} = K_c \int_0^{30} (750 - 25x) \, dx$$

$$= K_c \left[750(x \, |_0^{30}) - 25 \left(\frac{x^2}{2} \Big|_0^{30} \right) \right]$$

$$= K_c(11,250).$$

It follows that, when $K_c = 0.02$ (the cost of storing one case for 1 day is 2¢), the total monthly carrying cost is $(0.02)(11,250) = \$225.00$. If $K_c = 0.03$, the monthly carrying cost is $(0.03)(11,250) = \$337.50$.

We can generalize this example to other situations. Once $I(x)$, the inventory on hand as a function of time, and K_c, the unit holding cost, are known, the total carrying cost is simply

$$\text{Total carrying cost} = K_c \int_0^T I(x) \, dx, \qquad (12)$$

where T is the time period under consideration.

Example 4 Dataprocess Supply Company receives a shipment of 7200 boxes of line printer paper every 60 days. From past experience, it is known that the inventory on hand is related to the number of days x since the last shipment by the equation $I(x) = 7200 - 2x^2$. The daily holding cost for one box of paper is 0.2¢. Determine the total cost of maintaining inventory for 60 days.

Solution Using Eq. (12), we calculate

$$\text{Total carrying cost} = 0.2 \int_0^{60} (7200 - 2x^2) \, dx$$

$$= (0.2)(288,000)$$

$$= 57,600\text{¢ or } \$576.$$

Exercises

1. The marginal revenue function for a particular product is $5 + e^{-0.25x}$, where x denotes the number of units sold. Determine the total revenue obtained from the sale of 200 units.

2. Steelex, a manufacturer of steel springs, determines the price for each box of springs as a function of the number of boxes sold to date according to the formula

$$P = \frac{20}{x^2 + 1} + 10.$$

Set up but do not evaluate the required integral to determine the total revenue received in the sale of 1000 boxes of springs. (*Hint:* P is the manufacturer's marginal revenue.)

3. Determine the total cost of producing 1000 cartons of staples if the marginal cost function is given by $C(x) = (x^2/5000) + 0.50$, where $C(x)$ is in cents per carton. Find the additional cost incurred from an increase in production from 1000 to 2000 boxes.

4. Determine the total cost of producing 2500 boxes of pencils if the marginal cost of production is given by $C(x) = 0.25x^{1/2} + 0.10$ where $C(x)$ is in cents per box.

5. The marginal cost of producing pairs of shoes (x) is given by the function $C(x) = 5 + e^{0.01x}$. Calculate the cost of producing 100 pairs of shoes after an initial production run of 50 pairs.

6. The marginal revenue of selling x picture frames is $R(x) = 3.5$. The associated marginal cost function is $C(x) = 900e^{1.5x}$, where x is measured in thousands of frames. Determine the profit in producing and selling 1000 frames.

7. Mantrel Plastics receives a shipment of 8000 rolls of fiberglass every 30 days. The inventory function for this product (rolls on hand x as a function of time in days) is $I(x) = 9000 - 300x$. The cost of holding one roll of fiberglass for 1 day is 3¢. Find the total cost of maintaining inventory for 30 days. Determine the inventory carrying cost for 1 year under the assumption that a year has 12 months each of 30 days' duration.

8. Quantec Container receives a shipment of 63,000 steel drums every 30 days. The inventory function for this product is $I(t) = 63,000 - 70t^2$. Find the total cost of maintaining inventory for 30 days if the cost of holding one drum for 1 day is 2¢.

9. The demand per year D of a new brand of dog food (in thousands of cases) is given by the equation $D = 10 - 7e^{-x}$, where x denotes the number of years the product has been on the market. Determine the total sales for the first 4 years of the product's life.

10. A company's sales per year x (in thousands of boxes) is given as a function of the number of years the company has been in business by $S(x) = 3 + 5x$. Determine (a) the total sales for the first 8 years in business, (b) the total sales for the first 15 years in business, (c) a general expression for the total sales as a function of years in business, (d) the year when the total sales reaches 335.5 thousand boxes.

11. The company described in Exercise 10 earns $10 for each box sold. Using the results of parts (a) and (b) in Exercise 10 determine the income during the first 8 years of business and the first 15 years of business.

12. A company's sales revenue S as a function of years the company has been in business is given by $S = 70e^{0.15x}$. S is measured in thousands of dollars, while x is measured in years. Determine the total sales revenue earned over the first 10 years in business.

8.5 Substitution of Variables

The definite integral of $f(x)$ is easy to evaluate once the antiderivative of $f(x)$ is known. We simply use the fundamental theorem of integral calculus. In turn, antiderivatives can be obtained by the rules in Section 8.2 for many common functions. Unfortunately, these rules do not apply to all functions. For example, none of the rules are applicable to

$$\int_0^3 \frac{x}{\sqrt{x^2 + 16}} \, dx. \tag{13}$$

The method of substitution of variables is a four-step procedure for converting troublesome integrals such as Eq. (13) into the simple integral

$$\int_a^b z^n \, dz. \tag{14}$$

Integral (14) is particularly easy to evaluate, since one antiderivative of z^n is $z^{n+1}/(n+1)$. The procedure for substitution of variables is:

Step 1 Introduce a new variable z and set it equal to part of the function being integrated. Exactly which part is a matter of trial and error, experience, and luck.

Step 2 Differentiate z with respect to x and solve for dz. If dz does not appear in the original integral, stop; return to Step 1 and try another expression for z. The original choice of z will not work.

Step 3 Find the equivalent limits for the new z variable that correspond to the x limits of integration of the original integral.

Step 4 Substitute z, dz, and the new limits into the original integral.

Example 1 Use the method of substitution of variables to evaluate

$$\int_1^5 (1 + x)^3 \, dx.$$

Solution

Step 1: Set $z = (1 + x)$. \hfill (15)

Step 2: Differentiating z, we find $dz/dx = 1$. Treating dz/dx as a fraction, we solve for dz as $dz = dx$.

Step 3: The limits on the given integral are for x. Since the new integral will be in terms of z, we must determine suitable z limits corresponding to the known

x limits. It follows from Eq. (15) that, when $x = 1$ (the lower limit of integration), $z = 1 + 1 = 2$, and when $x = 5$ (the upper limit of integration), $z = 1 + 5 = 6$.

Step 4: Substituting z, dz, and the new limits into the original integral, we obtain

$$\int_1^5 (1 + x)^3 \, dx = \int_2^6 z^3 \, dz = \frac{z^4}{4}\bigg|_2^6 = \frac{(6)^4}{4} - \frac{(2)^4}{4} = 320.$$

Example 2 Evaluate $\int_0^2 12x^3(1 + 3x^4)^2 \, dx$.

Solution

Step 1: Set $z = (1 + 3x^4)$. (16)

Step 2: $dz/dx = 12x^3$. Solving for dz, we have $dz = 12x^3 \, dx$ which appears in the original integral.

Step 3: From Eq. (16), we conclude that, when $x = 0$ (the lower limit), $z = 1 + 3(0)^4 = 1$ and, when $x = 2$ (the upper limit), $z = 1 + 3(2)^4 = 49$.

Step 4: Making the following required substitutions,

$$\begin{array}{c} z=49 \to \\ z=1 \to \end{array} \int_{x=0}^{x=2} \underbrace{12x^3}_{dz}\underbrace{(1+3x^4)^2}_{z} \, dx$$

we have

$$\int_0^2 12x^3(1 + 3x^4)^2 \, dx = \int_1^{49} z^2 \, dz = \frac{z^3}{3}\bigg|_1^{49} = \frac{(49)^3}{3} - \frac{(1)^3}{3} = 39{,}216.$$

The choices of z in Examples 1 and 2 were largely a matter of experience. In each, we chose a z which on substitution resulted in an expression of the form z raised to a power.

In solving for dz, we treated dz/dx as a fraction. Although first derivatives can be treated in this fashion, it is not good practice to do so except when using the method of substitution of variables. Higher-order derivatives are never treated as fractions.

Example 3 Evaluate $\int_0^1 5e^{5x}(1 + e^{5x})^{10} \, dx$.

Solution We first inspect the integral for some function of x raised to a power. Noting that $1 + e^{5x}$ is such a function, we try

$$z = 1 + e^{5x}. \tag{17}$$

Differentiating z, we obtain $dz/dx = 5e^{5x}$ and $dz = 5e^{5x}\, dx$ which appears in the original integral. It follows from Eq. (17) that, when $x = 0$, then $z = 1 + e^0 = 2$ and, when $x = 1$, then $z = 1 + e^5 \approx 1 + 148.41 = 149.41$. These values of e^0 and e^5 are tabulated in Table C.2. Making the following required substitutions,

$$\begin{array}{c} z=149.41 \to x=1 \\ z=2 \to x=0 \end{array} \int 5e^{5x}\underbrace{(1 + e^{5x})^{10}}_{z}\underbrace{}_{dz}\, dx$$

we find

$$\int_0^1 5e^{5x}(1+e^{5x})^{10}\, dx = \int_2^{149.41} z^{10}\, dz = \left.\frac{z^{11}}{11}\right|_2^{149.41}$$

$$= \tfrac{1}{11}[(149.41)^{11} - (2)^{11}] \approx 7.53 \times 10^{22}.$$

As stipulated in Step 3 of the method of substitution of variables, the choice of z is useful only if the resultant expression for dz appears in the original integral. However, sometimes dz can be forced into the integral even if it is not there initially.

Example 4 Evaluate $\int_{-1}^1 x^3(1 + 5x^4)^7\, dx$.

Solution Looking for a function raised to a power, we find $(1 + 5x^4)$ raised to the seventh power and try

$$z = (1 + 5x^4). \tag{18}$$

Then, $dz/dx = 20x^3$ and $dz = 20x^3\, dx$ which is *not* in the original integral. However, the only part missing is the constant 20. Applying Rule 2 in Section 8.2 to definite integrals and noting that multiplication by $1 = 20/20$ will not affect the final answer, we obtain

$$\int_{-1}^1 x^3(1+5x^4)^7\, dx = \int_{-1}^1 \tfrac{20}{20}x^3(1+5x^4)^7\, dx = \tfrac{1}{20}\int_{-1}^1 20x^3(1+5x^4)^7\, dx.$$

Our substitution is applicable to the last integral. Using Eq. (18), we have $z = 1 + 5(-1)^4 = 6$ when $x = -1$, and $z = 1 + 5(1)^4 = 6$ when $x = 1$. Therefore

$$\int_{-1}^1 x^3(1+5x^4)^7\, dx = \tfrac{1}{20}\int_{-1}^1 20x^3(1+5x^4)\, dx$$

$$= \tfrac{1}{20}\int_6^6 z^7\, dz$$

$$= \frac{1}{20}\left(\frac{z^8}{8}\Big|_6^6\right)$$
$$= \frac{1}{20}\left[\frac{(6)^8}{8} - \frac{(6)^8}{8}\right] = 0.$$

Example 5 Evaluate $\int_0^3 \frac{x}{\sqrt{x^2+16}}\, dx$.

Solution We first look for a function raised to a power. Since

$$\frac{x}{\sqrt{x^2+16}} = \frac{x}{(x^2+16)^{1/2}} = x(x^2+16)^{-1/2},$$

such a function is

$$z = x^2 + 16. \tag{19}$$

Here $dz/dx = 2x$ and $dz = 2x\,dx$. The expression $2x\,dx$ is not in the given integral, but only the constant 2 is missing. We can rectify this omission. From Eq. (19), $z = 16$ when $x = 0$, and $z = 25$ when $x = 3$. Then,

$$\int_0^3 \frac{x}{\sqrt{x^2+16}}\, dx = \int_0^3 x(x^2+16)^{-1/2}\, dx$$
$$= \int_0^3 \tfrac{2}{2}x(x^2+16)^{-1/2}\, dx$$
$$= \tfrac{1}{2}\int_0^3 2x(x^2+16)^{-1/2}\, dx$$
$$= \tfrac{1}{2}\int_{16}^{25} z^{-1/2}\, dz$$
$$= \frac{1}{2}\left(\frac{z^{-(1/2)+1}}{-\tfrac{1}{2}+1}\Big|_{16}^{25}\right)$$
$$= \frac{1}{2}\left(\frac{\sqrt{z}}{\tfrac{1}{2}}\Big|_{16}^{25}\right)$$
$$= \sqrt{25} - \sqrt{16}$$
$$= 1.$$

Exercises

In Exercises 1 through 10, evaluate the given integrals.

1. $\int_0^1 4x^3(1+x^4)^6\, dx$ 2. $\int_0^1 4x^3(1+x^4)\, dx$

3. $\int_{-1}^{0} x^2(6 + 6x^3)^3 \, dx$

4. $\int_{0}^{1} 3e^{3x}(10 + e^{3x})^3 \, dx$

5. $\int_{0}^{0.05} (2x + 4e^{4x})(x^2 + e^{4x})^3 \, dx$

6. $\int_{0}^{0.05} (x^2 + e^{3x})(5 + x^3 + e^{3x})^2 \, dx$

7. $\int_{-1}^{1} (3x^2 + 3e^x)(6 + x^3 + 3e^x) \, dx$

8. $\int_{0}^{\sqrt{5}/2} 8x(4 + 4x^2)^{1/2} \, dx$

9. $\int_{1}^{3} (x^2 + 2x)(x^3 + 3x^2) \, dx$

10. $\int_{-2}^{1} (x + 1)\sqrt{x^2 + 2x} \, dx$

The method of substitution of variables can be used to reduce troublesome integrals to simpler integrals not of the form given in Eq. (14). In Exercises 11 through 14, use the indicated substitution to reduce the given integral to a simpler one and then evaluate, using Appendix C.2.

11. $\int_{0}^{1} 2xe^{x^2} \, dx, \ z = x^2$.

12. $\int_{1}^{2} x^2 e^{x^3} \, dx, \ z = x^3$.

13. $\int_{-1}^{1} (1 + x)e^{(x^2 + 2x)} \, dx, \ z = x^2 + 2x$.

14. $\int_{-1}^{0} xe^{-2x^2} \, dx, \ z = -2x^2$.

PROBABILITY

Part

Chapter

Sets

The foundation of all probability is sets. Therefore we begin by developing the basic definitions and algebraic operations pertinent to them. Most important for our purposes is the relationship between outcomes of a commercial process and a set, which is covered in Section 9.4. Once this relationship is understood, it can be used to determine the probabilities associated with the various possible outcomes.

9.1 Sets

A *set* is a collection of objects. Examples are the set of all clocks in the world, the set of all males in the United States, the set of all males and females in New Jersey, the set of all elephants in the San Diego Zoo, and the set of all real numbers between 2.05 and 7.13. Any collection, group, or class of well-defined objects is a set. The objects in a set are called *elements*.

One way to specify sets is to list all the individual elements. Traditionally, this listing is written in a row with the elements separated by commas and the entire list enclosed in braces. Using this notation, the set of all integers between

2 and 9, inclusive, is written as {2, 3, 4, 5, 6, 7, 8, 9} and is read "the set consisting of 2, 3, 4, 5, 6, 7, 8, and 9." The set of the first five presidents of the United States is {George Washington, John Adams, Thomas Jefferson, James Madison, James Monroe}, and the set of all outcomes of flipping a coin twice is {HH, HT, TH, TT}, where H denotes a head and T denotes a tail.

The order in which elements are listed in a set is irrelevant; only the elements themselves matter. Thus {1, 2, 3} is the same as the set {3, 2, 1}, since they contain the same elements; both represent the set of the first three positive integers. The set {John, Jim, Mary, Alice} is the same as the sets {Mary, Jim, Alice, John} and {Alice, Mary, Jim, John}.

Two sets are *equal* if and only if they contain the same elements regardless of the order in which the elements are listed. Accordingly, {1, 2, 3} = {3, 2, 1} and {red, blue, green, yellow} = {yellow, blue, red, green}. Note, however, that {1, 2, 3} is not equal to {1, 2, 3, 4}, since the second set contains an element not found in the first set.

An element is never listed twice in a set. The set of letters in the word "meet" is {m, e, t}. Even though meet has two e's, only one appears in set notation. The set of letters in the word "hoola hoop" is {h, l, a, p, o}. Again, order does not matter, and each element is listed only once.

Specifying a set by listing its elements is reasonable when there are a few elements in the set, but is not reasonable or practical when there are many elements in the set. A listing of the set of all United States citizens would be lengthy indeed, and trying to list the set of all positive integers would be futile, since there are infinitely many. Another method for specifying a set is to state a rule that all the elements satisfy. Thus the set of all integers between 1 and 5 can be specified as {all integers between 1 and 5}. This set of course is the set {2, 3, 4}. The set of all United States citizens is given by {all United States citizens}. Here the rule clearly states what elements are in the set. Whenever a rule is used to specify a set, the elements are all objects that satisfy this rule.

It is common to denote sets by capital letters. For example, if we let A be the set of all integers between 1 and 5, then $A = \{2, 3, 4\}$. Similarly we can let $B = \{red, green, blue, yellow\}$ and $C = \{all\ positive\ integers\}$. The symbol $\in$ is a shorthand mathematical symbol meaning "is an element of" or "belongs to" and is used to denote that an element belongs to a particular set. The statement "$2 \in A$" is read "2 belongs to the set A," or "2 is an element of A." The statement "Mary Jones $\in C$" is read "Mary Jones is an element of C" or "Mary Jones belongs to C." To indicate that an element does *not* belong to a set, a slash is drawn through the $\in$ symbol. The statement "$5 \notin A$" is read "5 does not belong to A" or "5 is not a member of A."

Just as a particular element is a member of a set, it frequently happens that all the members of a complete set are also elements of another set. If every element in a set A is also in B, the set A is said to be a *subset* of the set B. The notation for a subset is the symbol $\subset$, which is read "is a subset of" or "is contained in." To illustrate this standard notation, consider the sets

$A = \{1, 2\}$
$B = \{1, 2, 3, 4\}$
$C = \{1, 3, 4\}$

and

$D = \{2, 5\}$.

The set A is a subset of B, written $A \subset B$, since every element of A is also in B. The set C is a subset of B, written $C \subset B$, since every element in C is also in B. The set D is not a subset of B, since D contains an element not found in B, namely, 5. Similarly, B is not a subset of A, since B contains an element (actually two elements) not found in A. The notation for "is not a subset" is $\not\subset$. Therefore $D \not\subset C$, and $C \not\subset A$.

It follows (somewhat surprisingly) from our definition of a subset that a set is always a subset of itself. That is, $A \subset A$ for every set A. Note that $A \subset A$ if every element in the first set (here A) is also an element of the second set (also A). This is obviously true, so $A \subset A$ is correct.

A particularly interesting set is the *empty set* (sometimes called the *null* set or *void* set), denoted by $\emptyset$. The empty set is a set containing no elements. Two examples are

$\emptyset = \{\text{people living on the moon}\}$

and

$\emptyset = \{\text{pregnant males}\}$.

The empty set is a subset of every other set, since it satisfies our definition of a subset; that is, $\emptyset \subset A$ for every set A.

Exercises

1. The following sets represent stock portfolios of five different investors:
   $A = \{\text{Exxon, General Motors, RCA, IBM}\}$
   $B = \{\text{Exxon, IBM}\}$
   $C = \{\text{General Motors, RCA, Exxon}\}$
   $D = \{\text{General Motors, RCA, General Electric, IBM}\}$
   $E = \{\text{General Motors, Exxon, IBM, RCA}\}$.
   Determine whether the following are true or false.
   (a) $A \subset B$ (b) $B \subset A$ (c) $C \subset A$ (d) $E \not\subset A$
   (e) $A = E$ (f) $D = A$ (g) $D \subset E$ (h) $\emptyset \not\subset D$
   (i) $\text{IBM} \in A$ (j) $\text{RCA} \in B$ (k) $\text{RCA} \notin D$ (l) $\text{IBM} \in \emptyset$.

2. The following sets represent warehouse distribution locations for four different companies:
   $A = \{$Buffalo, Nashville, Tampa$\}$
   $B = \{$Philadelphia, Cleveland, Nashville, Tampa$\}$
   $C = \{$Buffalo, Cleveland$\}$
   $D = \{$Nashville, Tampa$\}$
   Determine whether the following are true or false:
   (a) $D \subset A$ (b) $D \subset B$ (c) $D \subset C$ (d) $A \subset B$
   (e) $\emptyset \subset A$ (f) $C = D$ (g) $D = A$ (h) $A \not\subset C$
   (i) Buffalo $\in B$ (j) Cleveland $\notin D$ (k) Boston $\notin B$.
3. The following sets represent sales goals for three different companies:
   $A = \{$sales between \$2,000,000 and \$5,000,000$\}$
   $B = \{$sales between \$3,000,000 and \$6,000,000$\}$
   $C = \{$sales between \$3,500,000 and \$4,700,000$\}$.
   Determine whether the following are true or false.
   (a) $A \subset B$ (b) $B \not\subset C$ (c) $C \subset B$ (d) $C \subset A$
   (e) $A \not\subset C$ (f) $\emptyset \subset A$ (g) $\emptyset \subset C$ (h) $C = B$
   (i) $\$5{,}290{,}315 \in B$ (j) $\$3{,}712{,}127 \in B$
   (k) $\$3{,}712{,}127 \in A$ (l) $\$3{,}712{,}127 \in C$.
4. List all subsets of $A = \{1, 2, 3, 4\}$.
5. List all subsets of $B = \{$Ford, Chevrolet, Pontiac$\}$.
6. An author defines two sets A and B to be equal if and only if both $A \subset B$ and $B \subset A$. Is this definition consistent with ours?
7. An author defines A to be a subset of B if and only if A contains some or all of the elements of B. Is this definition consistent with ours?

9.2 The Algebra of Sets

On occasion, it is useful to combine two sets into a third set. One such combination is the union.

Definition 9.1 The *union* of two sets A and B, denoted by $A \cup B$, is the set of all elements that are either in A or in B or in both A and B.

An element is in the union of two sets if it is in either one or both of the original sets. For example, if $A = \{1, 2, 3\}$ and $B = \{3, 4, 5\}$, then $A \cup B = \{1, 2, 3, 4, 5\}$. Note that, although $3 \in A$ and $3 \in B$, it is listed only once in $A \cup B$, since elements are never repeated in any one set. If $C = \{$green, red, blue$\}$ and $D = \{$blue, white, black, red$\}$, then $C \cup D = \{$green, red, blue, white, black$\}$. Finally, if $E = \{$John, Mary$\}$ and $F = \{$Alan, Carol, Mary, George, John$\}$, then $E \cup F = \{$Alan, Carol, Mary, George, John$\}$.

A second method for combining sets is the intersection.

Definition 9.2 The *intersection* of two sets A and B, denoted by $A \cap B$, is the set of all elements that are in both A and B simultaneously.

If an element is in one set A but not in the second set B, or vice versa, it is *not* in the intersection of the two sets. An element is in the intersection if and only if it is in both sets. For example, if $A = \{1, 2, 3\}$ and $B = \{3, 4, 5\}$ then $A \cap B = \{3\}$, since 3 is the only element in both A and B. If $C = \{$green, red, blue$\}$ and $D = \{$blue, white, black, red$\}$, then $C \cap D = \{$red, blue$\}$, since only red and blue belong to C and D simultaneously. Finally, if $E = \{$John, Mary$\}$ and $F = \{$Alan, Carol, Mary, George, John$\}$, then $E \cap F = \{$John, Mary$\}$.

Two sets are called *disjoint* if their intersection is empty, that is, if they have no elements in common. In particular, $\{1, 2\}$ and $\{3, 4\}$ are disjoint, since $\{1, 2\} \cap \{3, 4\} = \emptyset$. Also, $\{$George, John, Mary$\}$ and $\{$Alan, Carol$\}$ are disjoint, since $\{$George, John, Mary$\} \cap \{$Alan, Carol$\} = \emptyset$.

Example 1 A new car dealer categorizes his previous customers into the following sets:

$A = \{$people who ordered power steering$\}$
$B = \{$people who ordered larger engines$\}$
$C = \{$people who ordered air conditioning$\}$.

Determine the elements of (a) $A \cup B$, (b) $A \cap B$, (c) $B \cup C$, (d) $B \cap C$.

Solution (a) $A \cup B$ is the set of all people who ordered power steering or larger engines or both. Any customer who ordered either one or both is in this set.

(b) $A \cap B$ is the set of all people who ordered both power steering *and also* larger engines.

(c) $B \cup C$ is the set of all people who ordered larger engines or air conditioning or both. Any customer who ordered either one or both is in this set.

(d) $B \cap C$ is the set of all people who ordered both larger engines and also air conditioning.

A third method for combining sets is the difference.

Definition 9.3 The *difference* of two sets A and B, denoted by $A - B$, is the set of all elements that are in A and also not in B.

To be in the difference, an element must satisfy two conditions simultaneously: It must be in the first set and it must not be in the second set. For example, if $A = \{1, 2, 3\}$ and $B = \{3, 4, 5\}$, then $A - B = \{1, 2\}$. The element 3 is not in the difference, since it is in B. The elements 4 and 5 are not in the difference, since they are not in A. Only the elements 1 and 2 are both in A and not in B. If $C = \{$green, red, blue$\}$ and $D = \{$blue, white, black, red$\}$, then $C - D = \{$green$\}$, since green is the only element both in C and not in D simultaneously. Finally, if $E = \{$John, Mary$\}$ and $F = \{$Alan, Carol, Mary,

George, John}, then $E - F = \emptyset$, since there are no elements simultaneously in E and not in F. Every element in E is also in F.

Example 2 Using the sets defined in Example 1, determine the elements in $A - B$, $B - A$, $B - C$, and $C - A$.

Solution (a) $A - B$ is the set of all people who ordered power steering but did not order a larger engine.

(b) $B - A$ is the set of all people who ordered a larger engine but did not order power steering.

(c) $B - C$ is the set of all people who ordered a larger engine but did not order air conditioning.

(d) $C - A$ is the set of all people who ordered air conditioning but did not order power steering.

The three operations of union, intersection, and difference can be used together. Since each combines two sets, however, parentheses are important for indicating which operations are performed first.

Example 3 Let $A = \{1, 2, 3\}$, $B = \{3, 4, 5\}$, and $C = \{2, 4\}$. Find (a) $(A - B) \cup C$ and (b) $A - (B \cup C)$.

Solution (a) Here we first find $A - B$ and then take the union of this result with C. In particular, $A - B = \{1, 2\}$. Then, $(A - B) \cup C = \{1, 2\} \cup \{2, 4\} = \{1, 2, 4\}$.

(b) Here we first find $B \cup C$ and then take the difference of this result from A. In particular, $B \cup C = \{2, 3, 4, 5\}$ and $A - (B \cup C) = \{1, 2, 3\} - \{2, 3, 4, 5\} = \{1\}$.

It follows from Example 3 that $(A - B) \cup C \neq A - (B \cup C)$. Obviously, the order of the operations is important. So too are the parentheses. We cannot calculate $A - B \cup C$, since it might mean either $(A - B) \cup C$ or $A - (B \cup C)$, which are not the same.

Example 4 Using the sets defined in Example 3, find (a) $A \cup (B \cap C)$ and (b) $(A \cup B) \cap C$.

Solution (a) We first calculate $B \cap C$ and then take the union of this result with A. Here $B \cap C = \{4\}$, and then

$A \cup (B \cap C) = \{1, 2, 3\} \cup \{4\} = \{1, 2, 3, 4\}$.

(b) Now we need $A \cup B$, and then the intersection of this result with C. Here $A \cup B = \{1, 2, 3, 4, 5\}$, and

$(A \cup B) \cap C = \{1, 2, 3, 4, 5\} \cap \{2, 4\} = \{2, 4\}$.

Again note that the order in which the operations are performed is important. In particular, $A \cup (B \cap C) \neq (A \cup B) \cap C$.

Example 5 Let $A = \{\text{IBM, RCA, GM, GE}\}$, $B = \{\text{IBM, GM, ITT}\}$, and $C = \{\text{IBM, ITT, CBS}\}$. Find $(A - B) \cup (B \cap C)$.

Solution We first find $A - B$, then $B \cap C$, and finally the union of these two results. Here

$A - B = \{\text{IBM, RCA, GM, GE}\} - \{\text{IBM, GM, ITT}\} = \{\text{RCA, GE}\}$.
$B \cap C = \{\text{IBM, GM, ITT}\} \cap \{\text{IBM, ITT, CBS}\} = \{\text{IBM, ITT}\}$
$(A - B) \cup (B \cap C) = \{\text{RCA, GE}\} \cup \{\text{IBM, ITT}\}$
$\qquad\qquad\qquad\quad = \{\text{RCA, GE, IBM, ITT}\}$.

The last operation of interest is the complement. The complement of a set A is the set of all elements not in A. This, however, poses a problem: If the elements of the complement of A are not in A, where are they? Consider the set $A = \{1, 2, 3\}$. What is its complement? Is it all positive integers except 1, 2, and 3? Is it all integers both positive and negative except 1, 2, 3? Is it all real numbers except 1, 2, 3? Perhaps, being somewhat absurd, it is all the elephants in the San Diego Zoo. After all, there are no elephants in A, so the complement (those elements not in A) could include elephants. The difficulty in finding the complement is that we have not established a frame of reference within which to work.

The *universal set*, denoted by U, is the set of all elements of interest. Obviously such a set is problem-dependent. If we were interested in stocks, we might take U to be the set of all stocks on the New York Stock Exchange. If this is not large enough we might add to U all stocks on the American Stock Exchange. If this still were not large enough for our purposes, we could let U be all stocks listed on any exchange in the world. In any event, U is always taken to be a set large enough to contain all possible elements of interest in a particular situation. It is our frame of reference for a given problem.

Definition 9.4 The *complement* of a set A, denoted by A', is the difference $U - A$, where U is the universal set.

In other words, A' is the set of all elements in U that are not in A. If $A = \{1, 2, 3\}$ and $U = \{1, 2, 3, 4, 5\}$, then $A' = \{4, 5\}$, the set of elements not in A but still in U. Redefining $U = \{1, 2, 3, 4, 5, 6, 7, 8\}$ but keeping the same A, we obtain $A' = \{4, 5, 6, 7, 8\}$. Since the complement of A depends on both A and U, redefining either A or U results in a new A'.

As another example, consider the sets A, B, and C defined in Example 1. Let U be the set of all previous customers.

$A' = \{\text{previous customers who did not order power steering}\}$
$B' = \{\text{previous customers who did not order a larger engine}\}$
$C' = \{\text{previous customers who did not order air conditioning}\}$.

For any set U, it follows from the definition that $U' = \emptyset$ and $\emptyset' = U$.

Example 6 Let $U = \{a, b, c, d, e\}$, $A = \{a, b\}$, and $B = \{d, e\}$. Find $(A - B') \cap (A' \cup B)$.

Solution Proceeding step by step, we find sequentially

$B' = U - B = \{a, b, c, d, e\} - \{d, e\} = \{a, b, c\}$
$A - B' = \{a, b\} - \{a, b, c\} = \emptyset$
$A' = U - A = \{a, b, c, d, e\} - \{a, b\} = \{c, d, e\}$
$A' \cup B = \{c, d, e\} \cup \{d, e\} = \{c, d, e\}$

and

$(A - B') \cap (A' \cup B) = \emptyset \cap \{c, d, e\} = \emptyset.$

Exercises

1. Let $U = \{1, 2, 3, 4, 5, 6\}$, $A = \{1, 2\}$, $B = \{5, 6\}$ and $C = \{1, 4, 5\}$. Find
   (a) $A \cup B$ (b) $A - B$ (c) $B - A$ (d) A'
   (e) $B \cap C$ (f) $B \cap C'$ (g) $A' - B'$ (h) $B' - A'$
   (i) $A' - A$ (j) $A \cup C'$.

2. Let $U = \{$red, white, blue, green, black, orange$\}$, $A = \{$red, blue$\}$, $B = \{$red, green, black$\}$, and $C = \{$red, white, black, orange$\}$. Find
   (a) $B \cup C$ (b) $B - C$ (c) $C - B$ (d) $A \cap C$
   (e) $C \cap A$ (f) C' (g) $A - C'$ (h) $C' - A$
   (i) $A \cup B'$ (j) $B' \cup A$.

3. The following sets represent warehouse distribution locations for four different companies:

   $A = \{$Buffalo, Nashville, Tampa$\}$
   $B = \{$Philadelphia, Cleveland, Nashville, Tampa$\}$
   $C = \{$Buffalo, Cleveland$\}$
   $D = \{$Nashville, Tampa$\}$.

   Find
   (a) $A - B$ (b) $A \cup D$ (c) $C \cap D$ (d) $C - D$
   (e) $D - C$ (f) $B \cap C$ (g) A'.

4. A company categorizes its employees into the following sets:

   $A = \{$male employees$\}$
   $B = \{$female employees$\}$
   $C = \{$minority employees$\}$
   $D = \{$married employees$\}$.

   Determine the elements of the following sets if U is the set of all employees.
   (a) A' (b) $A \cup B$ (c) $A \cap B$ (d) $A \cap C$
   (e) $B \cup C$ (f) $A - D$ (g) $B - D'$ (h) $B \cap (C \cap D)$

5. Let $A = \{a, b, c\}$. Find appropriate universal sets such that (a) $A' = \{x, y, z\}$, (b) $A' = \{r, s, t, u, v\}$, and (c) $A' = \{a, b, c, d, e, f\}$.

6. Let $A = $ {people who smoke}, $B = $ {males}, $C = $ {people over 40}, and $U = $ {people in the world}. Write the following sets in terms of A, B, and C:
   (a) {females}
   (b) {females who do not smoke}
   (c) {people in the world}
   (d) {people in the world who smoke}
   (e) {males over 40}
   (f) {females not over 40}
   (g) {smokers over 40}
   (h) {males over 40 who do not smoke}.
7. Let $R = $ {defective merchandise}, $S = $ {sale merchandise}, $T = $ {advertised merchandise}, and $U = $ {all merchandise in the store}. Write the following sets in terms of R, S, and T.
   (a) {defective merchandise on sale}
   (b) {advertised merchandise not on sale}
   (c) {advertised merchandise that is not defective}
   (d) {unadvertised merchandise}
   (e) {advertised merchandise that is defective but not on sale}
8. Let $A = \{1, 2, 3, 4\}$, $B = \{3, 4, 5, 6\}$, $C = \{1, 2, 9, 10\}$, and $U = $ {the first 10 positive integers}. Find
   (a) $(A - B) \cup (B - A)$
   (b) $(A \cap C') - (B \cap C')$
   (c) $(A - B') \cup (B - A')$
   (d) $(C \cap A) \cup (A \cap B)$
   (e) $[(C - A)' - B']$.
9. Let $R = $ {IBM, GM, GE}, $S = $ {RCA, IBM, CBS}, $T = $ {Exxon, IBM, GM, CBS}, and $U = $ {IBM, GM, GE, RCA, CBS, Exxon, NBC}. Find
   (a) $(R - S) \cup (T \cap S)$
   (b) $(R \cap T') - (T \cap S')$
   (c) $(R - T) - S$
   (d) $(R' - T') \cup (T' - R')$
   (e) $[(R \cap S)' - T']'$.
10. The *symmetric difference* of two sets A and B denoted by $A \triangle B$ is $(A - B) \cup (B - A)$. Find $A \triangle B$ and $B \triangle C$ for the sets defined in Exercise 3. Find $R \triangle T$ for the sets defined in Exercise 9.

9.3 Venn Diagrams

Sets can be represented pictorially by drawings known as Venn diagrams. The universal set U is drawn as a rectangle, and individual sets are drawn as disks within U. Regions of interest are shaded. The Venn diagram for a single set A is shown in Figure 9.1. The Venn diagram for the complement of A is given in Figure 9.2. Since the complement of A is the set of elements not in A (but still in U), the shaded portion of its Venn diagram is all points outside the disk representing A.

Venn diagrams involving two sets A and B are drawn initially as Figure 9.3. The particular region of interest is then shaded. For example, $A \cap B$ is the set of all elements in both A and in B; its Venn diagram is given in Figure 9.4. The Venn diagram for $A - B$ is shown in Figure 9.5. Note that only the points in A but not in B are shaded. Since $A \cup B$ is the set of elements in either A or B or in both, its Venn diagram is given by Figure 9.6.

Venn diagrams for complicated sets often can be obtained by combining Venn diagrams for simple sets. The key is remembering that the elements in a

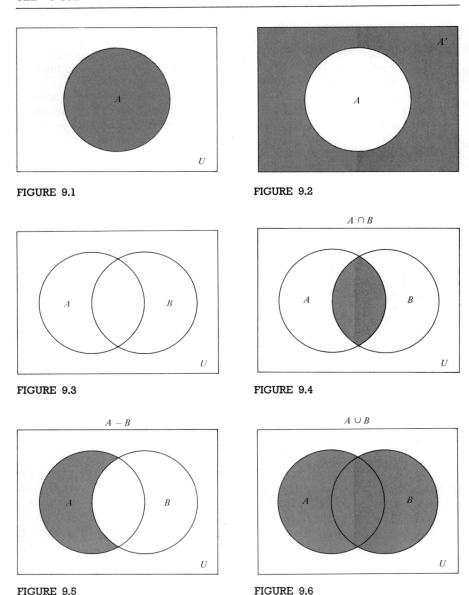

FIGURE 9.1

FIGURE 9.2

FIGURE 9.3

FIGURE 9.4

FIGURE 9.5

FIGURE 9.6

particular set are represented by the shaded region in the Venn diagram. As an example, let us draw the Venn diagram for $(A - B) \cup (B - A)$. This set is the union of $A - B$ and $B - A$, so we first draw the Venn diagrams for these sets, Figure 9.7A and B. The union of two sets is all elements in one set or the other set or in both. Pictorially, the union of $A - B$ with $B - A$ is all points shaded in either Figure 9.7A or B or in both, which is drawn in Figure 9.7C.

9.3 Venn Diagrams **323**

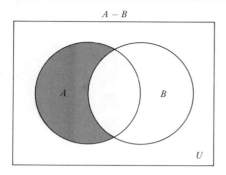

FIGURE 9.7A

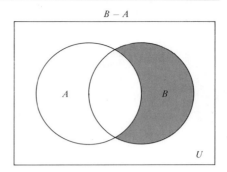

FIGURE 9.7B

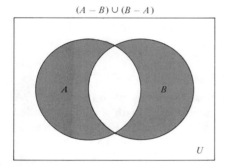

FIGURE 9.7C

Example 1 Draw the Venn diagram for $A - (A \cap B')$.

Solution We first need the Venn diagram for $A \cap B'$. Figure 9.8A is the Venn diagram for A, and Figure 9.8B is the Venn diagram for B'. Note that, since the set of interest $A - (A \cap B')$ involves A and B, all our intermediate diagrams also should involve these sets. Figure 9.8C is the diagram for $A \cap B'$. $A \cap B'$ is an intersection of two sets. The shaded region in its Venn diagram should be

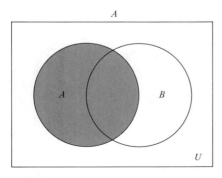

FIGURE 9.8A

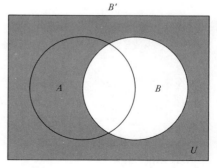

FIGURE 9.8B

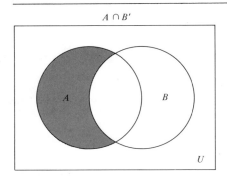

FIGURE 9.8C

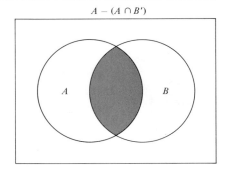

FIGURE 9.8D

the points that appear shaded in Figure 9.8A and B *simultaneously*. That is, the points in both A and B'. Having A and $A \cap B'$, we form their difference by using the points shaded in Figure 9.8A that are not shaded in Figure 9.8C. This results in Figure 9.8D.

Example 2 Draw the Venn diagram for $(A \cup B) - (B \cap A')$.

Solution The step-by-step construction of this diagram is given in Figure 9.9A through F. In particular, $B \cap A'$ is the intersection of B with A', so the shaded portion of its diagram (Figure 9.9E) is all points shaded in Figure 9.9B and D simultaneously. The required set is the difference of $A \cup B$ with $B \cap A'$. The shaded region of its Venn diagram, Figure 9.9F, is all points shaded in Figure 9.9C that are not shaded in Figure 9.9E.

Venn diagrams involving three sets, A, B, and C, are drawn initially as Figure 9.10. They are then manipulated exactly as Venn diagrams with only two sets.

Example 3 Draw the Venn diagram for $(A - C) \cup (C \cap B)$.

Solution Since we want a set that is a combination of A, B, and C, all our Venn diagrams will contain these sets. The step-by-step construction of this

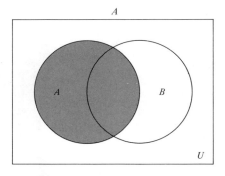

FIGURE 9.9A

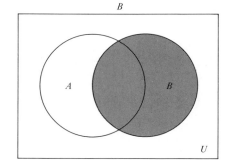

FIGURE 9.9B

9.3 Venn Diagrams **325**

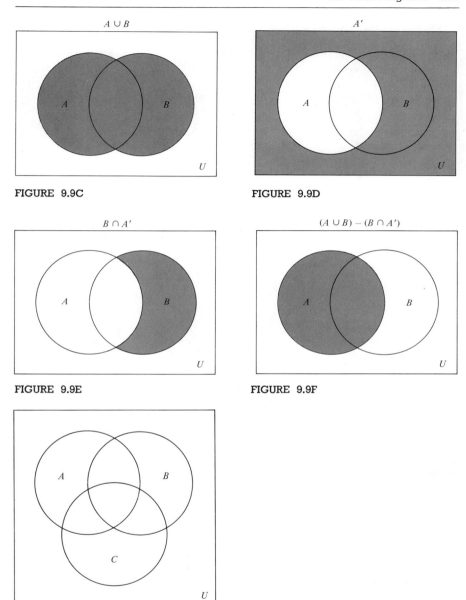

FIGURE 9.9C

FIGURE 9.9D

FIGURE 9.9E

FIGURE 9.9F

FIGURE 9.10

Venn diagram is given in Figure 9.11A through F. In particular, $A - C$ is the difference of A with C, so the shaded portion of its diagram is all points shaded in Figure 9.11A that are not shaded in Figure 9.11C. The set $C \cap B$ is the inter-

326 9 Sets

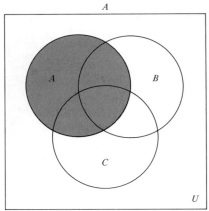

FIGURE 9.11A

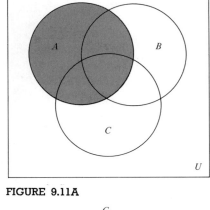

FIGURE 9.11B

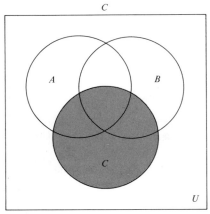

FIGURE 9.11C

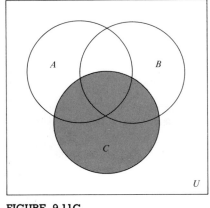

FIGURE 9.11D

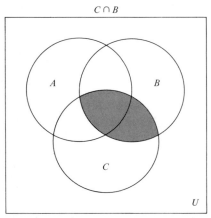

FIGURE 9.11E

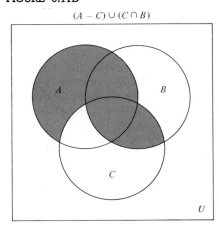

FIGURE 9.11F

section of C with B, so the shaded portion of its diagram is all points shaded in both Figure 9.11C and B simultaneously. Finally, the Venn diagram for $(A - C) \cup (C \cap B)$ is the points shaded either in Figure 9.11D or E or in both. It is drawn in Figure 9.11F.

One of the major uses of Venn diagrams is to verify pictorially whether or not two sets are equal. For example, is $A - (A \cap B')$ equal to $A \cap B$? We answer this question by constructing Venn diagrams for both sets. If the Venn diagrams are identical (i.e., the same region is shaded in both diagrams), the sets are equal. In particular, the Venn diagram for $A - (A \cap B')$ is drawn in Figure 9.8D, and the Venn diagram for $A \cap B$ is shown in Figure 9.4. They are indeed the same, so $A - (A \cap B') = A \cap B$.

Example 4 Use Venn diagrams to determine whether or not the equality $A \cap B' = A - B$ is valid.

Solution The Venn diagram for $A - B$ is drawn in Figure 9.5. The step-by-step construction of the Venn diagram for $A \cap B'$ is given in Figure 9.12A through C. Since Figure 9.12C is identical to Figure 9.5, the equality is valid.

Example 5 Use Venn diagrams to determine whether or not the equality $(A - C') - B = (A - C) \cup (C \cap B)$ is valid.

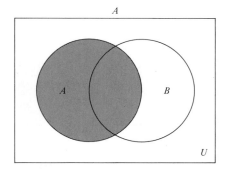

FIGURE 9.12A

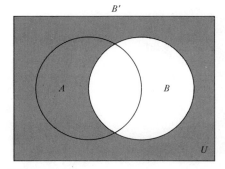

FIGURE 9.12B

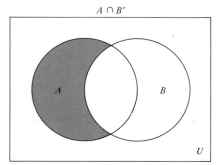

FIGURE 9.12C

328 9 Sets

Solution The step-by-step construction for $(A - C') - B$ is given in Figure 9.13A through F. The Venn diagram for $(A - C) \cup (C \cap B)$ is drawn in Figure 9.11F. Since Figures 9.13F and 9.11F are *not* identical, the equality is *not* valid. That is, $(A - C') - B \neq (A - C) \cup (C \cap B)$.

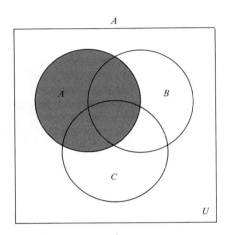

FIGURE 9.13A

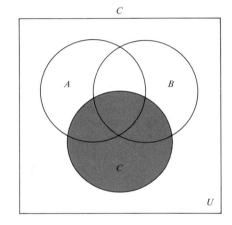

FIGURE 9.13B

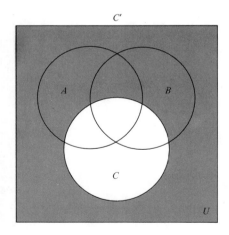

FIGURE 9.13C

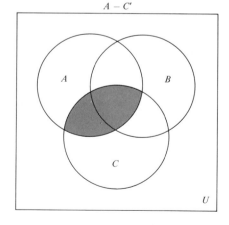

FIGURE 9.13D

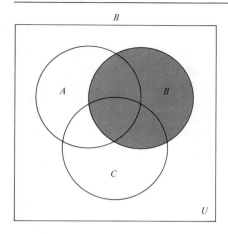

FIGURE 9.13E

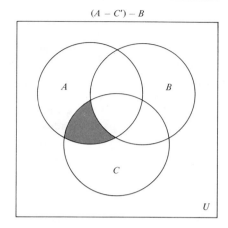

FIGURE 9.13F

Exercises

1. Draw Venn diagrams for the following sets:
   (a) $A \cup B'$ (b) $A' - B'$ (c) $A' \cap B'$.
2. Draw Venn diagrams for the following sets:
   (a) $(A - B)' \cap A$ (b) $(A - B) \cup A'$
   (c) $(A - B) \cup (B - A')$ (d) $(A \cup B) - A$.
3. Draw Venn diagrams for the following sets:
   (a) $(A - B) \cup (C \cap A)$ (b) $(A - B) - C$
   (c) $(A - B') \cup (C \cap A')$ (d) $(A - B)' \cap (C - A)'$.
4. DeMorgan's laws for sets are (a) $(A \cup B)' = A' \cap B'$ and (b) $(A \cap B)' = A' \cap B'$. Verify these laws with Venn diagrams.
5. Use Venn diagrams to verify the following equalities:
   (a) $A \cap (B \cap C) = (A \cap B) \cap C$
   (b) $A \cup (B \cup C) = (A \cup B) \cup C$
   (c) $A \cap (B \cup C) = (A \cap B) \cup (A \cap C)$
   (d) $A \cup (B \cap C) = (A \cup B) \cap (A \cup C)$.
6. Determine whether or not the equality $A - (B - C) = (A - B) - C$ is valid.
7. Draw the Venn diagram for the empty set $\emptyset$. Verify that $(A - B) \cap (B - A) = \emptyset$.

9.4 Sample Spaces

Whenever a company initiates a commercial process, several different outcomes are usually possible. The company expects a particular outcome, but other less desirable ones may occur. For example, an electronics firm may decide to initiate research for a more competitive calculator. Two outcomes are possible. Either the research will succeed or it will not. An automobile repair company may decide to hire two machinists from a list of three applicants, X, Y, and Z.

Now three outcomes are possible. Either X and Y are hired, or Y and Z are hired, or X and Z are hired.

Any action or decision that results in one or more possible outcomes is a *process*. A decision to initiate research is a process. Filling positions from a list of applicants is a process. Flipping a coin is a process.

The set of all possible outcomes of a particular process is called a *sample space* and is denoted by U, the same notation used earlier for the universal set. Since the sample space includes all possible outcomes, it is the universal set for a process. A sample space for the machinist problem described above is $U =$ {X and Y, Y and Z, X and Z}. A sample space for the electronic calculator research process is $U =$ {research succeeds, research fails}.

The sample space for any process can be subjective; a possible outcome to one person sometimes is not a possible outcome to a second person. Flipping a coin is such a situation. One person may give the sample space as $U_1 =$ {head, tail, side, coin lost}. A second person may not consider the coin landing on its side a realistic possibility (most people do not), so the sample space would be $U_2 =$ {head, tail, coin lost}. A third person may not consider losing the coin a realistic outcome, so the sample space may be the usual one, $U_3 =$ {head, tail}. The final choice of which outcomes are to be considered should be specified carefully by the decision maker. The importance of this selection will become evident when we encounter probabilities in Chapter 11.

We adopt the notation $n(U)$ to designate the number of elements or outcomes in a sample space. For the coin process mentioned above, $n(U_1) = 4$, $n(U_2) = 3$, and $n(U_3) = 2$.

Example 1 A barrel contains 10 balls numbered 1 through 10, inclusive. The odd-numbered balls are red, while the even-numbered balls are white. Determine a sample space for the process: Pick one ball and state its number.

Solution $U = \{1, 2, 3, 4, 5, 6, 7, 8, 9, 10\}$, and $n(U) = 10$.

Example 2 Determine a sample space for the process: Pick one ball from the barrel described in Example 1 and state its color.

Solution $U = \{$red, white$\}$, and $n(U) = 2$.

Example 3 Determine a sample space for the process: Pick 1 card from a regular deck of 52 playing cards and state its suit.

Solution $U = \{$club, diamond, heart, spade$\}$, and $n(U) = 4$.

As long as the number of possible outcomes remains small, there is little difficulty in listing the elements of a sample space and then counting them. One aspect, however, can be troublesome: Does the order in which the events occur matter? Consider again the problem of hiring two machinists from three applicants, X, Y, and Z. If there is a seniority system in effect at the company whereby the person hired first is fired last even if the time difference in hiring is a few seconds, the outcome XY, signifying X was hired first and Y second, is different from the outcome YX, signifying Y was hired first and X second. If order matters, there are six different outcomes, and $U = \{XY, YX, XZ, ZX, YZ, ZY\}$. Whether or not order matters is usually clear from the physical situation at hand. A problem should not be attempted unless this point is clarified, since the answer usually is different for each case.

Example 4 Falk Industries, with one plant each in Buffalo, Milwaukee, Cleveland, and Nashville, will close two of these plants, one each year, over a 2-year period. A vice-president will schedule the closings. Determine a sample space for all possible schedules.

Solution Obviously the order of plant closings is important, if not to the company, then certainly to the employees. Listing outcomes in the order they will occur and using only the first initial of each city, we have $U = \{BM, MB, CB, BC, BN, NB, MC, CM, MN, NM, CN, NC\}$, and $n(U) = 12$.

If both plants were to be closed at the same time in Example 4, order would not matter. Now, $U = \{BM, BC, BN, MC, MN, CN\}$, and $n(U) = 6$. The outcome MB is the same as BM, so they are not listed separately; both denote closings at Buffalo and Milwaukee.

Real difficulties occur when the number of outcomes is large. If, in Example 4, 3 plants will be closed 1 each year, there are 24 possible schedules, and it is not as simple to list all of them. The case of scheduling 5 closings from 7 plants, 1 each year, is even harder. We devote all of Chapter 10 to listing and counting large sample spaces. It will then follow that there are exactly 2520 ways to schedule 5 closings from 7 plants, 1 each year, and we also will describe a technique for listing each one of these schedules.

Exercises

1. Three balls, one green, one red, and one white, are placed in a barrel. Determine the sample space for the process: Pick one ball and state its color.
2. Four balls, numbered successively 1 through 4, are placed in a barrel. Determine a sample space for the process: Pick two balls and state their sum. Does order matter?
3. Determine a sample space for the process: Pick two balls from the barrel described in Exercise 2 and subtract the number on the second ball from the number on the first ball. Does order matter?
4. Determine $n(U)$ for the process: Pick 1 card from a regular deck of 52 cards and state what it is.
5. Determine the sample space for the process: Pick 1 card from a regular deck of 52 cards and state its rank (ace, two, three, etc.).
6. The American Citrus Corporation asks three of the country's best architects to submit designs for its new home office. Determine a sample space for the number of responses to these requests.

7. Determine a sample space for the process: Keep track of the order in which designs (as described in Exercise 6) arrive, assuming that each architect submits a design.
8. Redo Exercise 7 without the assumption.
9. Lincoln Hamburger Company institutes a psychological testing program for prospective management. It is hoped that this program will pass ideal managers and fail individuals unsuited for management. Determine the sample space for the results of a test when it is taken by an individual applicant.
10. Four people, John, Mary, Paula, and Ted, have applied for a position as salesperson in a department store. There are two openings, both in the jewelry department. Determine a sample space for the process: Fill both openings from the list of four applicants if order does not matter.
11. Redo Exercise 10 if order matters.
12. Redo Exercise 10 if one opening is in the jewelry department and the second vacancy is in the hardware department.
13. Redo Exercise 10 if there are four openings in the jewelry department.
14. Redo Exercise 10 if one of the positions must be filled by a female.

Chapter

Permutations and Combinations

Counting is fundamental to all businesses. The amount of merchandise on hand, the number of orders received, the attendance at a stockholder's meeting, and the number of sales necessary for a new product to break even are but a few examples. The bottom line of an annual fiscal report, the amount of money made or lost, is the ultimate example.

The number of elements in a sample space is also of great interest, especially in commercial processes. If the numbers are small, the sample space can be counted by first listing all possible outcomes. Unfortunately, this is not always feasible. The situation described at the end of Section 9.4 involving five closings from seven plants is such a case. In this chapter, we develop methods for systematically listing and counting processes with large sample spaces.

10.1 Trees

One method of counting the elements in a sample space is to first list all possible outcomes and then count the number of entries on the list. The problem then reduces to constructing a complete list. One approach is to write entries as they

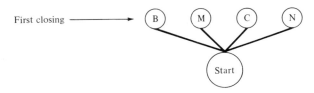

FIGURE 10.1

come to mind, but this is usually inadequate. It is too easy to overlook possible outcomes, especially if the list is large. A more systematic approach is required.

Consider the problem described in Example 4 in Section 9.4, where two plants must be closed, one each year over the next 2 years. The company has one plant each in Buffalo, Milwaukee, Cleveland, and Nashville. To see how we can generate a systematic technique for listing all possible closing schedules, let us put ourselves in the position of the vice-president responsible for these schedules. To start the scheduling process, we must decide which plant will be closed first. Do we have a choice? Obviously we do, since we can close any one of the four. Therefore we begin graphically with Figure 10.1.

Figure 10.1 is an example of a one-level tree. It is one level since only one decision has been reached, which plant will be closed first, and it is called a tree since it vaguely resembles one. The "start" is the trunk, and the four appendages are the branches. Care must be taken, however, in reading a tree. Each branch represents *one* possible outcome. In Figure 10.1 we have four branches, so there are four possibilities for scheduling the first closing. We can close the first plant in Buffalo (B) or Milwaukee (M) or Cleveland (C) or Nashville (N).

Once we have decided which plant will be closed first, we then have a second decision: Which plant will be closed next. Do we have a choice? Again we do, but the choice depends on which plant we closed first. Obviously, if we close the Buffalo plant first, we cannot schedule it for the second closing too. If we decide to close Buffalo first, we have only three possible closings the second year—Milwaukee, Cleveland, or Nashville. This is shown in Figure 10.2.

If, however, we decide to close the Nashville plant first, we again have three choices for the second closing, but a different set is involved. Now the

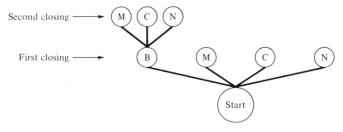

FIGURE 10.2

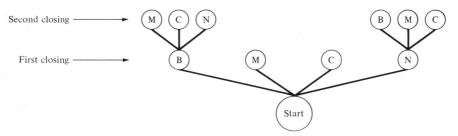

FIGURE 10.3

choices are Buffalo, Milwaukee, or Cleveland. Adding these choices to Figure 10.2, we obtain Figure 10.3.

We complete the entire tree by adding the possible second choices to the first choices of both Milwaukee and Cleveland. We then obtain Figure 10.4, a two-level tree which includes all possible outcomes (schedules) for the process at hand.

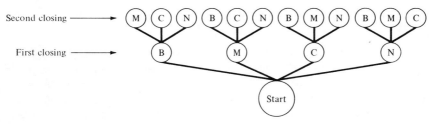

FIGURE 10.4

Each complete branch is one possible outcome. The leftmost branch is Buffalo first and Milwaukee second. The next branch is Buffalo first and Cleveland second. Using this tree, we can list all 12 possible schedules starting from the leftmost branch and moving to the rightmost branch, as $U = \{$BM, BC, BN, MB, MC, MN, CB, CM, CN, NB, NM, NC$\}$.

Trees are a straightforward technique for listing possibilities. They also are systematic, thereby minimizing the chances of omitting possibilities.

Example 1 A barrel contains three balls, two red and one white. Construct a tree for the sample space of the process: Pick two balls from the barrel, one at a time, without any replacement, and state their colors as they are picked.

Solution To start this process we must first pick a ball. Do we have a choice? Clearly we do and, since our first selection is either red or white, the first level of the tree is given in Figure 10.5.

If our first pick is a red ball, then our second pick can be either red or white, since one ball of each color remains in the barrel. If, however, our first pick is a white ball, our second pick must be red; there are no other possibilities. Figure 10.6 is the complete tree.

336 10 Permutations and Combinations

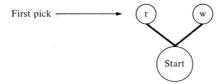

FIGURE 10.5

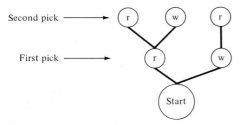

FIGURE 10.6

Each branch represents one possible outcome. Therefore $U = \{rr, rw, wr\}$ and $n(U) = 3$.

When listing by trees, *order always matters*. In the example dealing with closing two plants, one each year, a schedule of Buffalo first and Nashville second is different from Nashville first and Buffalo second. Although the same two cities are involved in both outcomes, the order in which they are chosen is crucial. The same is true of Example 1; the order in which individual selections is made is important, and different orderings of the same distinguishable objects are counted as different outcomes. A red ball first and a white ball second is different from a white ball first and a red ball second.

In some processes, order does not matter. That is, different arrangements of the same distinguishable objects are not counted as different outcomes. As an example, consider the process of making a sandwich from two of four possible ingredients, salami, ham, American cheese, and provolone. A sandwich made of ham and American cheese is *not* different from one made of American cheese and ham. The order in which the ingredients are used does not matter, only the ingredients themselves.

Trees always list outcomes as if order matters. We show in Section 10.4 how to modify the counting procedure when different orderings of the same distinguishable objects are irrelevant.

The process of constructing a tree can be simplified if a few general procedures are followed. First, always place yourself in the position of the person involved in the process. In Example 1, pretend you are the person picking the balls. In the process leading to Figure 10.4 pretend you are the vice-president actually making a schedule for closing two plants. Then, mentally perform the required process, always asking two questions: (1) What must I do next? and (2) Do I have a choice?

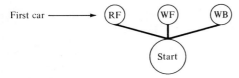

FIGURE 10.7

Example 2 A used car dealer has two Fords, one red and one white, and one white Buick which he wishes to display in a showroom next to each other. Construct a tree which lists all possible arrangements of the three cars.

Solution We place ourselves in the position of the dealer and ask, "What do we do next?" Our first step must be to drive in a car. Note that we cannot drive all the cars into the showroom simultaneously. Even with three drivers, one car must enter first. Do we have a choice? Yes. The first car can be either a red Ford (RF) or a white Ford (WF) or a white Buick (WB). Therefore the first level of the tree is as given in Figure 10.7.

What do we do next? Obviously, drive in the second car. Do we have a choice? Yes, but the choice depends on which car was driven in first. If the first car was the red Ford, the second car must be either the white Ford or the white Buick. If the first car was the white Ford, the second car must be either the red Ford or the white Buick. If the first car was the white Buick, the second car must be either the red Ford or the white Ford. The first two levels of the tree are given in Figure 10.8.

What do we do next? Drive in the third car. Do we have a choice? Not really. After two cars have been positioned, only one other car remains to be placed, even though its type will depend on which cars were positioned previously. The complete tree is given in Figure 10.9.

Each complete branch represents one possible outcome. Since there are six branches, there are six outcomes ranging from the red Ford first, the white Ford second, and the white Buick third (the leftmost branch) to the white Buick first, the white Ford second, and the red Ford third (the rightmost branch).

Example 3 Associated Caterers, specializing in box lunches, receives an order

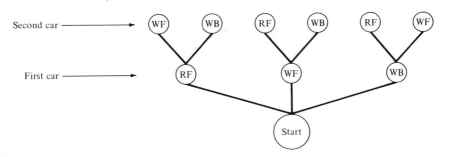

FIGURE 10.8

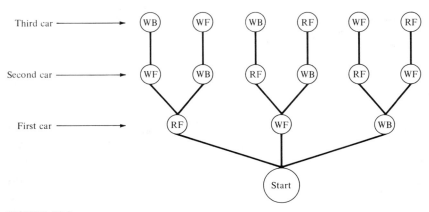

FIGURE 10.9

for 240 lunches each containing an entree item, a side dish, and dessert. Each lunch must cost under $2, and the customer wants as much variety among the lunches as possible. Currently Associated has two types of entrees, fried chicken ($1.50) and ham sandwiches (90¢), two different side dishes, potato salad (20¢) and deluxe salad (55¢), and three types of desserts, chocolate bars (15¢), cake (65¢), and apples (25¢), where the prices in parentheses denote the individual cost of each item. Construct a tree indicating the number of different lunches the caterer can prepare. Two lunches are different if any item is different.

Solution We place ourselves in the position of the caterer and ask, "What do we do next?" Since each lunch must have an entree, we first decide which entree to use. Do we have a choice? Yes, both entrees are under $2; we can use either. The first level of the tree is given in Figure 10.10.

What do we do next? After the entree, we must add the side dish. Do we have a choice? Yes, although the choices depend on the entree. If the entree is fried chicken, the side dish cannot be the deluxe salad, since the cost of fried chicken plus the deluxe salad is greater than the $2 maximum. Only the potato salad is possible. If the entree is a ham sandwich, both side dishes are acceptable. Adding these possibilities to Figure 10.10, we obtain Figure 10.11.

What do we do next? Add the dessert. Do we have a choice? Yes, but again the choice depends on what has happened previously. If the box lunch already contains chicken and potato salad, the cake is not a possibility, since the

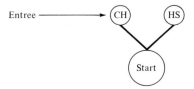

FIGURE 10.10

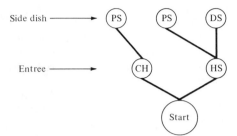

FIGURE 10.11

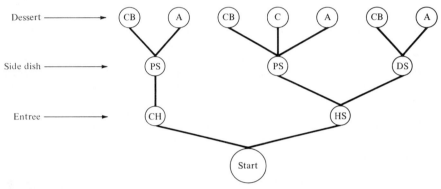

FIGURE 10.12

cost would exceed $2. The other two desserts, however, are possible. If the box lunch contains a ham sandwich and the deluxe salad, again the cake is too expensive, and the only possible desserts are the chocolate bar or the apple. The complete tree is given in Figure 10.12.

Since each branch denotes one possible box lunch, seven different lunches can be prepared ranging from fried chicken–potato salad–chocolate bar (the leftmost branch) to a ham sandwich–deluxe salad–apple (the rightmost branch).

Exercises

Construct a tree and then count the number of possible outcomes for the processes described in Examples 1 through 15.

1. A luxury car manufacturer offers three body colors, silver, black, and green, and three interior colors, black, brown, and red. How many different color combinations can prospective buyers choose?
2. Frozen dinners consist of one entree, either chicken, salisbury steak, or ham steak, one vegetable, either peas or corn, and one dessert, apple pie or a brownie. How many different dinners can be offered? Two dinners are different if any one item is different.
3. A builder has plans for three different house models, a ranch, a colonial, and a split level. He will build three houses in a row, but he does not want to build two

identical models side by side. Two identical models separated by a different model is acceptable. How many different ways can he construct the houses?

4. Messages are relayed to passing trains by hoisting two flags up a flagpole, one under the other. Flags are red, white, or striped, and each different combination signifies a different message. A red flag under a white flag is different from two red flags. How many different two-flag messages are possible?

5. Using the flags described in Exercise 4, determine how many different three-flag messages are possible if two flags of the same color are not allowed adjacent to each other. For instance, two red flags *separated* by a white flag is acceptable, but two red flags and then a striped flag is not.

6. A local department store has two vacancies, one in the jewelry department and one in the housewares department. There are four applicants for these positions, John, Mary, Laura, and William. How many different ways can the two positions be filled if *at least one* of the jobs must go to a female?

7. A well-known fruit drink is a combination of orange, grape, and apple ingredients. The manufacturer can use either real fruit juice or artificial flavoring for each ingredient, but *at least one* of the ingredients must be real juice, since the product will be advertised as a "real fruit drink." How many different combinations of artificial and real ingredients can be used in the production of the drink?

8. How many different three-letter combinations can be made from the four letters a, b, c, and e if no letter can be used twice in the same arrangement?

9. Redo Exercise 8 for two-letter combinations if letters can be repeated.

10. A drug chain with five stores in Tampa decides to close one store every Saturday and another store every Sunday. The store chosen for the Saturday closing will close every Saturday, and the same is true for the store that closes on Sunday. How many different ways can the choice of closings be made?

11. A bank gives free gifts to all new depositors. Those who deposit $50 can choose one gift from a set of bridge cards and a lighter. Those who deposit $100 can choose one gift from a radio, a silver-plated baby spoon, or a stuffed animal. Those who deposit $500 can choose one gift from a toaster and a camera. A $1000 depositor can choose one item from each of the three categories for a total of three gifts. How many different sets of gifts can the bank offer to $1000 depositors?

12. A condominium developer offers four different types of lighting to buyers for both their foyer and living room. How many different combinations are available to each buyer?

13. Redo Exercise 12 if buyers cannot choose the same fixture for both rooms.

14. Redo Example 3 if chicken is $1.25 and potato salad is 40¢.

15. A Los Angeles distributor imports merchandise from Italy by way of New York. The merchandise is shipped from Italy to New York by either plane or ship, and from there to Los Angeles by plane, train, or truck. How many different ways can merchandise be shipped from Italy to Los Angeles?

10.2 Fundamental Theorem of Counting

Trees are an efficient means of listing, but they are still cumbersome when the list is large. A process with 1000 possible outcomes requires a tree with 1000 branches. If, however, we are interested in just the *number* of outcomes rather than the outcomes themselves, we may be able to count without first listing. If so, we avoid the necessity of constructing a tree.

Let us return to Figure 10.4, which is the tree for all possible closings of two plants, one each year, from the individual plants in Buffalo, Milwaukee, Cleveland, and Nashville. It is a two-level tree with the first level representing the first closing and the second level representing the second closing. Note that there are four ways to reach the first level (i.e., four different possibilities), and each first-level entry has three second-level entries associated with it. This is important. Although each first-level entry has a different *set* of second-level entries, the total *number* of second-level entries associated with each first-level entry is the same, 3. The total number of different branches is 4 (the number of first-level entries) times 3 (the number of second-level entries associated with each first-level entry), or 12. This is an example of the following general rule.

Theorem 10.1 (The Fundamental Theorem of Counting) If the first level of a tree has r different entries and, if every first-level entry has exactly s second-level entries associated with it, the two-level tree has exactly $r \times s$ different branches.

Example 1 A chain with seven stores wishes to close two stores, one each year over the next 2 years. How many different closing schedules can be constructed?

Solution Without drawing the tree, we visualize it as having two levels, the first level representing the first closing and the second level representing the second closing. There are seven entries on the first level, since there are seven possibilities for the first closing. Regardless of which store is closed first, there are six possibilities for the second closing. Therefore each first-level entry has six second-level entries. Based on the fundamental theorem of counting, it follows that there are $(7)(6) = 42$ branches on the two-level tree, each branch representing one possible outcome.

Example 2 An airline has 10 different flights from Boston to Chicago daily and 7 flights each day from Chicago to Boston. How many different roundtrip schedules can it offer Bostonians who plan to leave and return on different days?

Solution If we were to construct a tree (which we try to avoid), it would have two levels, the first representing a Boston–Chicago flight and the second representing the return flight. There would be 10 entries on the first level. Each first level would have the same 7 second-level entries associated with it. Therefore there are $(10)(7) = 70$ different roundtrip schedules.

The fundamental theorem of counting is easily generalized to trees with three or more levels. The only condition is that each first-level entry have the same number of second-level entries associated with it, that each second-level entry have the same number of third-level entries associated with it, and so on. The entries themselves can be different; the number of entries *must* be the same.

In particular, if the first level of a three-level tree has r different entries, if each first-level entry has exactly s second-level entries associated with it, and if each second-level entry has exactly t third-level entries associated with it, the entire tree has exactly $r \times s \times t$ different branches. Further generalizations are similar.

Figure 10.9 is a three-level tree. There are three first-level entries. Each first-level entry has two second-level entries. Each second-level entry has one third-level entry. It follows from the generalized fundamental theorem of counting that there are $(3)(2)(1) = 6$ branches, which is the case.

Example 3 How many different combinations of 3-letter initials exist using the 26 letters of the English alphabet?

Solution We first visualize (but do not construct) a tree for the process of combining initials. It is a three-level tree. The first level represents the first initial, and there are 26 different possibilities. The second level represents the second initial. For each first initial, there are again 26 possible second initials. The third level represents the third initial, and again there are 26 possibilities for it associated with each second initial. Therefore we have $(26)(26)(26) = 17{,}576$ different combinations.

Certainly, we solved Example 3 faster with the fundamental theorem of counting than would have been possible by first drawing a tree and then counting the branches. A tree with 17,576 different branches is not constructed quickly. But we did not abandon trees completely. In fact, we actually visualized the tree construction in our minds, to guarantee that the conditions of Theorem 10.1 were met, before we used the theorem.

We are ready to tackle the problem proposed at the end of Chapter 9.

Example 4 A company has seven plants, one each in seven different cities. It wishes to close five of these plants, one every year, over the next 5 years. How many different closing schedules can be assembled?

Solution A tree for this process has five levels, the first level representing the first closing, and the fifth level representing the fifth closing. There are seven possibilities for the first closing. Associated with each first closing, there are six possible second closings. For each second closing, there are five possible third closings. Continuing, we find exactly $(7)(6)(5)(4)(3) = 2520$ different closing schedules.

Example 5 A dress manufacturer produces one style in basic white and then trims each dress differently. There are 14 styles of buttons, 8 colors for lace trim, and 9 types of emblems which can be sewn on the single pocket. How many different finished designs containing all three trims can be advertised?

Solution $(14)(8)(9) = 1008$.

Obviously, Theorem 10.1 is a useful theorem when it applies. One must be careful, however, not to apply it incorrectly. Examples 1 and 3 in Section 10.1 are two situations in which the fundamental theorem cannot be applied. In both examples, one first-level entry has two second-level entries associated with it, while the other first-level entry has only one second-level entry. The theorem is not applicable.

Trees have one advantage over the fundamental theorem of counting: They exhibit all possible outcomes. If one is interested in the actual outcomes rather than just the number of outcomes, a tree must be used.

Finally, since Theorem 10.1 deals with trees, and since trees deal with processes in which the order that events occur matters, it follows that the fundamental theorem of counting can be used only if order matters. If order does not matter, the theorem is not applicable in its present form. We consider such situations in Section 10.4.

Exercises

1. A department store advertises a watch sale offering a watch face and a band for $30. They have 15 different faces and 20 different bands from which to choose. How many different sets can they offer?
2. A furniture store advertises ensembles, three different rooms of furniture, for $2000. Buyers can choose from 7 different living-room sets, 5 different dining-room sets, and 10 different bedroom sets. How many different ensembles can the store advertise?
3. How many different license plates are available if each plate has five places, the first two being reserved for numbers and the last three for letters?
4. An executive wishes to fly to Chicago on Monday, to San Francisco on Tuesday, and return to Boston on Wednesday. There are 15 different flights from Boston to Chicago, 21 different flights from Chicago to San Francisco, and 5 different flights from San Francisco to Boston, all available on a daily basis. From how many different itineraries can the executive choose?
5. An automobile manufacturer offers 6 different models, with the choice of 10 different exterior colors and 7 different interior colors. How many different cars does it offer?
6. A buyer of a particular car model is offered the following options: automatic or standard transmission; air conditioner or none; choice of seven different tire models; choice of AM radio, AM-FM radio, stereo radio, or none; and choice of front window defroster, front and rear window defrosters, or none. How many different ways can a buyer equip a car?
7. A restaurant offers a price-fixed dinner for $19.95. A diner may choose one each from 7 appetizers, 8 entrees, 4 vegetables, 3 potatoes, 8 desserts, and 12 wines. How many different complete dinners does the restaurant offer?
8. A motel has 31 rooms and 6 reservations. How many different ways can the 6 parties be assigned rooms?
9. A rent-a-car agency has 14 available cars and 4 customers in the office. How many different ways can the 4 customers be assigned cars?
10. Messages are relayed to passing trains by displaying two single-colored flags, one

under the other. Each combination of colors represents a different message. How many messages are possible from a set of
- (a) 8 flags each of a different color?
- (b) 16 flags each of a different color?
- (c) 16 flags 2 each of 8 different colors? Assume that 2 flags of the same color cannot be used together.
- (d) Same as part (c) with the assumption removed.

11. A baseball team has three pitchers, five outfielders, six infielders, and two catchers. Each outfielder can play each of the three outfield positions, and each infielder can play each of the four infield positions. How many different rosters can be assembled from this team? A roster consists of one player assigned to each one of the nine positions. Two rosters are different if any one assignment is different.
12. Redo Exercise 5 if the pitchers can also play the outfield.
13. A telephone repair depot has nine crews and five requests for repairs. How many different ways can repairs be scheduled if no crew is assigned more than one repair?
14. A computer is given the names of five females and eight males. How many different ways can the computer match each of the five females with a male?
15. Can the fundamental theorem of counting be applied to Exercise 15 in Section 10.1?

10.3 Permutations

Certain sample spaces can be counted quickly if the possible outcomes are known to be permutations. A *permutation* of a set of elements is one arrangement of all the elements of that set. Three permutations of the set {A, B, C, D, E} are ABCDE, ABDEC, and CEADB. Two permutations of the set {red, green, blue, yellow} are the arrangements red-blue-green-yellow and green-yellow-blue-red. All possible permutations of the set {1, 2, 3} are 123, 132, 213, 231, 312, and 321.

Suppose we are given a set of five objects. Can we determine the number of all possible permutations of this set? Yes, with the fundamental theorem of counting. In a five-element set, there are five choices for the first entry of a permutation. For each possible first entry, there are four possible second entries. For each second entry, there are three third-entry possibilities. Continuing, we find exactly $(5)(4)(3)(2)(1) = 120$ different permutations.

Can we count the total number of permutations of a set containing 11 elements? Yes, and the same procedure is applicable. Each permutation contains 11 entries. There are 11 choices for first entry, 10 choices for the second entry, 9 choices for the third entry, and so on. It follows from the fundamental theorem of counting that there are exactly $(11)(10)(9)(8)(7)(6)(5)(4)(3)(2)(1) = 39,916,800$ permutations.

A pattern is developing, and it involves the product of successive integers.

Such products are called factorials which are denoted by an integer followed by an exclamation point. For the first six integers, we define

One factorial: $1! = 1$
Two factorial: $2! = (2)(1) = 2$
Three factorial: $3! = (3)(2)(1) = 6$
Four factorial: $4! = (4)(3)(2)(1) = 24$
Five factorial: $5! = (5)(4)(3)(2)(1) = 120$
Six factorial: $6! = (6)(5)(4)(3)(2)(1) = 720.$

In general, for any *positive* integer n, n factorial, denoted $n!$, is defined as the product $n! = (n)(n-1)(n-2)\cdots(3)(2)(1)$.

Factorials become very large very quickly. Six factorial is 720, 8 factorial is 40,320, 11 factorial is nearly 40 million, and 13 factorial is over 6 billion. Factorials are a simple notation for indicating certain large numbers which occur frequently. But it is only a notation. To compute a given factorial, say 7!, the product $(7)(6)(5)(4)(3)(2)(1)$ must be determined. The only shortcut is to know 6 factorial ($6! = 720$) and realize that $7! = (7)(6!) = 7(720) = 5040$. Furthermore, factorials have very few properties in common with the usual algebraic properties of numbers. In particular, $(3!)(2!) \neq 6!$, as one can verify easily by computing both sides.

For later work, we also will have need of *zero factorial*. We define it as $0! = 1$.* Therefore quotients having zero factorial in the denominator, like $720/0!$, are arithmetically valid.

The following result is now a simple consequence of the fundamental theorem of counting.

Theorem 10.2 A set of n elements has exactly $n!$ possible permutations.

It follows that there are exactly 5! permutations of a 5-element set, 11! permutations of an 11-element set, and 25! (a very large number) permutations of a 25-element set.

Example 1 A new car dealer has four cars of different colors, red, white, green, and yellow, which are to be arranged in a row in a showroom. How many different arrangements can be made?

Solution Each arrangement is a permutation of the four colors red, white, green, and yellow. There are as many different arrangements as there are different permutations, exactly $4! = 24$. The situation is shown schematically in Figure 10.13.

---

* In more advanced mathematics, it is shown that this definition is consistent with the other factorials.

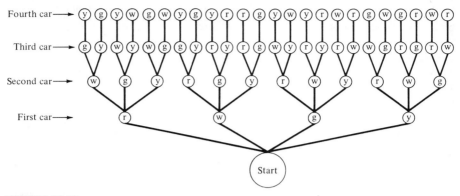

FIGURE 10.13

Example 2 A seven-store chain plans to go out of business by closing one store every month for the next 7 months. How many different ways can the closings be scheduled?

Solution If we label the stores 1 through 7, respectively, each closing schedule is a permutation of the seven integers 1 through 7. The permutation 2715634 denotes closing store 2 first, then store 7, then store 1, until finally store 4 is closed last. There are as many different closing schedules as there are different permutations, exactly $7! = 5040$.

At times, we do not want to arrange all the elements of a particular set but only some. Perhaps we wish to close only five of the seven stores in Example 2, or the showroom has room for only three of the four cars in Example 1. Although Theorem 10.2 is not applicable, factorials still can be useful.

A permutation of r objects from a set of n objects, denoted $P(n, r)$, is any one arrangement of r different objects. One permutation of three objects from the five-object set {A, B, C, D, E} is AED. Two other permutations are ABE and DBC. One permutation of four objects from the six-object set {red, blue, green, white, black, orange} is white-black-green-red. All possible permutations of two objects from the three-object set {1, 2, 3} are 12, 13, 21, 23, 31, and 32.

Let us return to the problem of scheduling five closings from seven stores. Each closing schedule is one permutation of five stores from the list of seven stores, and we know (Example 4 in Section 10.2) that there are exactly (7)(6)(5)(4)(3) such arrangements. If we are interested in the number of ways three cars can be arranged from a set of four cars, we use the fundamental theorem of counting to obtain (4)(3)(2). Again a pattern is emerging.

Theorem 10.3 The number of permutations of r objects selected from n objects is

$$P(n, r) = (n)(n - 1)(n - 2) \cdots (n - r + 1).$$

The formula is not as forbidding as it appears. It is simply the product of successive integers beginning at n (the number of elements in the set) and continuing downward until $n - r + 1$. If $n = 7$ and $r = 5$, then $n - r + 1 = 7 - 5 + 1 = 3$. If $n = 4$ and $r = 3$, then $n - r + 1 = 4 - 3 + 1 = 2$. In particular, $P(n = 7, r = 5) = (7)(6)(5)(4)(3)$, and $P(n = 4, r = 3) = (4)(3)(2)$. More commonly, $P(n = 7, r = 5)$ is written as $P(7, 5)$, and $P(n = 4, r = 3)$ as $P(4, 3)$. The first integer denotes n, and the second integer denotes r. The notation $P(8, 4)$ has $n = 8$ and $r = 4$ and is the product $(8)(7)(6)(5) = 1680$. The notation $P(11, 3)$ has $n = 11$ and $r = 3$ and is the product $(11)(10)(9) = 990$.

Example 3 An organization of 50 members meets to elect 3 officers, president, vice-president, and secretary, from its ranks. How many different ways can this executive slate be formed?

Solution Each arrangement of 3 persons from the 50-person membership constitutes one slate. The arrangement Abrams-Taletti-Johnson for example, denotes Abrams as president, Taletti as vice-president, and Johnson as secretary. There are as many different slates as there are permutations of 3 people from a 50-person set, which is exactly

$P(50, 3) = (50)(49)(48) = 117,600.$

Example 4 The judges at a beauty contest with 22 contestants will designate a winner and a runner-up. How many different ways can the contest end?

Solution Each arrangement of 2 people from the 22-person field constitutes one possible outcome. The arrangement Motley-Jones, for instance, denotes Motley the winner and Jones the runner-up, whereas the arrangement McNutley-Motley denotes McNutley the winner and Motley the runner-up. There are as many different endings as there are permutations of 2 people from a 22-person set, which is exactly

$P(22, 2) = (22)(21) = 462.$

Example 5 A developer plans to build 7 houses on a particular block from a list of 42 different designs. How many different ways can the block be built if no 2 houses are of the same design?

Solution Each permutation of 7 designs from the set of 42 designates one possible block arrangement. There are

$P(42, 7) = (42)(41)(40)(39)(38)(37)(36) = 135,970,773,120.$

In conclusion, we stress that nothing new has been developed in this section other than notation. We have introduced both the factorial notation and $P(n, r)$, and we have used both in several examples. However, all these examples could have been done with the fundamental theorem of counting. The advantage of this new notation is that it provides a compact form for representing large numbers. But that is all. It is very well to give an answer as $P(50, 3)$ and have everyone nod in agreement, but to obtain the numerical answer one must still compute $P(50, 3) = (50)(49)(48)$.

Exercises

In Exercises 1 through 7 leave the answers in either factorial or $P(n, r)$ form.
1. Exercise 8 in Section 10.2
2. Exercise 9 in Section 10.2
3. Exercise 13 in Section 10.2
4. Exercise 14 in Section 10.2
5. Exercise 8 in Section 10.1
6. Exercise 10 in Section 10.1
7. Exercise 13 in Section 10.1
8. How many different 4-letter arrangements can be made from the 26 letters of the alphabet if no letter can be repeated in the same arrangement?
9. A hotel with 141 single rooms receives 141 single reservations. How many different ways can the rooms be assigned?
10. How many different seating charts are there for
    (a) A class of 25 students and 25 desks?
    (b) A class of 25 students and 30 desks?
    (c) A class of 30 students and 25 desks?
11. A rent-a-car agency has 95 cars and 80 reservations. How many different ways can the reservations be filled?
12. The Army has 182 new second lieutenants and 4251 companies. How many different ways can it assign lieutenants to companies if no company is to receive more than one new lieutenant?
13. How many different five-digit numbers can be made from the numbers 1, 2, 3, 4, 5 if no number is used twice in any one arrangement?
14. Show that $P(10, 3)$ can be written as $10!/7!$ and that $P(7, 5)$ equals $7!/2!$.
15. Generalize Exercise 14 to show that $P(n, r) = n!/(n - r)!$.

10.4 Combinations

To this point, all processes have had one characteristic in common: Different orderings resulted in different outcomes. For example, the initials AJB are different from the initials BJA, although both contain the same letters. In the plant-closing example in Section 10.3, closing the Buffalo plant first and the Nashville plant a year later is a different outcome than first closing the Nashville plant and later closing the Buffalo plant, even though both outcomes involve the same two plants. Different arrangements of the same distinguishable objects constitute different outcomes. But this is not always the case. In some processes, different arrangements of the same objects do *not* constitute different

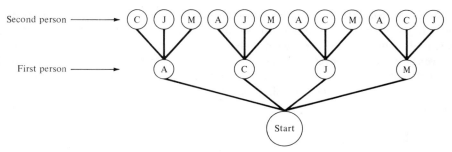

FIGURE 10.14

outcomes; sometimes two outcomes are different if and only if they contain different elements.

Consider a telephone repair unit consisting of four people, Allan, Carl, John, and Mary. A request for service is received which requires two individuals. How many different ways can this service call be handled? If we construct a tree listing all possible outcomes, we obtain Figure 10.14. Apparently, there are 12 possible outcomes, with $U = \{$AC, AJ, AM, CA, CJ, CM, JA, JC, JM, MA, MC, MJ$\}$. But this sample space differentiates between different orderings of the same people. It should not. The team of Allan and Carl (AC) is the same team as Carl and Allan (CA). The team of Carl and Mary (CM) is the same team as Mary and Carl (MC). Since each team is listed twice (the number of different ways two people can be rearranged among themselves) the true answer is 12 divided by 2, or 6 different ways to answer the service call.

As a second example, consider a chain of hero sandwich stores which stock four types of ingredients for their sandwiches: salami, ham, American cheese, and provolone. A regular hero contains three of these ingredients, chosen by the customer. How many different regular heros does the chain offer? Figure 10.15 is a tree for the process of making a hero, and it appears that there are 24

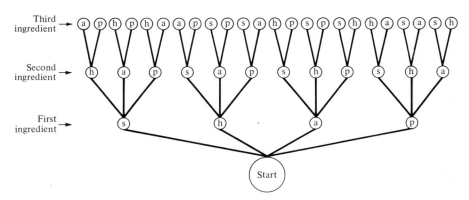

FIGURE 10.15

different possibilities (one for each branch). But this is not correct. The order in which ingredients are placed in the sandwich is irrelevant; only the ingredients themselves matter. The sandwiches sha (first branch), sah (third branch), hsa (seventh branch), has (ninth branch), ash (thirteenth branch), and ahs (fifteenth branch) all contain ham, American cheese, and salami and represent the same hero. Since each hero is listed six times (the number of different ways the same three ingredients can be rearranged among themselves), the true answer is 24 divided by 6, or 4 different regular heros. They are salami–ham–American cheese, ham–American cheese–provolone, salami–ham–provolone, and salami–American cheese–provolone.

These two examples suggest a procedure for counting sample spaces in which order does not matter. First, count the number of possibilities assuming that order does matter. This step is accomplished by any of the methods given in Sections 10.1 through 10.3. Then divide this answer by the number of ways the objects in each outcome can be rearranged *among themselves*.

Applying this procedure to the telephone repair team just considered, where we must choose two people from a set of four, we first determine the number of ways to do the problem if order mattered, which is $P(4, 2) = (4)(3) = 12$. We then divide this number by the number of ways two people (the number in each team) can be arranged among themselves. But two people can be arranged $2! = 2$ ways among themselves. Therefore the final answer is $12/2 = 6$.

If we reconsider the hero sandwich problem in which three ingredients are selected from a set of four, we first determine the number of ways to make the selection if order matters, namely, $P(4, 3) = (4)(3)(2) = 24$. We then divide this number by the number of ways the same three ingredients can be arranged among themselves, which is $3! = 6$. The final answer is $24/6 = 4$.

Let us consider the general case of selecting r different objects from a set of n objects when *order does not matter*. First we count the number of different selections assuming that order does matter. From Theorem 10.3, this is $P(n, r) = n(n - 1)(n - 2) \cdots (n - r + 1)$. We then divide this result by the number of ways the same r objects can be arranged among themselves, namely, $r!$ The final answer is

$$\frac{P(n, r)}{r!} = \frac{n(n - 1)(n - 2) \cdots (n - r + 1)}{r!}.$$

Example 1 How many ways can a nominating committee of 4 members be selected from a club membership of 80 people?

Solution The order in which committee members are selected is irrelevant; only the people themselves are important. Here $n = 80$ and $r = 4$; therefore $P(80, 4) = (80)(79)(78)(77) = 37,957,920$, $r! = 4! = (4)(3)(2)(1) = 24$, and the answer is

$$\frac{P(80, 4)}{4!} = \frac{37,957,920}{24} = 1,581,580 \text{ different possible committees.}$$

Example 2 There are 25 people in a typing pool. Next Monday, 4 of these people will be selected to fill vacant secretarial positions. How many different ways can these selections be made assuming no seniority system?

Solution Here we select 4 objects (secretaries) from a 25-object set (typists). The order of selection is not important, only the people. Therefore $n = 25$, $r = 4$, and there are

$$\frac{P(25, 4)}{4!} = \frac{(25)(24)(23)(22)}{24} = 12{,}650 \text{ different ways.}$$

The quantity $P(n, r)/r!$ occurs so frequently in counting problems that we denote it simply by $C(n, r)$. Thus

$$C(n, r) = \frac{n(n - 1)(n - 2) \cdots (n - r + 1)}{r!}.$$

The notation $C(n, r)$ is read "a *combination* of r objects selected from a set of n objects" or, more concisely, "n things taken r at a time." Whenever the word "combination" is used, it is understood that order does not matter. If order matters, the word "permutation" is used.

Example 3 A bakery receives requests for new accounts from 20 different supermarkets, but it has the capacity to service only 3. How many different ways can the bakery choose its new accounts?

Solution The order in which the 3 accounts to be serviced are selected is immaterial. Here $n = 20$ (the number of accounts from which to choose), $r = 3$ (the number to be chosen), and order does not matter. The answer is

$$C(20, 3) = \frac{(20)(19)(18)}{3!} = 1140.$$

Example 4 A jai alai game involves eight teams, numbered 1 through 8, in competition. A quinella wager is a bet on the two teams that will place first and second regardless of order. A 5-6 quinella wager wins if team 5 finishes first and team 6 finishes second, or vice versa. How many different quinella wagers are possible?

Solution There are as many wagers as there are ways to select two teams from a set of eight without distinguishing between different orderings of the same teams. Therefore $n = 8$ (the number of teams from which to choose), $r = 2$ (the number of teams to be chosen), and order does not matter. The answer is

$$C(8, 2) = \frac{(8)(7)}{2!} = 28.$$

Example 5 A hotel with 141 rooms receives a reservation from an organization for 5 rooms. How many different ways can the reservation be filled?

Solution Since all 5 rooms are reserved under the organization name, the order in which the rooms are assigned is irrelevant; only the room numbers matter. Here $n = 141$, $r = 5$, and the answer is

$$C(141, 5) = \frac{(141)(140)(139)(138)(137)}{5!} = 432{,}295{,}143.$$

Often the number $C(n, r)$ for particular values of n and r is very large and difficult to calculate. If so, it is sometimes left in the form $C(n, r)$ and later calculated with the aid of a computer. If, in Example 5, the organization requests 37 rooms instead of 5, the number of ways to make the necessary room assignments is easily given by $C(141, 37)$. One can verify, however, that it is not easy to compute the actual number.

Exercises

1. Find
   (a) $C(6, 2)$
   (b) $C(10, 4)$
   (c) $C(19, 2)$
   (d) $C(15, 3)$
   (e) $C(7, 5)$
   (f) $C(9, 0)$
   (g) $C(5, 5)$
   (h) $C(5, 8)$
2. A rent-a-car agency with 20 available cars receives an order from a corporate client for 7 cars. How many ways can this order be filled?
3. An ice cream concern stocks 30 flavors and offers a rainbow banana split consisting of 3 scoops of ice cream, each of a different flavor. How many different rainbow splits can it advertise?
4. A department store chain has options to buy 20 different parcels of land in various sections of the country for new stores. It decides to build 3 stores. How many different ways can it select the required locations?
5. A candy manufacturer produces Halloween surprise bags by filling bags with 5 different pieces of penny candy. How many different surprise bags can it produce if it stocks 14 kinds of penny candy?
6. The maintenance department of an industrial plant receives calls for new neon bulbs from 17 offices. How many ways can it respond if it has only 8 bulbs in stock?
7. A telephone survey is to be conducted with 50 members of a 1000-member organization. How many different ways can the organization be surveyed?
8. How many different ways can two horses tie for the win position in an eight-horse race?
9. Seventeen colonels will be promoted to generals. How many ways can the promotion list be prepared if 645 colonels are eligible for promotion?
10. A jury call in a county of 97,000 people involves 25 people for 2 weeks. How many different lists can be prepared for each call?
11. Show that $C(n, r)$ can be given compactly by $n!/(n - r)!\, r!$.
12. Use Exercise 11 to prove that $C(n, r) = C(n, n - r)$.
13. Use Exercise 11 to prove that $C(n - 1, r - 1) + C(n - 1, r) = C(n, r)$.

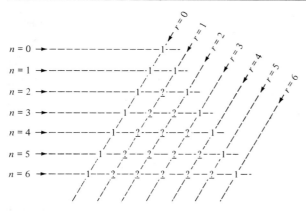

FIGURE 10.16

14. Use the result from Exercise 13 to fill in the interior of Figure 10.16. The complete result is known as *Pascal's triangle*. The rows correspond to successive values of n (beginning at $n = 0$), while the diagonals correspond to successive values of r (beginning at $r = 0$).

10.5 Counting Complex Processes*

The previous material on the fundamental theorem of counting, factorials, permutations, and combinations covers the basic methods used for counting sample spaces. Often two or more of these methods are used together to count the outcomes of complex processes.

As an example of such a situation, consider a rent-a-car agency which has 10 luxury cars, 52 standard cars, and 35 compacts available for service. The agency receives an order from an organization for 2 luxury cars, 15 standard cars, and 21 compacts. How many different ways can this order be filled? Since none of the standard procedures by themselves is sufficient, we break the problem into manageable parts.

We first consider the luxury cars. We have 10, we must select 2, and order does not matter. There are $C(10, 2)$ different ways to make the selection. Having chosen the luxury cars, we then turn our attention to the standard models. We have 52, we want 15, and order does not matter. There are $C(52, 15)$ different ways to make this selection. Finally, we consider the compacts. We have 35, we want 21, and again order does not matter. There are $C(35, 21)$ ways to fill this part of the order.

To determine the different ways to fill the entire order we use the fundamental theorem of counting. There are $C(10, 2)$ ways to select the luxury cars. For each selection, there are $C(52, 15)$ ways to select the standard cars. For

---

* This section contains more advanced optional material.

each one of these possibilities, there are $C(35, 21)$ ways to select compact cars. The product $C(10, 2)C(52, 15)C(35, 21)$ is the different number of ways to fill the entire order.

The procedure of first finding the number of ways individual parts can be selected, and then using the fundamental theorem of counting to determine the number of possibilities for the entire process, is useful whenever the processes can be broken into sequential parts.

Example 1 A Chinese dinner for 6 consists of 1 selection from group A, 3 selections from group B, and 2 selections from group C. There are 8, 20, and 13 entrees listed under groups A, B, and C, respectively. How many different dinners for 6 does the restaurant offer?

Solution This problem can be broken into sequential parts by ordering from the individual groups one at a time. Since there are 8 entrees available in group A, and 1 selection is permitted, there are 8 possibilities. There are 20 entrees available in group B, and 3 selections are required. Since the order in which entrees are selected is irrelevant, there are $C(20, 3)$ ways to order from group B. Similarly, there are $C(13, 2)$ ways to order the 2 dishes from the 13 listed under group C.

To count the sample space of the entire process (ordering dinner for 6), we note that there are 8 ways to order from group A. For each order, there are $C(20, 3)$ ways to order from group B. For each of these selections, there are $C(13, 2)$ ways to order from group C. Using the fundamental theorem of counting, we obtain the total number of possible dinners as the product

$8C(20, 3)C(13, 2)$.

Example 2 The residence office of a university has three dormitory rooms unassigned, two doubles and one triple, and requests for rooms from 10 female students. How many different ways can it assign the available space?

Solution For convenience, we number the rooms 101, 102, and 103 for the triple and the two doubles, respectively. This problem can be broken into sequential parts if we fill the rooms one at a time. Filling the triple first, we must select three students from the 10 applying, and order does not matter. The assignment Mary-Joan-Paula to room 101 is the same as the assignment Joan-Paula-Mary. There are $C(10, 3)$ ways to assign students to room 101. We now turn our attention to room 102, a double. Since three students have been assigned to 101, there are only 7 students left without a room. We must select 2 of these students for 102, and again order does not matter. There are $C(7, 2)$ possible selections. There are now 5 students without a room and one double still unassigned. Accordingly, we have $C(5, 2)$ ways to fill room 103.

To count the sample space of the entire process, we note that there are $C(10, 3)$ ways to fill room 101. For each of these possibilities there are $C(7, 2)$ ways to fill room 102, and for each of these possibilities there are $C(5, 2)$ ways to fill room 103. Using the fundamental theorem of counting, we obtain the total number of ways to fill all three rooms as the product $C(10, 3)C(7, 2)C(5, 2)$.

Example 3 How many different ways can a horse race with eight entries end if two horses tie for first and one horse finishes third?

Solution We consider first place first. We want two horses from a set of eight, and order does not matter. Horses 1 and 5 tied for first is the same outcome as horses 5 and 1 tied for first. There are $C(8, 2)$ possibilities. Once two horses have won, there are only six candidates for third place, hence six possibilities. Since there are $C(8, 2)$ selections for first place, and each of these has exactly six possibilities for third, the product $6C(8, 2)$ represents the number of ways the entire race can end.

We have left the answers to Examples 1 through 3 in combination notation as a matter of convenience. If we want the actual number, we must calculate each $C(n, r)$ separately. In Example 1, we first calculate $C(20, 3) = 1140$ and $C(13, 2) = 78$, and then the product $8C(20, 3)C(13, 2) = 8(1140)(78) = 711{,}360$.

A second type of process that cannot be solved by only one of the methods described in Sections 10.1 through 10.4 involves outcomes in which certain orderings matter but others do not. The general procedure is similar to the one given in Section 10.4. Do the problem assuming that order matters and then divide by the number of ways the indistinguishable elements can be arranged among themselves.

As an example, let us determine the number of different four-letter arrangements that can be made from the letters of the word *meet*. Two arrangements are emet and eemt. There are more. We begin by assuming that all the letters are distinguishable and then calculate the number of different arrangements that can be made from four distinguishable letters. There are 4! such arrangements, representing all possible permutations of the four letters. But some of these permutations do not yield different arrangements. The permutation meet and meet, in which the two e's have been rearranged, are not different. Neither are mtee and mtee, where again we have rearranged the two e's. Obviously, any rearrangement of the e's does not lead to a different combination. And for each permutation there are exactly 2! such rearrangements, the number of ways the two e's can be rearranged among themselves. Accordingly, there are only $4!/2! = 12$ distinguishable combinations.

Example 4 Determine the number of different seven-letter arrangements that can be formed from the letters of the word "tweezer."

Solution If we assume momentarily that all seven letters are distinguishable, we find 7! possible arrangements, representing all possible permutations of the seven letters. But three of the letters, the three e's are not distinguishable, and any rearrangement of just these three letters alone does not yield a new word. Since there are 3! ways the three e's can be rearranged among themselves, there are only $7!/3! = 840$ distinguishable arrangements of the letters in "tweezer."

Example 5 Determine the number of different six-letter arrangements that can be formed from the letters in the word "banana."

Solution There are 6! possible arrangements if all the letters are distinguishable, which they are not. There are three a's which can be rearranged among themselves in 3! ways. Therefore 6!/3! is a better answer, although it too is not correct. The word "banana" also has two n's which can be rearranged among themselves in 2! ways. The correct answer is

$$\frac{6!}{3!\,2!} = 120.$$

A situation that occurs often in probability problems (see Section 11.6) is the following.

Example 6 Determine the number of different n-letter arrangements that can be formed with r S's and $(n - r)$ F's.

Solution There are $n!$ possible arrangements if all the letters are distinguishable, which they are not. There are r S's which can be rearranged among themselves in $r!$ ways, and there are $(n - r)$ F's which can be rearranged among themselves in $(n - r)!$ ways. Accordingly, there are $n!/r!(n - r)!$ different arrangements of all n letters that can be written succinctly as $C(n, r)$ (see Exercise 11 in Section 10.4).

A different but similar process is the following.

Example 7 A used furniture dealer has a dining-room set of 1 table and 6 chairs, 2 matching love seats, and 3 identical lamps to be aligned in a row for a promotional picture. How many different pictures can be taken?

Solution There are 12 pieces of furniture for the picture, of which the 6 chairs, 2 love seats, and 3 lamps are indistinguishable among themselves. There are 12!/6! 2! 3! different possible arrangements. Each arrangement results in a different picture.

A third type of process involves constraints, that is, restrictive conditions which must be met. A good procedure for counting the possible outcomes of such processes is first to satisfy the given constraints.

Example 8 A new car dealer receives an order from one corporation for 12 cars and another order from a second corporation for 5 black cars. How many ways can the orders be filled if the dealer has 8 black cars, 7 white cars, 3 red cars, and 2 blue cars in stock.

Solution The problem involves one constraint: All cars ordered by the second corporation must be black. Following the recommended procedure, we fill the second order first. There are 8 black cars available, we want 5, and order does not matter. There are $C(8, 5)$ ways to fill the order. We now handle the needs of the first corporation. We have 15 cars available (3 remaining black, 7 white, 3 red, and 2 blue), we want 12, and order does not matter. There are $C(15, 12)$ ways to fill the first order. Since there are $C(8, 5)$ ways to fill the second order, and for each selection there are $C(15, 12)$ ways to fill the first order, it follows that there are $C(8, 5)C(15, 12)$ ways to fill both orders.

Example 9 How many ways can a 5-card hand be dealt from a regular deck of 52 cards if the hand must contain exactly 2 kings.

Solution We first agree that order does not matter. The order in which cards are dealt is immaterial; only the cards themselves are important. Since we have a constraint for exactly 2 kings, we satisfy this requirement first. There are 4 kings in a deck, we want to pick 2, and order does not matter. There are $C(4, 2)$ ways to make the selection. The remaining 3 cards in the hand must be selected from the 48 cards in the deck that are not kings. Since order does not matter, there are $C(48, 3)$ ways to choose them. The product $C(4, 2)C(48, 3)$ is the number of different ways the hand can be dealt.

Example 10 An airline employs 50 captains, 30 copilots, and 40 engineers, from which it must assemble 28 flight crews for the next day's schedule. A flight crew consists of 1 captain, 1 copilot, and 1 engineer. How many ways can the assignments be made if captains can act as copilots but copilots cannot act as captains?

Solution The constraints are that a captain's position can be filled only by a captain and an engineer's position can be filled only by an engineer. Therefore we make these selections before choosing the copilots. We need 28 pilots, and 50 are available. If we assume that order matters (which is the case if the first captain is assigned the New York–Miami flight, while the second captain is assigned the New York–Boston flight), there are $P(50, 28)$ ways to make the selection. Similarly, there are $P(40, 28)$ ways to fill the engineer slots. We also need 28 copilots, but they can be chosen from the 30 available copilots *and* the 22 (50 − 28) captains still unassigned. We want 28 selections from a set of 52 candidates (30 + 22), which can be made in $P(52, 28)$ different ways. The product $P(50, 28)P(40, 28)P(52, 28)$ represents the total number of different ways the 28 crews can be assembled assuming that order matters.

Exercises

1. A dinner consists of 1 appetizer, 1 entree, 2 different vegetables, and 1 dessert to be chosen from a menu containing 5 appetizers, 10 entrees, 5 vegetables, and 6 desserts. How many different dinners can be ordered? Two dinners are different if any item is different.
2. A 9-person bargaining committee and a 3-person grievance committee are to

be formed from the membership of a local union. No member may serve on both committees. How many different ways can the committees be formed if 25 people belong to the union?

3. The head of the legal department of a large corporation plans to send 2 lawyers to Arizona and 3 lawyers to Boston to conclude stock merger plans. How many different ways can the department head make these assignments if she employs 10 lawyers?

4. Seventeen colonels will be promoted to general, 15 majors will be promoted to colonel, and 10 captains will be promoted to major. How many different ways can the promotions be made if there are 100 colonels, 215 majors, and 125 captains eligible for promotion?

5. The manager of a new store must hire 20 sales people from 52 applicants, 3 bookkeepers from 7 applicants, 5 cashiers from 31 applicants, and 3 stock people from 11 applicants. Determine the number of different ways the manager can fill his personnel openings.

6. A football team plays nine different opponents. How many different ways can it end the season with six wins, two losses, and one tie. *Note:* One way is to win the first six games, lose the next two, and tie the last. A second way is to win the first two, lose the next two, tie the fifth, and win the last four.

7. A man owns 3 suits, 20 ties, and 22 shirts. How many ways can he select 2 suits, 4 ties, and 4 shirts to take with him on a vacation?

8. How many different five-letter arrangements can be made from the letters in "bollo"?

9. How many different 11-letter arrangements can be made from the letters in "Mississippi"?

10. A new car dealer plans to arrange two white cars, three red cars, and four blue cars on his lot in a row. Determine the number of different arrangements if the cars are all the same model and indistinguishable except for color.

11. Redo Exercise 10 if all the red cars must be placed next to each other at the beginning of the row.

12. A man flips a coin nine times. How many different ways can he obtain five heads and four tails? *Note:* One way is a head on the first flip, a tail on the next four flips, and a head on the last four flips.

13. A woman picks one ball from each of 10 identical barrels standing in a row. Each barrel contains one red ball and one white ball. How many different ways can she select six red balls and four white balls? *Note:* One way is to pick a red ball from the first two barrels, then a white ball from the next four barrels, and then a red ball from the last four barrels.

14. Redo Exercise 13 if each barrel contains two red balls and one white ball.

15. A quality-control engineer selects 12 radios from the production line one at a time. How many different ways can she select 6 good radios and 6 defective radios? *Note:* One way is to select 2 good radios first, then 6 defective radios, and then 4 good radios. In how many ways can the selections be made if no two consecutive radios are both good or both defective?

16. Fifty tubes of the same toothpaste, 40 boxes of the same detergent, and 10 similar packages of bacon are to be given free to the first 100 people to patronize a particular supermarket next Monday. How many different ways can the free gifts be distributed?

17. How many ways can a five-card hand containing exactly two aces and three jacks be dealt from a regular deck of cards?
18. How many ways can a 13-card hand containing 1 heart, 3 clubs, 4 diamonds, and 5 spades be dealt from a regular deck of cards?
19. How many ways can a five-card hand containing exactly two aces, two kings, and one other card be dealt from a regular deck of cards?
20. How many ways can a 13-card hand containing 4 hearts, 4 spades, 4 diamonds, and 1 club be dealt from a regular deck of cards?

Exercises 21 through 25 involve processes slightly more involved than those considered to date. Nonetheless, they can be done with the methods developed in this section.

21. How many ways can a 13-card hand containing a 4-4-4-1 suit distribution (4 of one suit, 4 of a second suit, 4 of a third suit, and 1 of the remaining suit) be dealt from a regular deck of cards?
22. The French club at a local high school will stand in a row for a group picture. The club consists of four males and three females.
    (a) How many different pictures can be taken?
    (b) How many different pictures can be taken if the males and females must be grouped together by sex?
    (c) How many different pictures can be taken if the sexes alternate?
23. Redo Example 6 if objects of the same type must be grouped together.
24. How many different ways can four red books, three blue books, and two green books (all different titles) be arranged on a shelf grouped together by color?
25. A car dealer has a white Chevrolet, a red Chevrolet, a white Ford, a red Ford, a black Ford, a red Dodge, a white Pontiac, a red Pontiac, a black Pontiac, and a green Pontiac. He wishes to arrange all the cars in a row on his lot.
    (a) How many different arrangements can he make?
    (b) How many different arrangements can he make if the cars must be grouped together by make?
    (c) How many different arrangements can he make if the cars must be grouped together by color?

Chapter

Probability

Very few things in business are guaranteed; uncertainties almost always exist. A corporation may anticipate a profit of $6 million for the next fiscal year, but no one really believes that the profit will be exactly $6 million to the penny. There is a chance that the company may make $7 million, and there is a chance that the company may make $5 million. There is even a chance that the company may lose money.

Since most decisions, actions, or projections allow for several possible outcomes, not all of which are desirable, management is interested in both the possible outcomes and the probabilities that these outcomes will materialize. That is, management needs to know what is possible and what is probable. It is one thing to know that a suggested research project can either succeed or fail. It is quite another thing to know that the chances of success are only 10% and the chances for failure are 90%.

In Chapter 10, we developed methods for listing and counting sample spaces, those outcomes that are possible. We now turn our attention to determining the probability, or likelihood, of each outcome actually occurring.

11.1 Equal Probability—Simple Processes

The simplest type of probability problems occurs when a sample space for a process is chosen in which all the outcomes have the same chance of happening. We call such sets *equally likely sample spaces*. As an example, consider the process of throwing an ordinary die (the singular of dice) and announcing the number that turns up. An equally likely sample space for this process is $U = \{1, 2, 3, 4, 5, 6\}$, since each outcome has the same chance of happening.

Often the same process can have several different sample spaces associated with it, some of which are equally likely and some of which are not. Probability problems are easier to solve if we use the equally likely sample space when it is available.

Consider a barrel containing four red balls, numbered from 1 through 4, and one white ball. One sample space for the process: Pick a ball from the barrel. is $U = \{\text{red, white}\}$. Unfortunately, these two outcomes are not equally likely to occur. Since there are four red balls and only one white ball, there is a greater chance of picking a red than there is of picking a white. To construct an equally likely sample space for this process, we take $U = \{\text{red 1, red 2, red 3, red 4, white}\}$, where each red ball is distinguished by both its color and its number. Now each outcome has the same likelihood of happening.

Example 1 Determine an equally likely sample space for the process: Flip a coin twice and describe the results.

Solution We first try $U = \{\text{two heads, two tails, one of each}\}$. This is a sample space for the process, but unfortunately the outcomes are *not* equally likely. The outcome "one of each" is twice as likely to occur as the outcome "two heads." "One of each" can happen in two ways: a head first and a tail second, or a tail first and a head second, whereas "two heads" can occur in only one way.

A second sample space for this process is $U = \{\text{HH, HT, TH, TT}\}$, where the first and second letter of each outcome represent the results of the first and second toss, respectively. This sample space is an equally likely one, since each outcome has the same chance of occurring.

Example 2 Determine an equally likely sample space for the process: Throw two dice and announce the sum of the faces that turn up.

Solution When two dice are thrown, any sum between 2 and 12 can occur, so we are tempted to take $U = \{2, 3, 4, 5, 6, 7, 8, 9, 10, 11, 12\}$. Unfortunately the outcomes of this sample space are *not* equally likely. In particular, the outcome 7, which can be rolled as 6 plus 1 or 5 plus 2 or 3 plus 4, is more likely to occur than the outcome 2, which can be rolled only as 1 plus 1.

To construct an equally likely sample space for this process we imagine that the two dice are distinguishable. Perhaps one is red and the other is green. We then list outcomes in sum form with the first number representing the face of the red die and the second number representing the face of the green die. Now,

$$\begin{aligned}U = \{&1+1, 1+2, 1+3, 1+4, 1+5, 1+6,\\&2+1, 2+2, 2+3, 2+4, 2+5, 2+6,\\&3+1, 3+2, 3+3, 3+4, 3+5, 3+6,\\&4+1, 4+2, 4+3, 4+4, 4+5, 4+6,\\&5+1, 5+2, 5+3, 5+4, 5+5, 5+6,\\&6+1, 6+2, 6+3, 6+4, 6+5, 6+6\}.\end{aligned}$$

Here each outcome has the same chance of occurring. (Note that we have listed the most elementary outcomes possible.)

More often than not, one is interested in the probability that a set of outcomes will occur rather than the probability of only one outcome occurring. Consider again the process of picking one ball from a barrel containing four red balls and one white ball as discussed at the beginning of this section. An equally likely sample space for this process is $U = \{$red 1, red 2, red 3, red 4, white$\}$. If we were interested in the event of choosing a red ball, this would be satisfied by any one of the four equally likely outcomes, red 1, red 2, red 3, and red 4.

We define an *event*, denoted by E, to be a subset of the sample space U. Two different subsets represent two different events. In general, we are asked to find the probability of an event such as picking a red ball from a barrel containing four red balls and one white ball. Verbally, the event is picking a red ball from a barrel. If we use the equally likely sample space $U = \{$red 1, red 2, red 3, red 4, white$\}$, the subset that defines this event is $E = \{$red 1, red 2, red 3, red 4$\}$.

What we have done is first to construct an equally likely set of outcomes for the given process and then select a subset of these outcomes to define the event of interest.

Example 3 The same coin is flipped twice. Find the subset that defines the event. Both tosses are different.

Solution We first need a sample space for the process of flipping a coin twice. From Example 1, we take $U = \{$HH, HT, TH, TT$\}$. Then $E = \{$HT, TH$\}$. Note that the outcome HT is different from the outcome TH. Since both outcomes HT and TH satisfy the event that both tosses are different, they must both be included in E.

Example 4 Two dice are rolled. Find the subset that defines the event that the sum of the dice is 7.

Solution We take U to be the second sample space described in Example 2, since the outcomes are all equally likely. Then $E = \{6+1, 5+2, 4+3,$

3 + 4, 2 + 5, 1 + 6} is the subset of all outcomes in this sample space that sum to 7.

We are now ready to find probabilities associated with processes in which all outcomes are equally likely. For notation, we again let U denote the sample space, E an event, $n(U)$ the number of elements in U, and $n(E)$ the number of elements in E. We denote the probability that E will happen by $P(E)$ and define this probability as

$$P(E) = \frac{n(E)}{n(U)}. \tag{1}$$

That is, the probability that an event will happen is the number of elements in E divided by the number of elements in U. It is important to note, however, that *the formula is valid if and only if all the outcomes in U are equally likely to occur.*

Example 5 A coin is flipped twice. Find the probability that both tosses are different.

Solution The process is to flip a coin twice while the event is that both tosses are different. From Example 1, we have $U = \{HH, HT, TH, TT\}$ as an equally likely sample space. Clearly $n(U) = 4$. From Example 3, we take $E = \{HT, TH\}$, so $n(E) = 2$. Using Eq. (1), we conclude

$$P(E) = \frac{n(E)}{n(U)} = \frac{2}{4} = \frac{1}{2}.$$

The probability of both tosses being different is $\frac{1}{2}$.

Example 6 Two dice are thrown. Determine the probability that the sum of the faces that turn up is 7.

Solution The process is to throw two dice while the event is that the sum of the faces that turn up is 7. Using Example 2 and counting the elements of the second sample space (the equally likely one), we find $n(U) = 36$. From Example 4, we take $E = \{6 + 1, 5 + 2, 4 + 3, 3 + 4, 2 + 5, 1 + 6\}$. Since there are exactly six outcomes in E, $n(E) = 6$. Using Eq. (1), we calculate

$$P(E) = \frac{n(E)}{n(U)} = \frac{6}{36} = \frac{1}{6}.$$

The probability of the sum of the faces being 7 is $\frac{1}{6}$.

Example 7 The numbers 3 through 11, inclusive, are written on separate but identical sheets of paper and placed in a bag. One number is chosen from the bag. Determine the probability that the number will be even.

Solution The process is to pick a number, and the event is that the number is even. We take $U = \{3, 4, 5, 6, 7, 8, 9, 10, 11\}$, which represents all possible outcomes. The outcomes are all equally likely, and $n(U) = 9$. The subset of U that defines the event of interest is $E = \{4, 6, 8, 10\}$. Accordingly, $n(E) = 4$. Using Eq. (1), we calculate

$$P(E) = \frac{n(E)}{n(U)} = \frac{4}{9}.$$

Example 8 Pick one card from a regular deck. Determine the probability that the card is a heart.

Solution The process is to pick a card, while the event is that the card is a heart. We can do this problem in two different ways, depending on how we initially choose the sample space.

Method 1: Take U to be the set of all 52 cards in the deck. Since each card is as likely to be picked as any other card, the outcomes are all equally likely and $n(U) = 52$. With this sample space, the event E is the set of all hearts ranging from the ace of hearts through the king of hearts. There are 13 hearts in the deck, so $n(E) = 13$. Then, using Eq. (1),

$$P(E) = \frac{n(E)}{n(U)} = \frac{13}{52} = \frac{1}{4}.$$

Method 2: Take $U = \{\text{club, diamond, heart, spade}\}$. Since all suits are equally likely to be chosen, the $n(U) = 4$ outcomes of this sample space are all equally likely. The subset of this U that defines E is simply $E = \{\text{heart}\}$, hence $n(E) = 1$. Using Eq. (1), we again obtain $PE = n(E)/n(U) = \frac{1}{4}$.

By either method, the probability of picking a heart is $\frac{1}{4}$.

Exercises

1. Two identical bags each contain the numbers 1 through 4 written on separate but identical slips of paper. One number is drawn from each bag, and their sum is announced. Determine which of the following sample spaces for this process contain equally likely outcomes.
   (a) {odd, even}
   (b) {1, 2, 3, 4, 5, 6, 7, 8}
   (c) {1 + 1, 1 + 2, 1 + 3, 1 + 4, 2 + 1, 2 + 2, 2 + 3, 2 + 4, 3 + 1, 3 + 2, 3 + 3, 3 + 4, 4 + 1, 4 + 2, 4 + 3, 4 + 4}
   (d) {1 + 1, 1 + 2, 1 + 3, 1 + 4, 2 + 2, 2 + 3, 2 + 4, 3 + 3, 3 + 4, 4 + 4}.
   Note: In this sample space the outcomes are given by listing the smaller number first.
   (e) {less than 5, equal to 5, greater than 5}.

2. Four cards consisting of a king of clubs, a king of hearts, a king of diamonds, and a king of spades are shuffled and placed face down on a table. One card is then selected. Determine which of the following sample spaces for this process contain equally likely outcomes.
   (a) {red king, black king}
   (b) {king of clubs, not the king of clubs}
   (c) {red king, not a red king}
   (d) {king of clubs, king of diamonds, king of hearts, king of spades}.
3. Consider the second sample space (the one with equally likely outcomes) given in Example 2. Describe in words the events defined by the following subsets:
   (a) {1 + 1}
   (b) {1 + 1, 6 + 6}
   (c) {1 + 1, 1 + 2, 2 + 1, 2 + 2}
   (d) {1 + 1, 1 + 2, 1 + 3, 1 + 4, 1 + 5, 1 + 6, 2 + 1, 3 + 1, 4 + 1, 5 + 1, 6 + 1}
   (e) {4 + 6, 5 + 5, 5 + 6, 6 + 4, 6 + 5, 6 + 6}.
4. Consider the sample space consisting of each one of the 52 cards for the process of choosing one card from a regular deck. Describe in words the events defined by the following subsets. In each outcome the rank is listed before the suit.
   (a) $A = \{2H, 2C, 2S, 2D\}$
   (b) $B = \{AH, 2H, 3H, 4H, 5H, 6H, 7H, 8H, 9H, 10H, JH, QH, KH\}$
   (c) $C = \{JC, JD, JH, JS, QC, QD, QH, QS, KC, KD, KH, KS\}$
   (d) $A \cap B$
   (e) B'
   (f) $B \cap C$.
5. A coin is flipped three times in succession.
   (a) Determine a sample space of equally likely outcomes for this process.
   (b) Find the subset of this sample space that defines the event of obtaining exactly two heads.
   (c) Find the subset of this sample space that defines the event of obtaining at least two tails.
   (d) Determine the probability of flipping a coin three times and obtaining exactly two heads.
   (e) Determine the probability of flipping a coin three times and obtaining at least two tails.
6. A bag contains two spoons and three forks. A magnet is placed in the bag; it attaches itself to one of the utensils, and then it is pulled out. The utensil drawn from the bag is announecd.
   (a) Determine a sample space of equally likely outcomes for this process.
   (b) Find the subset of this sample space that defines the event of drawing a spoon.
   (c) Determine the probability of drawing a spoon from the bag.
7. A woman has three green dresses, two red dresses, and two white dresses. Her maid, who is color blind, packs two of these dresses in a suitcase for a business trip.
   (a) Determine a sample space of equally likely outcomes for the process of packing two dresses in the suitcase.
   (b) Find the subset of this sample space that defines the event that both dresses are a different color.

(c) Determine the probability that both dresses are a different color.
8. One card is drawn from a regular deck. Determine the probability that it will be a king.
9. One die is thrown and the number that turns up is announced. What is the probability that the number is less than 5?
10. One ball is selected from a barrel containing two red balls, three white balls, and four green balls. What is the probability that the ball selected is not green?
11. A jury list of 100 names contains 55 men and 45 women. Classified by education, 42 men and 31 women have college degrees, while 13 men and 14 women do not. Determine the probability that
    (a) The first person selected from the list has a college degree
    (b) The first person selected from the list is female
    (c) The first person selected from the list is a female with a college degree.
12. A candy manufacturer knows that 2% of his chocolate bars with almonds and 1% of his chocolate bars without almonds spoil during shipping. The manfacturer mails one gift package of 200 bars with almonds and 100 chocolate bars without almonds to an orphanage. What is the probability that the first candy bar picked is spoiled? What is the probability that the first candy bar picked contains almonds?
13. The editor of the annual report of a corporation is informed by mail that a photographer has taken a picture of two lawyers in the firm's Tuscon office for the report. The editor knows that the Tuscon office employs five lawyers, four males and one female. What is the probability that the female lawyer is in the picture?
14. A man has three suits, each a different color, and three ties, one for each suit. Without looking he quickly grabs one suit and one tie. What is the probability that the suit and tie match?
15. The menu at a small restaurant lists three entrees, sliced beef, chef's salad, and broiled halibut, and four desserts, pie, cake, ice cream, and fruit. A customer tells the waiter to bring whatever the waiter thinks is best. Determine the probability that the customer will have sliced beef for his entree. What is the probability that he will have sliced beef and ice cream?
16. Mr. Jones purchases two cans of chicken soup and three cans of vegetable soup at a bargain price. The cans have no labels, and all look identical. On the way home, the cans spill out of their bags and become thoroughly mixed. Once home, Mr. Jones arbitrarily opens two cans. Determine the probability that the contents of both cans are the same.

11.2 Equal Probability—Complex Processes*

The problems considered in Section 11.1 were impractical in one sense: The sample spaces involved were small. We were able to first list the sets of all possible outcomes and then count U and E. In contrast, most commercial

---

* This section contains advanced material requiring a knowledge of the material covered in Section 10.5.

processes involve a large number of possible outcomes, usually too many to list conveniently.

In Chapter 10, we developed methods for counting sets without first listing. These methods can be employed to determine probabilities associated with many business problems. The first step is to construct an appropriate sample space. If the formula

$$P(E) = \frac{n(E)}{n(U)} \qquad [1]$$

is to be used, all outcomes in the sample space must be equally likely to occur. This is crucial. When we apply Eq. (1) to a sample space in which the outcomes are not equally likely, we invariably obtain incorrect answers. Having an appropriate U, we then determine E, count the elements in both U and E, and finally use Eq. (1) to calculate the desired probability.

Several methods are available for counting U and E. Sometimes we use the fundamental theorem of counting (Section 10.2). Other times permutations (Section 10.3) or combinations (Section 10.4) may be appropriate. Occasionally we need two or more of these methods together (Section 10.5). As a last resort we can revert to trees. There is no formula to tell us which method of counting should be used. Each problem must be handled separately.

Example 1 The manager of the Portland branch of a department chain employing 100 men and 35 women will choose 3 people at random to represent the store at a regional sales meeting. Determine the probability that the 3 people selected will all be men.

Solution We take the sample space to be all possible groups of 3 people that can be formed from the 135 employees. Since no group is any more likely to be chosen than any other group, this is an equally likely sample space. To compute $n(U)$, we realize that we have 135 people, we want 3, and order does not matter. Therefore $n(U) = C(135, 3)$.

The event E is that subset of U containing all possible groups of 3 people involving only men. Such groups can be selected only from the 100 available men. We want 3, and order does not matter, so $n(E) = C(100, 3)$.

Using Eq. (1), we have

$$P(E) = \frac{C(100, 3)}{C(135, 3)}.$$

Since

$$C(100, 3) = \frac{(100)(99)(98)}{3!} = 161,700$$

and

$$C(135, 3) = \frac{(135)(134)(133)}{3!} = 400{,}995,$$

it follows that

$$P(E) = \frac{161{,}700}{400{,}995} = .40 \text{ rounded to two decimal places.}$$

Example 2 A new car dealer receives 10 cars, 5 red and 5 white, from a manufacturer. An employee is instructed to drive them into the lot and park them in a row. What is the probability that the cars will be parked together by color?

Solution We take the sample space to be all possible groupings of the 10 cars. Clearly all the outcomes are equally likely. We have 10 choices for the first car, 9 choices for the second car, 8 choices for the third car, and so on. Accordingly, $n(U) = 10! = 3{,}628{,}800$.

The event E is the subset of all outcomes in which the red cars are grouped together and the white cars are grouped together. To count E, we note that we have 10 choices for the first car. Once it has been parked, we have only 4 choices for the second car, since it must match the first car in color. Similarly the third, fourth, and fifth cars also must match the first car in color, so we have 3, 2, and 1 choice, respectively, for them. The sixth car can be any one of the 5 remaining cars. Once it is placed, the seventh car can be any one of the 4 remaining cars, and so on. Using the fundamental theorem of counting, we obtain

$$n(E) = (10)(4)(3)(2)(1)(5)(4)(3)(2)(1) = 28{,}800.$$

It follows from Eq. (1) that

$$P(E) = \frac{28{,}800}{3{,}628{,}800} = .008 \text{ rounded to three decimal places.}$$

Example 3 A caterer prepares 100 box lunches for an organization's picnic. Each box lunch is to contain 1 sandwich—ham, roast beef, salami, or corned beef and the caterer has prepared 25 of each type of sandwich. A family of 4 each takes 1 box lunch. What is the probability that each member of the family receives a different sandwich?

Solution We take U to be the set of all possible selections for the family. One possible outcome is father–ham, mother–ham, brother–salami, and sister–roast beef. A second possible outcome is father–salami, mother–corned beef, brother–ham, and sister–roast beef. To count U, we note that the father can receive any one of 4 different sandwiches. Once he has chosen, the mother selects, and she too can receive any one of 4 different sandwiches. The same is true for the brother and sister. Using the fundamental theorem of counting, we obtain $n(U) = (4)(4)(4)(4) = 256$, and we are on our way to the wrong answer.

We have determined a valid sample space but *not* an equally likely one. If the father selects a box lunch with a roast beef sandwich, then there will be only 24 roast beef sandwiches available to the mother. But there are 25 ham, 25 salami, and 25 corned beef sandwiches still available. Although the mother can choose any one of the 4 types of sandwiches, she has a greater chance of selecting ham than roast beef. The outcome father–roast beef, mother–roast beef, brother–salami, and sister–corned beef is not as likely to occur as the outcome father–roast beef, mother–ham, brother–salami, and sister–corned beef.

To generate an equally likely sample space, we pretend that the different types of sandwiches are all numbered from 1 to 25, so that all sandwiches are distinguishable. Again we take U to be the set of all possible selections for the family, but now an outcome has the form father–roast beef 10, mother–roast beef 3, brother–ham 22, and sister–salami 17. Accordingly, the father has 100 possible sandwiches available to him, the mother then has 99 sandwiches available to her (all except the one chosen by the father), the brother has 98 possible choices, and the sister then has 97 possible choices. Now $n(U) = P(100, 4) = (100)(99)(98)(97)$. All outcomes are equally likely.

The event E is that subset of U in which each member of the family has a different type of sandwich. To count E, we note that the father can pick any one of the 100 available sandwiches. Once he has chosen, the mother can select only one of 75 sandwiches. She cannot have the same type of sandwich as the father. The brother can only select from 50 sandwiches; he cannot have the same type as either the father or the mother. Similarly, the sister can have only 1 of 25 sandwiches, any of the type not yet taken by the rest of the family. Using the fundamental theorem of counting, we have $n(E) = (100)(75)(50)(25)$.

It now follows from Eq. (1), that

$$P(E) = \frac{(100)(75)(50)(25)}{(100)(99)(98)(97)} = .10 \text{ rounded to two decimal places.}$$

Example 4 Determine the probability that the first 4 digits of a telephone number will all be the same if the first digit cannot be either 0 or 1.

Solution We let U be the set of all allowable 4-digit numbers. There are 8 choices for the first digit (2 through 9) and 10 choices (0 through 9) for each of the other 3 digits. Therefore $n(U) = (8)(10)(10)(10) = 8000$.

The event E is the subset of U in which all 4 digits match. There are only 8 possibilities: 2222, 3333, 4444, ..., 9999. Using Eq. (1), we obtain

$$P(E) = \frac{8}{8000} = .001.$$

Example 5 Determine the probability of being dealt 5 cards from a regular deck and receiving exactly 2 kings.

Solution We define U to be the set of all possible 5-card hands. The outcomes are all equally likely. There are 52 available cards, we want 5, and order does not matter; $n(U) = C(52, 5)$.

The event E is the subset of U in which each hand contains exactly 2 kings. From Example 9 in Section 10.5, we have $n(E) = C(4, 2)C(48, 3)$.

It now follows from Eq. (1) that

$$P(E) = \frac{C(4, 2)C(48, 3)}{C(52, 5)}.$$

But

$$C(4, 2) = \frac{(4)(3)}{2!} = 6$$

$$C(48, 3) = \frac{(48)(47)(46)}{3!} = 17{,}296$$

and

$$C(52, 5) = \frac{(52)(51)(50)(49)(48)}{5!} = 2{,}598{,}960,$$

so

$$P(E) = \frac{(6)(17{,}296)}{2{,}598{,}960} = .04 \text{ rounded to two decimal places.}$$

Example 6 A manufacturer of ball bearings knows that exactly 1% of all bearings produced will be defective. A quality-control engineer samples 10 bearings from a recent production run of 5000 bearings. Determine the probability that exactly 2 of the 10 bearings sampled are defective.

Solution Since 1% of the bearings are defective, there are 1% of 5000 or 50 defective bearings in the production run being sampled. We take U to be the sample space of all possible groups of bearings that can be taken from the 5000 available bearings. This is an equally likely sample space. To count U, we note that 5000 bearings are available, we want 10, and order does not matter. Therefore $n(U) = C(5000, 10)$.

Let E be the subset of all groups of 10 bearings containing exactly 2 defective ones. To count E we must count all possible ways of picking 2 bad bearings and 8 good bearings. There are 50 bad bearings available, we want 2, and order does not matter; the bad bearings can be chosen in $C(50, 2)$ different ways. Once they are chosen, we then need 8 good bearings which must be chosen from 4950 available good bearings. Since order does not matter, there are $C(4950, 8)$. Therefore, $n(E) = C(50, 2)C(4950, 8)$.

Having $n(E)$ and $n(U)$, we now use Eq. (1) to calculate

$$P(E) = \frac{n(E)}{n(U)} = \frac{C(50, 2)C(4950, 8)}{C(5000, 10)} = .004 \text{ rounded to three decimal places.}$$

Exercises

1. A jury of 12 people is to be selected from a group of 80 people of whom 65 have college degrees and 15 do not. Determine the probability that everyone on the jury will have a college degree, assuming that each person has an equal chance of being selected.
2. A candle manufacturer produces 6000 candles daily in equal amounts of red, white, green, and blue. The candles are then randomly packed in boxes of 6. Determine the probability that the first box will contain no red candles.
3. For the process described in Example 2, determine the probability of having the cars parked in an alternating color sequence.
4. For the process described in Example 3, determine the probability of each member of the family receiving the same type of sandwich.
5. A florist has 12 red roses, 12 white roses, and 24 yellow roses. A customer picks 6 roses at random.
   (a) What is the probability of the customer selecting all red roses?
   (b) What is the probability of the customer selecting no red roses?
   (c) What is the probability of the customer selecting 2 roses of each color?
6. A candy manufacturer knows that exactly 2% of all candy shipped cross-country spoils during transit. A delivery of 10,000 bars is shipped from New York to Los Angeles. Determine the probability that the first 7 bars sampled are all spoiled.
7. A manufacturer of portable radios knows that 1 out of every 500 radios manufactured is defective. A quality-control engineer samples 10 radios from a production run of 3000. What is the probability that (a) none of the radios are defective? (b) all of the radios are defective? (c) 2 radios are defective?
8. A car washer for a rent-a-car agency washes 6 of 10 cars available for rental before leaving for lunch. On his return, he learns that 3 cars have been rented while he was out. What is the probability that all 3 cars rented had been washed?
9. The manager of a rent-a-car company knows that one out of every 20 cars rented will experience some kind of mechanical difficulty while being driven by the customer. At present there are 80 cars on the lot. Determine the probability that none of the next 10 cars rented will experience difficulty. What is the probability that only 1 will experience difficulty?
10. Determine the probability of being dealt five cards from a regular deck and receiving exactly two of one rank and three of a second rank (a full house).

11. Determine the probability of being dealt five cards from a regular deck and receiving two of one rank and three other cards that do not match (a pair).
12. Determine the probability of being dealt a bridge hand (13 cards) from a **regular** deck and receiving a 5-4-3-1 suit distribution.
13. A barrel is known to contain 80 good apples and 20 bad apples. What is the probability that exactly 2 of the first 7 apples taken from the barrel are defective?
14. Two hundred lottery tickets are sold for a fund-raising dinner. Two numbers are to be selected at random for prizes. What is the probability that a person will win one prize if he or she buys five tickets?
15. Assume that a year has exactly 365 days. Determine the probability that 22 people selected at random all have different birthdays (i.e., the month and date on which they were born are different).
16. It is known that a delivery of 1000 light bulbs contains 5 defective bulbs. Bulbs are used one at a time. What is the probability that none of the first 100 bulbs used will be defective?
17. A barrel contains six red balls, three green balls, and one white ball. Three balls are picked from the barrel without replacement. Determine the probability that the selected balls will all differ in color.
18. Referring to the process in Exercise 17, determine the probability that all the colors will be different if each time a ball is chosen it is returned to the barrel before the next selection is made.
19. An airport limousine service has 20 cars on the road. Five of the cars can accommodate 5 people each, 6 of the cars can accommodate 8 people each, and 9 of the cars can accommodate 11 people each. What is the probability that exactly 2 of the next 3 cars that arrive at the airport seat 11 people?

11.3 The Rules of Probability

Most probability problems involve sample spaces in which all the outcomes are *not* equally likely to occur and for which the formula

$$P(E) = \frac{n(E)}{n(U)} \qquad [1]$$

is *not* applicable. Other formulas must be developed in these cases. To do so, however, we first must clearly understand the rules and properties we want probabilities to possess.

All of us have an intuitive feeling for probabilities. We have not, however, formally defined a probability. That is, we have not provided an answer to the question, What is a probability? Let us use our intuition and experience with equal probability to develop such a definition. As a consequence, we will develop a set of rules that can be applied to all probability problems with or without equally likely sample spaces.

The first thing to note is that probabilities are numbers. In Section 11.2 we found that the probability that three people selected for a sales meeting will all be men (Example 1) was .40, the probability that all members of the same family will receive different sandwiches (Example 3) was .10, and that the probability of being dealt five cards and receiving exactly two kings (Example 5) was .04. Probabilities are numbers associated with sets of outcomes called events. Each event has a number associated with it that measures the chance or likelihood of that particular event actually happening. Furthermore, no event has two different numbers associated with it. In other words, probabilities are functions.

Requirement 1 A probability is a function. The domain is subsets of the sample space called events, and the range is real numbers.

Observe that our notation in Eq. (1) is consistent with this function interpretation of probability. The probability that an event will occur is denoted by $P(E)$. The event E is the input, and $P(E)$, a real number, is the output.

Let us consider the range a little closer to see whether or not we can limit the allowable real numbers a probability function assumes. In the examples just considered, we had probabilities of .40, .10, and .04. Intuitively, we expect probabilities always to range between 0 and 100% or, in decimal form, between 0 and 1. We do not expect the probability of an event to be less than 0%(0) or greater than 100% (1).

Requirement 2 The range of a probability function is (in decimals) between 0 and 1.

The sample space U is the set of *all* possible outcomes. Once a process is initiated, the result must be an element of U. If not, we do not have a valid sample space, since by definition a sample space contains *all* outcomes. Accordingly, the probability that something in the sample space will occur is 100%.

From our definition of a process (Section 9.4), we know that it is a decision or an action that results in one or more possible outcomes. Once the process is initiated, something must happen. The probability that nothing will happen after a process has started is zero.

These conclusions are inherent to any process and therefore are requirements for all probabilities.

Requirement 3 The probability that something in the sample space will occur is 1, and the probability that nothing will occur is 0. That is, $P(U) = 1$ and $P(\emptyset) = 0$.

There is one more requirement that probabilities possess, but it is not as obvious as the previous three. Consider the process of picking one card from a regular deck. Let E be the event that the card is a 2, and let F be the event that the card is a picture card (jack, queen, or king). Using the techniques developed in Sections 11.1 and 11.2, we find $P(E) = \frac{4}{52}$ and $P(F) = \frac{12}{52}$. What is the probability of $E \cup F$? The set $E \cup F$ is the event that the card is either a 2 or a picture card. We easily calculate $P(E \cup F) = \frac{16}{52}$ and then note that

$$P(E \cup F) = P(E) + P(F).$$

To see whether or not this formula, $P(E \cup F) = P(E) + P(F)$, is always valid let us use the same process, picking one card from a deck, with two other events. Let E be the event that the card is a heart, and let F be the event that the card is red. Now, $P(E) = \frac{13}{52}$, and $P(F) = \frac{26}{52}$. The set $E \cup F$ is the event that either the card is a heart or the card is red, which is the same event as the card being red. That is, $E \cup F = F$ and $P(E \cup F) = P(F) = \frac{26}{52}$. Then $P(E \cup F) \neq P(E) + P(F)$.

The difference between the two examples is that, in the first E and F had nothing in common whereas, in the second E and F had several elements in common. In the first example $E \cap F = \emptyset$, while in the second example $E \cap F = F$.

Requirement 4 If $E \cap F = \emptyset$, then $P(E \cup F) = P(E) + P(F)$.

The important feature in Requirement 4 is that the events E and F have nothing in common. In particular, if the probability that an investment will make money (event E) is .42 and the probability that it will lose money (event F) is .35, then the probability that it will either make money or lose money (event $E \cup F$) is .42 + .35 = .77, since $E \cap F = \emptyset$.

Requirement 4 can be generalized to more than two sets. As an example, again consider the process of picking one card from a regular deck. Let E_1, E_2, and E_3 be the events that the card is, respectively, a jack, a queen, and a king. Then $P(E_1) = P(E_2) = P(E_3) = \frac{4}{52}$. Note that E_1 has nothing in common with E_2 or E_3, that E_2 has nothing in common with either E_1 or E_3, and that E_3 has nothing in common with either E_1 or E_2. The set $E_1 \cup E_2 \cup E_3$ is the event that the card is either a jack or a queen or a king or, in other words, the event that the card is a picture card. Therefore

$$P(E_1 \cup E_2 \cup E_3) = \frac{12}{52},$$

and

$$P(E_1 \cup E_2 \cup E_3) = P(E_1) + P(E_2) + P(E_3).$$

Requirement 4 (Generalized) If $E_1, E_2, \ldots, E_n$ are *mutually exclusive* (i.e., each pair of sets has nothing in common), $P(E_1 \cup E_2 \cup \cdots \cup E_n) = P(E_1) + P(E_2) + \cdots + P(E_n)$.

Combining Requirements 1 through 4, we obtain the mathematical definition of a probability.

Definition 11.1 Let a process have a finite number of possible outcomes. A *probability* is a function defined on all subsets of the sample space (the events) satisfying the following properties:

a. $P(U) = 1$.
b. $P(\emptyset) = 0$.
c. For any event E, $0 \leq P(E) \leq 1$.
d. If $E_1, E_2, \ldots, E_n$ are mutually exclusive, $P(E_1 \cup E_2 \cup \cdots \cup E_n) = P(E_1) + P(E_2) + \cdots + P(E_n)$.

Any function defined on subsets of a sample space which satisfies properties (a) through (d) is called a probability. We verify in Exercise 17 that the probability function defined by Eq. (1) for equally likely sample spaces satisfies Definition 11.2.

Example 1 Ms. Tilson buys 100 shares of stock. Her broker assures her that there is a 75% chance the stock will go up and only a 10% chance that the stock will go down. Determine the probability that the stock will remain unchanged.

Solution We take the sample space to be $U = \{\text{stock goes up, stock goes down, stock remains unchanged}\}$. The events of interest are $E_1 = \{\text{stock goes up}\}$, $E_2 = \{\text{stock goes down}\}$, and $E_3 = \{\text{stock remains unchanged}\}$. We are given $P(E_1) = .75$ and $P(E_2) = .10$, and we want $P(E_3)$.

For this particular sample space, $U = E_1 \cup E_2 \cup E_3$, hence

$$P(U) = P(E_1 \cup E_2 \cup E_3). \tag{2}$$

But E_1, E_2, and E_3 are mutually exclusive, so (Requirement 4)

$$\begin{aligned} P(E_1 \cup E_2 \cup E_3) &= P(E_1) + P(E_2) + P(E_3) \\ &= .75 + .10 + P(E_3) \end{aligned}$$

or

$$P(E_1 \cup E_2 \cup E_3) = .85 + P(E_3). \tag{3}$$

Equations (2) and (3) together imply

$$P(U) = .85 + P(E_3). \tag{4}$$

Since $P(U) = 1$ (Requirement 3), Eq. (4) becomes

$$1 = .85 + P(E_3),$$

and we conclude that $P(E_3) = 1 - .85 = .15$.

Example 2 Rough Rider Airways advertises that the probability of its aircraft leaving and arriving on time is .8, while the probability of its aircraft leaving late but still arriving on time is .1. It is also known that the probability of an aircraft leaving late and arriving late is .07. What is the probability of a plane leaving on time but arriving late?

Solution We let $E_1 = $ {leave on time and arrive on time}, $E_2 = $ {leave late and arrive on time}, $E_3 = $ {leave late and arrive late}, and $E_4 = $ {leave on time and arrive late}, and take $U = E_1 \cup E_2 \cup E_3 \cup E_4$. Therefore

$$P(U) = P(E_1 \cup E_2 \cup E_3 \cup E_4). \tag{5}$$

But E_1, E_2, E_3, and E_4 are mutually exclusive, so

$$P(E_1 \cup E_2 \cup E_3 \cup E_4) = P(E_1) + P(E_2) + P(E_3) + P(E_4). \tag{6}$$

Together Eqs. (5) and (6) imply

$$P(U) = P(E_1) + P(E_2) + P(E_3) + P(E_4)$$
$$= .8 + .1 + .07 + P(E_4)$$

or

$$P(U) = .97 + P(E_4). \tag{7}$$

Since $P(U) = 1$ (always), Eq. (7) becomes

$$1 = .97 + P(E_4),$$

and it follows that $P(E_4) = .03$.

Using Definition 11.1 and the algebraic properties of sets, we can develop (see Exercises 19 and 21) two other useful properties of probabilities. They are

$$P(E') = 1 - P(E) \tag{8}$$

and

$$P(E \cup F) = P(E) + P(F) - P(E \cap F). \tag{9}$$

Recall that E' is the complement of E. Therefore $P(E')$ is the probability that E' will happen or, equivalently, the probability that E will not happen. Equation (8) formulates the property that the probability that an event E will *not* happen is equal to 1 minus the probability that the event will happen. Equation (9) is a formula for finding $P(E \cup F)$ for any subsets E and F. It is therefore a generalization of Requirement 4 to arbitrary events. In particular, if E and F are disjoint $E \cap F = \emptyset$, $P(E \cap F) = P(\emptyset) = 0$, and Eq. (9) reduces to Requirement 4 exactly.

Example 3 Determine the probability that in a group of 23 people at least 2 people have the same birthday (day and month). Assume that a year has 365 days.

Solution Let E be the event that at least 2 people have the same birthday. Unfortunately, E can happen in several ways: 2 people can have the same birthday, or 3 people can have the same birthday. It is easier to find E', the event that none of the people in the group have the same birthday. We then use Eq. (8) to compute $P(E)$.

We let U be all possible birthday arrangements for 23 people. To count U, we note that the first person can have a birthday on any one of 365 possible days. Regardless of the day, the second person also can have a birthday on any one of 365 possible days. Regardless of the days chosen for the first 2 people, the third person can have a birthday on any one of 365 days, and so on. Using the fundamental theorem of counting, we have

$$n(U) = \underbrace{(365)(365)(365) \cdots (365)}_{\text{23 times}} = (365)^{23}.$$

To count E', the event that no 2 people in the group have the same birthday, we note that the first person can have a birthday on any one of 365 days. Once this birthday is chosen, however, the second person can have a birthday on any one of 364 days, any day except the birthday of the first person. Once this birthday is selected, the third person can have a birthday on any one of 363 days, any day except the birthdays of the first 2 people. We continue this analysis through all 23 people and obtain $n(E') = P(365, 23) = (365)(364)(363) \cdots (343)$.

Since U is an equally likely sample space,

$$P(E') = \frac{n(E')}{n(U)} = \frac{P(365, 23)}{(365)^{23}} = .493.$$

It follows from Eq. (8) that $P(E) = 1 - P(E') = 1 - .493 = .507$. The probability that 2 people have the same birthday is .507 or over 50%. Is this surprising?

Example 4 A department store survey shows that, of all the women entering the store on the day of the survey, 22% made a purchase in the jewelry department, 17% made a purchase in the cosmetics department, and 8% made purchases in both departments. What is the probability that a woman selected at random has made a purchase in at least one of the two departments?

Solution Let E be the event that a woman makes a purchase in the jewelry department and F the event that a woman makes a purchase in the cosmetics department. We are given $P(E) = .22$, $P(F) = .17$, and $P(E \cap F) = .08$. The event of a woman making a purchase in at least one of the two departments is $E \cup F$. Using Eq. (9), we calculate

$$P(E \cup F) = P(E) + P(F) - P(E \cap F) = .22 + .17 - .08 = .31.$$

Example 5 A political party is running two candidates for two different positions in the state government. The probability that candidate A will win is .74, and the probability that candidate B will win is .55. The probability that both will win is .38. What is the probability that at least one candidate will win?

Solution Let E be the event that A will win and F be the event that B will win. Then $P(E) = .74$, $P(F) = .55$, and $P(E \cap F) = .38$. The event that at least one candidate will win is the event that either A will win or B will win or both will win, which is $E \cup F$. It follows from Eq. (9) that

$$P(E \cup F) = P(E) + P(F) - P(E \cap F) = .74 + .55 - .38 = .91.$$

Note that had we used the formula in Requirement 4 directly, we would have calculated $P(E \cup F) = P(E) + P(F) = .74 + .55 = 1.29$, which is wrong and absurd. A probability can never be greater than 1. The error is in applying the formula $P(E \cup F) = P(E) + P(F)$ which is valid only if E and F have nothing in common. But we are told that $P(E \cap F) = .38$, which implies that $E \cap F$ contains some outcomes.

Exercises

1. A government spokesperson claims that there is a 25% chance that the rate of inflation will decrease and a 37% chance that it will remain constant. What is the probability that the rate will increase?
2. Four candidates are running for mayor. The first three commission pollsters to determine their chances of winning. Candidate A is given a 15% chance of winning, candidate B is given a 37% chance of winning, while candidate C is given a 29% chance of winning. Determine the probability that candidate D will win.
3. A shoe company commissions a survey on its advertising effectiveness. The results are as follows: (1) The probability of a person hearing the advertisement is .23; (2) the probability of a person not hearing the advertisement but being told about it is .37; and (3) the probability of a person not hearing the advertisement and not being told about it is .63. What is wrong with the survey?
4. Let E be the event that a company earns a profit for the first quarter and let F be

the event that a company declares a dividend for the first quarter. Determine (in words) the events defined by

(a) E' (b) $E \cup F$ (c) $E \cap F$
(d) $E' \cap F$ (e) $E \cup F'$ (f) $E' \cap F'$.

5. Let E be the event that a customer is female and let F be the event that a customer makes a purchase. Determine (in words) the meaning of

(a) $P(E')$ (b) $P(E \cup F)$ (c) $P(E \cap F)$
(d) $P(E \cap F')$ (e) $P(E' \cap F)$ (f) $P(E' \cup F')$.

6. All females on a jury list are either college-educated or not. The probability of having a college-educated woman on the jury is .59. The probability of having a woman on the jury who is not college-educated is .20. Determine the probability of having a woman on the jury. Determine the probability of having a male on the jury.

7. With reference to the jury list in Exercise 6, it is also known that the probability of having a college-educated male on the jury is .10. Determine the probability of having a male on the jury who is not college-educated.

8. A bridge toll plaza has two booths. The probability that booth 1 has a line of waiting cars at any point in time is .87. The probability for booth 2 is .62. The probability that both booths have a line at the same time is .55. Determine the probability that there is a line in front of at least one booth.

9. Redo Example 5 if the probability that both candidates will win is only .28. What conclusions can you draw from your answer?

10. A survey taken in a particular city showed that 43% of all wives and 51% of all husbands had a college degree. The probability that both a husband and a wife had a college degree was .37. Determine the probability that at least one individual of a randomly chosen couple had a degree.

11. From experience, an automobile company knows that 14% of its cars will require mechanical repairs and 9% of its cars will require electrical repairs within the first year of operation. In addition, 4% of all new cars sold will require both types of repairs during the first year. Determine the probability that a car recently sold to Mr. Jones will require mechanical repairs, electrical repairs, or both during its first year of operation.

12. Redo Exercise 11 if 27% of all new cars require both types of repairs within the first year. What conclusions can you draw from your answer?

13. Two outcomes are equally likely to occur if the probabilities of each occurring are equal. Show that, if a total of 14 equally likely possible outcomes composes the complete sample space for a process, the probability of any one outcome occurring is $\frac{1}{14}$.

14. A department head wants you to fund a new project. He claims that the project can result in only one of three outcomes and that the probability of each outcome occurring is .25. What is wrong with his analysis?

15. A coin is weighted so that a head comes up twice as often as a tail. Determine the probability of flipping this coin and obtaining a tail.

16. A school gives out grades of A, B, C, D, and F. A student determines that he has probability of .9 of passing history and a probability .6 of receiving a grade lower than a B. What is the probability that he will fail history? What is the probability that he will receive either a C or a D in history?

17. Show that the probability function given by $P(E) = n(E)/n(U)$ for sample spaces with equally likely outcomes satisfies all the properties of Definition 11.2.
18. Let A and B be two events, with $A \subset B$. Show that $P(B - A) = P(B) - P(A)$. *Hint*: Use Requirement 4 and the identity $B = (B - A) \cup A$.
19. Verify Eq. (8) using the properties of Definition 11.1. *Hint*: $A \cup A' = U$.
20. Show that for any two events E and F the formula $P(E - F) = P(E) - P(E \cap F)$ is valid. *Hint*: Write $E = (E - F) \cup (E \cap F)$. Show that $(E - F)$ and $(E \cap F)$ are mutually exclusive and then use Requirement 4.
21. Prove Eq. (9). First use Venn diagrams to show that $E \cup F = (E - F) \cup (F - E) \cup (E \cap F)$. Then show that $E - F$, $F - E$, and $E \cap F$ are mutually exclusive, which in turn implies that $P(E \cup F) = P(E - F) + P(F - E) + P(E \cap F)$. Finally, use the results from Exercise 20 to deduce $P(E - F) = P(E) - P(E \cap F)$ and $P(F - E) = P(F) - P(E \cap F)$.

11.4 Conditional Probability

Probabilities can change as additional information becomes available. A case in point involves an investor who is considering a purchase of 100 shares of XYZ stock. Initially, the investor feels that the probabilities of the stock going up, going down, and remaining unchanged are all $\frac{1}{3}$. After an in-depth analysis of XYZ, the investor feels that the company will lose $10 million during the first quarter. Obviously, this new information will change the initial estimates of the probabilities.

As another example, consider the process of throwing one die and the event E of having a number greater than 3 turn up. A sample space for this process is $U = \{1, 2, 3, 4, 5, 6\}$, from which we deduce $P(E) = n(E)/n(U) = \frac{3}{6} = \frac{1}{2}$. Suppose we are told that the die is loaded so that only even numbers will turn up. We may wish to change our answer for $P(E)$. With this additional information, the only possible outcomes are $U_1 = \{2, 4, 6\}$. Of these, the outcomes 4 and 6 are greater than 3 so the probability of rolling a number greater than 3 is $\frac{2}{3}$. Not surprisingly, the answer for $P(E)$ has changed.

Conditional probabilities are probabilities determined with prior knowledge of additional information. In effect, the probabilities are conditioned or affected by the new information. The probability that XYZ stock will go up, *knowing that* the company will declare a loss, is a conditional probability. The probability is determined with the additional information that XYZ will announce a loss. Similarly, the probability of rolling a number greater than 3 with a die, *knowing that* only even numbers will turn up, is also a conditional probability. The probability is determined with the additional information that only even numbers can occur.

Notationally, we let $E \mid F$ denote the event E, knowing the event F. Then $P(E \mid F)$ is the probability that E will occur, knowing the event F. If E is the event of rolling a number greater than 3 with one die, and F is the event that the number is even, $P(E \mid F)$ is the probability of rolling a number greater than 3, knowing that the number will be even.

In almost all business situations, market and production conditions constantly change. New data are produced, sometimes from day to day, sometimes from minute to minute. As this new information becomes available, previous estimates and forecasts must be reassessed; probabilities for success and failure must be updated. These updates based on new information are conditional probabilities.

Example 1 A motel has 31 single rooms. Some rooms have color television sets, some have waterbeds, and some have both. The actual breakdown is given in Table 11.1. Determine (a) the probability of being assigned a single room with a color television set, and (b) the conditional probability of getting a color television set if the reservation guarantees a waterbed.

TABLE 11.1

	Regular Bed	Waterbed
Color television set	7	3
Black and white television set	13	8

Solution (a) There are 31 single rooms each of which is equally likely to be assigned. Of these, 10 rooms have color television sets. Therefore $n(U) = 31$, $n(E) = 10$, and $P(E) = \frac{10}{31}$, where E is the event of getting a color television set.

(b) If the reservation guarantees a waterbed, the sample space of possible rooms U_1 is reduced to only those having waterbeds. Now, $n(U_1) = 11$. Of these, only 3 have color television sets, so the probability of getting a color television set, knowing that the room has a waterbed, is $\frac{3}{11}$.

One way of attacking conditional probability problems is to modify the sample space each time new information is introduced. We did this in both of the last two problems. In Example 1, we reduced the sample space from 31 possible outcomes to 11 possible outcomes once we knew the room had a waterbed. In the die problem, we modified the sample space from $\{1, 2, 3, 4, 5, 6\}$ to $\{2, 4, 6\}$ once we knew that only even numbers could occur.

If a sample space can be easily modified, this is usually the most efficient way to find conditional probabilities. Sometimes, however, the sample space cannot be modified (see Example 4). Other times, so many initial calculations have been made with respect to the original sample space that it is not feasible to change sample spaces. In both situations, it is useful to have a formula that allows one to calculate conditional probabilities using the original sample space.

Let us consider a bit closer the procedure for modifying a sample space by returning to Example 1. We retain our designation of E as the event of getting a color television, and we define F to be the known event, in this example the event of getting a waterbed. Our interest remains $P(E \mid F)$. In the solution to Example 1(b), we first modified U; U_1 became the set of all rooms that had a waterbed. That is, $U_1 = F$. We then restricted ourselves to U_1 and found all elements in the modified sample space which also satisfied E. These were the rooms which had both a waterbed and also a color television, the set $E \cap F$. We counted $n(E \cap F)$ as 3 and obtained the answer $\frac{3}{11}$ by calculating

$$P(E \mid F) = \frac{n(E \cap F)}{n(F)}.$$

Interestingly, this relationship holds for all equally likely sample spaces, and it can be used to generate a useful formula for conditional probabilities. The modified sample space U_1 is always F, and those elements of the modified sample space which also satisfy E is the set $E \cap F$. Now, if we take the above relationship and divide both the numerator and denominator of the right side of the equality by $n(U)$, the number of elements in the *original*, unmodified sample space, we obtain

$$P(E \mid F) = \frac{n(E \cap F)/n(U)}{n(F)/n(U)}.$$

For equally likely sample spaces, we recognize

$$\frac{n(E \cap F)}{n(U)} = P(E \cap F) \quad \text{and} \quad \frac{n(F)}{n(U)} = P(F).$$

Combining our results, we have the useful equation

$$P(E \mid F) = \frac{P(E \cap F)}{P(F)}$$

which can be extended to conditional probabilities for all sample spaces.

Definition 11.2 If $P(F) \neq 0$, the conditional probability that an event E will occur, knowing the event F, is

$$P(E \mid F) = \frac{P(E \cap F)}{P(F)}. \tag{10}$$

The important feature of Definition 11.3 is that both probabilities on the right side of Eq. *(10)* are calculated *with respect to the original sample space assuming no additional information.*

Example 2 The following data pertaining to the highest educational degree held by each employee of a small company was obtained from personnel files: 4 people held graduate college degrees, 8 people held undergraduate college degrees, 39 people held high school degrees, and 6 people held no degree. Determine the probability that a company employee selected at random has a college degree if it is known that the person does not have a graduate degree.

Solution This is a conditional probability problem since we are asked to find the probability of one event, the person has a college degree, *knowing* a second event, the person does not have a graduate degree. We do this problem two ways, first with Eq. (10) and second by modifying the sample space.

(a) *With Eq. (10):* An equally likely sample space is the set of all 57 (4 + 8 + 39 + 6) employees. Clearly $n(U) = 57$. Let E be the event of the person having a college degree; E is the subset of U consisting of all people whose highest degree is a graduate college degree or an undergraduate college degree. Therefore, $n(E) = 4 + 8 = 12$, and $P(E) = n(E)/n(U) = \frac{12}{57}$. Let F be the event of a person not having a graduate degree; F is all elements of U without the 4 people who have a graduate degree. Therefore, $n(F) = 57 - 4 = 53$ and $P(F) = \frac{53}{57}$. Finally, $E \cap F$ is all people who have a college degree and do not have a graduate degree. These are the 8 people who have the undergraduate college degree as their highest degree, so $n(E \cap F) = 8$ and $P(E \cap F) = n(E \cap F)/n(U) = \frac{8}{57}$. Using Eq. (10), we easily calculate without any modification in U

$$P(E \mid F) = \frac{P(E \cap F)}{P(F)} = \frac{\frac{8}{57}}{\frac{53}{57}} = \frac{8}{53}.$$

(b) *Modifying the sample space:* Once we know that the person does not have a graduate degree, the set of possible outcomes is those people in U whose highest degree is either an undergraduate college degree or a high school degree, or those people who do not hold any degree. Of these $8 + 39 + 6 = 53$ possible outcomes, only the 8 people with an undergraduate college degree satisfy the event of having a college degree, so again $P(E \mid F) = \frac{8}{53}$.

Example 3 Referring to the data given in Example 2, determine the probability that a person does not have a graduate degree if it is known that the person does have a college degree.

Solution Using the probabilities calculated in Example 2(a) and Eq. (10) with

E and F interchanged, we find with very little extra work that

$$P(F \mid E) = \frac{P(F \cap E)}{P(E)} = \frac{\frac{8}{57}}{\frac{12}{57}} = \frac{8}{12} = \frac{2}{3}.$$

Recall that $E \cap F = F \cap E$.

Example 3 illustrates one of the advantages of Eq. (10). If $P(E \cap F)$ and $P(F)$ are known from previous calculations, $P(E \mid F)$ is obtained with one division. No new sample spaces or probabilities need be computed. The other advantage of Eq. (10) occurs when the sample space cannot be modified, which is the case in the following example.

Example 4 A recent survey of new car buyers who currently own a car resulted in the following data: 40% had two or more children and owned a station wagon; 44% owned a station wagon; and 80% had two or more children. A person walks into the showroom and claims to own a station wagon. Determine the probability that she has two or more children.

Solution Let E be the event of having two or more children and let F be the event of owning a station wagon. We want $P(E \mid F)$. Since we are not given the sample space specifically, we cannot modify it. We are given certain probabilities with respect to this sample space, however, so Eq. (10) may be useful. In particular, we know from the data that $P(E \cap F) = .40$, $P(F) = .44$, and $P(E) = .80$. It follows from Eq. (10) that

$$P(E \mid F) = \frac{P(E \cap F)}{P(F)} = \frac{.40}{.44} = .91 \text{ rounded to two decimal places.}$$

To generate another useful formula, we first interchange the events E and F in Eq. (10), obtaining

$$P(F \mid E) = \frac{P(F \cap E)}{P(E)}.$$

But $F \cap E = E \cap F$, hence

$$P(F \mid E) = \frac{P(E \cap F)}{P(E)}.$$

Finally, multiplying both sides of the last equation by $P(E)$, we obtain

$$\boxed{P(E \cap F) = P(F \mid E)P(E).} \qquad (11)$$

That is, the probability that both E and F will occur is the probability that F will occur, *knowing* that E has already happened *times* the probability that E will occur.

Example 5 The probability that the Lincoln Hamburger Company will report a profit this year is .63. If it reports a profit, there is a 70% chance its stock will rise and a 25% chance its stock will drop. What is the probability that the company will report a profit and its stock will rise?

Solution Let E be the event that the company will report a profit and let F be the event that its stock will rise. We are given $P(E) = .63$ and $P(F \mid E) = .70$, and we want $P(E \cap F)$. It follows from Eq. (11) that $P(E \cap F) = P(F \mid E)P(E) = (.70)(.63) = .44$ rounded to two decimal places.

Example 6 Determine the probability of being dealt 2 cards from a regular deck and receiving 2 kings.

Solution We let E be the event that the first card dealt is a king and let F be the event that the second card dealt is a king. We want $P(E \cap F)$. To determine $P(E)$, we reason that there are 52 cards in the deck, all equally likely to be dealt, of which 4 are kings. Therefore $P(E) = \frac{4}{52}$. To determine $P(F \mid E)$, the probability that the second card will be a king, knowing the first card is a king, we reason as follows: Once a king has been dealt E, there are only 51 cards left, all equally likely to be dealt, of which only 3 are kings. Therefore $P(F \mid E) = \frac{3}{51}$. It follows from Eq. (11) that $P(E \cap F) = P(F \mid E)P(E) = (\frac{3}{51})(\frac{4}{52}) = .005$ rounded to three decimal places.

Exercises

1. Determine the probability of rolling two dice and having the sum of their faces be 4. What is the probability if we know that both dice will not have the same face value?
2. Table 11.2 lists the results of a recent survey of people entering a supermarket. From these data, determine the probability that a person selected at random was a woman. What was the probability that a person spending over $35 was a woman. Do this problem two ways, first by modifying the sample space and second by using formula (10).

TABLE 11.2

	Spends under $35	Spends over $35
Male	125	50
Female	150	175

3. Table 11.3 lists the results of a questionnaire mailed to subscribers to a national magazine. From these data determine the probability that
   (a) A reader is under 30.
   (b) A reader's income is greater than or equal to $20,000.
   (c) A reader under 30 has an income of less than $20,000.

TABLE 11.3

	Income Less Than $20,000	Income of $20,000 or More
Under 30	8000	3,000
30 or over	2000	12,000

(d) A reader with an income of $20,000 or more is under 30.
(e) A reader is at least 30 and earns $20,000 or more.

4. Table 11.4 lists the results of a random telephone survey commissioned by Rough Rider Airways. People were asked whether or not they had flown with Rough Rider Airways during the last year and what they thought of a recent advertisement by the company. From these data, determine the probability that
   (a) A person found the advertisement effective.
   (b) A person had flown Rough Rider Airways during the last year.
   (c) A person who had flown Rough Rider Airways during the last year found the advertisement effective.
   (d) A person who did not find the advertisement effective had flown Rough Rider Airways during the last year.
   (e) A person who had not flown Rough Rider Airways during the last year had no opinion.
   (f) A person who had no opinion had not flown Rough Rider Airways during the last year.

TABLE 11.4

	Felt the Ad Was Effective	Felt the Ad Was Not Effective	No Opinion
Those who had flown within the year	50	25	25
Those who had not flown within the year	400	300	200

5. During a test program for a new drug, the following data were gathered. The drug was effective on 73% of those tested; 45% of those tested experienced after effects; on 5% of those tested the drug was not effective but after effects were still experienced. Determine the probability that (a) the drug was not effective on a person who experienced after effects, and (b) a person would experience after effects if the drug were not effective.

6. In a small town market tests for a new bleach resulted in the following data: 10% of the households received a free sample of the bleach, 3% of the household shoppers bought the bleach at least once, and 25% of the people who received a free sample later bought the bleach. What is the probability that a household shopper received a free sample and later bought the bleach?

7. A health service knows that 70% of all adults in one area of the country smoke. Additionally, 17% of the deaths in that area are due to cancer. If a person smokes,

however, he stands a 20% chance of dying of cancer. What is the probability that a person selected at random will be a smoker and will die from cancer?

8. A survey commissioned by a meat packer resulted in the following data: The probability that a person eats meat for dinner at least three times a week is .65. The probability that a person eats fish for dinner at least once a week is .45. The probability that a person does both is .20. Determine the probability that (a) a person who eats meat at least three times a week will have fish at least once a week, and (b) a person who eats fish at least once a week will have meat at least three times a week.

9. An insurance company knows that 19% of all policy holders will have one accident in a year, and 11% of *all* policy holders will have two accidents in a year. What is the probability that a person who has had one accident will have a second in the same year?

11.5 Independent Events

Although probabilities can change as additional information becomes available, there are times when they do not. Such situations occur when the additional information is either irrelevant or unrelated to the event of interest.

As an example of irrelevant additional information, consider again the process of throwing one die. We easily determine that the probability of rolling a number greater than 3 (either 4, 5, or 6) is $\frac{1}{2}$. Suppose we are given the additional information that the die is red. Obviously this new information does not change our previous answer, $\frac{1}{2}$; the color of the die has no bearing on which number turns up. Letting E be the event that a number greater than three occurs and F be the event that the die is red, we have $P(E \mid F) = P(E)$. The probability that E will occur is the same *whether or not* we know F.

In general, whenever we have two events E and F such that $P(E \mid F) = P(E)$, we say that the events are *independent*. Knowledge of F does not affect the chances of E occurring. The events are unrelated.

Usually, it is clear from the context of a problem whether or not two events are independent. The die problem just considered is one example. Occasionally, however, it is not easy to determine independence. In such cases, one must calculate $P(E)$ and also $P(E \mid F)$. If they are equal, the events E and F are independent; if they are not equal, the two events are not independent.

Example 1 Consider all families with two children and pick one family at random. Let E be the event that the family has children of both sexes and let F be the event that the family has at most one girl. Are E and F independent?

Solution A sample space for this event is $U = \{bb, bg, gb, gg\}$, where the first initial indicates the sex of the oldest child and the second initial indicates the sex of the youngest child. If we assume that there is an equal chance for a child to be of either sex, this is a set of equally likely outcomes. Here $E = \{bg, gb\}$, $F = \{bb, bg, gb\}$, and $E \cap F = \{bg, gb\} \cap \{bb, bg, gb\} = \{bg, gb\}$. Therefore $P(E) = \frac{2}{4} = \frac{1}{2}$, $P(F) = \frac{3}{4}$, and $P(E \cap F) = \frac{2}{4}$, from which it follows that

$$P(E \mid F) = \frac{P(E \cap F)}{P(F)} = \frac{\frac{1}{2}}{\frac{3}{4}} = \frac{2}{3}.$$

Since $P(E \mid F)$ does *not* equal $P(E)$, the events are *not* independent.

Example 2 Redo Example 1 with families having three children.

Solution Now $U = \{$bbb, bbg, bgb, bgg, gbb, gbg, ggb, ggg$\}$, where the first, second, and third initials designate the sex of the first, second, and third child, respectively. This is an equally likely sample space. With this U, $E = \{$bbg, bgb, bgg, gbb, gbg, ggb$\}$, $F = \{$bbb, bbg, bgb, gbb$\}$, and $E \cap F = \{$bbg, bgb, gbb$\}$, from which $P(E) = \frac{6}{8} = \frac{3}{4}$, $P(F) = \frac{4}{8} = \frac{1}{2}$, and $P(E \cap F) = \frac{3}{8}$. Then

$$P(E \mid F) = \frac{P(E \cap F)}{P(F)} = \frac{\frac{3}{8}}{\frac{1}{2}} = \left(\frac{3}{8}\right)\left(\frac{2}{1}\right) = \frac{6}{8} = \frac{3}{4}.$$

Since $P(E \mid F) = P(E)$, the events are independent.

Examples 1 and 2 illustrate the difficulty in determining the independence of certain events. The same two events, in words, can be independent with respect to one sample space and not independent with respect to a second sample space. Luckily, such situations do not occur frequently in the business world. It is generally clear from context whether or not two events are independent.

A more interesting problem for commercial processes arises when we have two events that we know are independent and we want the probability that both will happen. For example, suppose we know that the probability of a washing machine needing repairs during its first year is .13 and that the probability of a dryer needing repairs during its first year is .09. Can we determine the probability that a customer who purchases both machines will need repairs on both during the first year?

We let E be the event that the washing machine will need repairs during its first year and F be the event that the dryer will need repairs during its first year. We know $P(E) = .13$ and $P(F) = .09$, and we want $P(E \cap F)$. Furthermore, we assume that E and F are independent. The washer and the dryer are different machines, so a breakdown in one should not affect the performance of the other.

To solve this problem, we recall two facts. For any two events E and F,

$$P(E \mid F) = \frac{P(E \cap F)}{P(F)}. \qquad [10]$$

and if the two events are independent,

$$P(E \mid F) = P(E).$$

Substituting the last equation into the left side of Eq. (10), we have

$$P(E) = \frac{P(E \cap F)}{P(F)}.$$

Multiplying both sides of this equation by $P(F)$ and rearranging, we obtain $P(E \cap F) = P(E)P(F)$, which proves the following useful result.

Theorem 11.1 If E and F are independent events,

$$P(E \cap F) = P(E)P(F). \tag{12}$$

Formula (12) is applicable if and only if the events under consideration are independent. If the events are not independent, the formula is not valid. If it is unclear whether or not two events are independent, as was the case in both Examples 1 and 2, it is safer not to use Eq. (12). For the washing machine–dryer problem, where we know both events are independent, we can use Eq. (12) to obtain the probability of both machines needing repairs during the first year. It is

$$P(E \cap F) = P(E)P(F) = (.13)(.09) = .01 \text{ rounded to two decimal places.}$$

Example 3 A card is picked from a regular deck and then put back. The deck is shuffled, and a second card is drawn. What is the probability that both cards will be kings?

Solution Let E be the event that the first card is a king and let F be the event that the second card is a king. Since the first card is returned to the deck before the second card is drawn, it has no effect on the second pick. As such, $P(E) = \frac{4}{52}$, $P(F) = \frac{4}{52}$, and E and F are independent. The event $E \cap F$ is the subset of two kings. Using Eq. (12), we find

$$P(E \cap F) = P(E)P(F) = \left(\tfrac{4}{52}\right)\left(\tfrac{4}{52}\right) = .006 \text{ rounded to three decimal places.}$$

If we change the problem slightly by not replacing the first card after it is picked, Eq. (12) is no longer applicable. Now E and F are not independent. The first pick affects the probabilities associated with the second pick. If the first card is a king, and it is not replaced, the probability that the second card will also be a king is $\frac{3}{51}$. If the first card is not a king, the probability that the second card will be a king is $\frac{4}{51}$. The solution to this problem is discussed in Example 6 in Section 11.4.

Example 4 A cough syrup will suppress a cough for 4 hours with a probability of .85. Two people take the cough syrup. Determine the probability that neither will have a coughing attack in the next 4 hours.

Solution Let E be the event that the first person will not have a coughing attack in the next 4 hours and let F be the same event for the second person. We know $P(E) = .85$ and $P(F) = .85$, and we want $P(E \cap F)$. The two events are independent; the effect of the syrup on one person has no bearing on the syrup's effect on the other person. Using Eq. (12), we calculate

$P(E \cap F) = P(E)P(F) = (.85)(.85) = .72$ rounded to two decimal places.

Theorem 11.1 can be generalized to more than two events. If E, F, and G are all independent of each other, that is, if each event has no bearing on the other two, then

$$P(E \cap F \cap G) = P(E)P(F)P(G). \tag{13}$$

If E, F, G, and H are all independent,

$$P(E \cap F \cap G \cap H) = P(E)P(F)P(G)P(H). \tag{14}$$

Five or more events are handled analogously. The probability that the intersections of many independent events will occur together is the product of the individual probabilities.

Example 5 The probability that an orange tree will bear fruit once planted is .95. What is the probability that four orange trees planted at the same time will all bear fruit?

Solution Let E, F, G, and H be the events that the first, second, third, and fourth trees planted will all bear fruit, respectively. We want $P(E \cap F \cap G \cap H)$. The events are independent, since the productivity of one tree has no bearing on the productivity of the other three. Using Eq. (14), we have

$$P(E \cap F \cap G \cap H) = P(E)P(F)P(G)P(H)$$
$$= (.95)(.95)(.95)(.95)$$
$$= .81 \text{ rounded to two decimal places.}$$

Example 6 The probability that an alarm received at a particular fire house is legitimate (not a false alarm) is .8. What is the probability that the next seven calls will all be legitimate?

Solution

$P(\text{all seven calls are legitimate}) = (.8)(.8)(.8)(.8)(.8)(.8)(.8)$
$\phantom{P(\text{all seven calls are legitimate}) } = .21$ rounded to two decimal places.

Exercises

1. Find $P(E \cap F)$ if E and F are independent events and
   (a) $P(E) = P(F) = \frac{1}{2}$
   (b) $P(E) = P(F) = .1$
   (c) $P(E) = .8$ and $P(F) = .9$
   (d) $P(E) = .6$ and $P(F) = .8$.

2. Find $P(E \cap F \cap G)$ if E, F, and G are all independent of each other and
   (a) $P(E) = P(F) = P(G) = .2$
   (b) $P(E) = P(F) = P(G) = .8$
   (c) $P(E) = P(F) = .6$, and $P(G) = .4$
   (d) $P(E) = .3$, $P(F) = .8$, and $P(G) = .5$.

3. Find $P(E)$ if $P(E \cap F) = .4$, $P(F) = .3$, and E and F are independent events. What conclusions can you draw from your answer?

4. Find $P(E)$ if $P(F) = .7$, $P(G) = .3$, $P(E \cap F \cap G) = .1$, and E, F, and G are all independent of each other.

5. A coin is tossed three times. Let E be the event that at least one toss is a head and let F be the event that the three tosses result in both a head and a tail occurring at least once. Are these events independent?

6. Redo Example 1 for families with four children.

7. Prove that the two events in Example 3 are independent. *Hint:* Show directly that $P(E \mid F) = P(E)$.

8. Five people, Anne, Bruce, Carl, Doris, and Edith, are arranged in a row for a picture. Let E be the event that Carl is first and Edith is last and let F be the event that Bruce is in the middle. Are these events independent?

9. A barrel contains three red balls and four green balls. Two balls are selected at random one after the other. Determine the probability that both balls will be red if
   (a) The first ball is returned to the barrel before the second ball is selected.
   (b) The first ball is not returned to the barrel.

10. A die is thrown, and then a coin is flipped. What is the probability that the die will turn up 6 and the coin will turn up tails?

11. A vaccine is effective on 97% of all people inoculated. Last week a doctor inoculated three people. What is the probability that the vaccine will be effective on all three people?

12. The probability that XYZ's stock will go up is $\frac{1}{3}$. The probability that ABC's stock will rise is $\frac{4}{10}$. Determine the probability that both stocks will rise.

13. The probability of a licensed driver having an accident within the year is .21. Four licensed drivers are selected at random. What is the probability that none of them will have an accident within the year?

14. An aptitude test given to all applicants for jobs at the Chubby Cat Food Corporation is 90% effective. That is, the test accurately predicts the aptitudes of 90% of those tested. Determine the probability that the test will correctly predict the aptitudes of the next five people who take it.

15. The American Citrus Corporation employs 30 lawyers in its legal department. The probability that a lawyer selected at random will be a woman is .2. The company picks two lawyers to handle a new case. What is the probability that both lawyers will be women?

16. A dress manufacturer knows that 10% of all the dresses he produces are defective.

What is the probability that the first two dresses he checks will be perfect?

17. A plumbing manufacturer guarantees the probability of a new faucet working properly when installed correctly is .99. Determine the probability that
   (a) The next two faucets installed correctly by a plumber will work properly.
   (b) The next two faucets installed will work properly if the plumber knows that the last faucet installed worked properly.

11.6 Bernoulli Trials

A particularly interesting class of problems involving independent events occurs when the same process is repeated a fixed number of times. Flipping a coin 3 times is such a situation. A commercial example is a quality-control expert making tests for purity on 10 different bottles of beer randomly sampled from the production line. This process can be decomposed into 1 smaller process, that of testing 1 bottle of beer, repeated 10 times.

We call each repeat of a process a *run*. Flipping a coin 3 times is 3 runs of flipping a coin once. Testing 10 bottles of beer is 10 runs of testing 1 bottle of beer.

A Bernoulli trial is a set of repeated runs of the same process. The process is assumed to have only two outcomes, one called a *success* and the other called a *failure*, and the outcome of each run is assumed to be independent of the outcome of every other run. In particular, each outcome of one flip of a coin (either heads or tails) has no bearing on the outcomes of subsequent flips. Furthermore, in Bernoulli trials, the probability of a success is known as is the probability of a failure. The problem is to determine the probability that a prescribed number of runs will end in a given number of successes.

The process of flipping a coin three times is a set of Bernoulli trials. The two outcomes are heads (which we may call a success) and tails (which we may call a failure). The probability of each is $\frac{1}{2}$. A typical problem is to determine the probability of obtaining two successes (heads) in three flips.

Any process can be viewed as a two-outcome process, namely, (1) a particular event happens or (2) a particular event does not happen. Consider the process of closing one plant out of four existing plants in Buffalo, Cleveland, Milwaukee, and Nashville. An employee at the Buffalo plant can view the possible closings as the four-element sample space $U = \{$Buffalo, Cleveland, Milwaukee, Nashville$\}$. More than likely, however, he views it as the two-element sample space $U = \{$Buffalo is closed, Buffalo is not closed$\}$.

To see how we can attack a general Bernoulli trials problem, we consider the following situation. Suppose the probability that an alarm received at a fire house is legitimate (not a false alarm) is .8. What is the probability that two of the next three alarms will be legitimate. To solve this problem, we first note that the process of receiving three alarms can be broken down into one smaller process, that of receiving one alarm, repeated three times. Furthermore, the outcomes of individual alarms are independent. Regardless of the legitimacy of the first alarm, the second call still has a probability of .8 of being legitimate and a probability of .2 of being false. No matter how the second alarm turns out, the probability that the third call is legitimate remains .8. One alarm has no effect on any other alarm.

For convenience, we designate a legitimate alarm by S (for success) and a false alarm by F (for failure). How many different ways can the fire house receive two out of three legitimate alarms? We easily list all possibilities as {SSF, SFS, FSS} and see that there are exactly three ways.

Next we determine the probability that each of these three possibilities (two S's and one F) will occur. Since the outcomes of each run are independent, it follows from the Section 11.5 that

$P(\text{SSF}) = P(S)P(S)P(F) = (.8)(.8)(.2) = (.8)^2(.2)$
$P(\text{SFS}) = P(S)P(F)P(S) = (.8)(.2)(.8) = (.8)^2(.2)$

and

$P(\text{FSS}) = P(F)P(S)P(S) = (.2)(.8)(.8) = (.8)^2(.2)$.

Since each possibility contains exactly two S's and one F, they all have the same probability, $(.8)^2(.2)$.

There are three ways to receive exactly two legitimate alarms in the next three calls, and each way has a probability of $(.8)^2(.2)$ actually to happen. It follows from Requirement 4 (Generalized) of Section 11.3 that the probability of receiving two out of three legitimate alarms in any order is

$P(\text{SSF} \cup \text{SFS} \cup \text{FSS}) = P(\text{SSF}) + P(\text{SFS}) + P(\text{FSS}) = 3[(.8)^2(.2)] = .384$

Before generalizing the solution of this fire alarm problem to other Bernoulli trials problems, we summarize our steps. We first counted the different ways of obtaining the exact number of successes required. We then calculated the probability of each of these ways occurring and found, since each way had the same number of S's and F's, that all the probabilities were equal. Last, we multiplied the number of ways by the probability of any one way occurring to obtain the final answer.

In the general problem, a process is repeated n times, and we want the probability of obtaining r successes and $n - r$ failures. Every outcome of this type can be considered an n-letter arrangement of r S's and $(n - r)$ F's. There are as many different ways to obtain r successes and $n - r$ failures as there are different arrangements of r S's and $(n - r)$ F's. After all, each possibility in the fire alarm problem, either SSF, SFS, or FSS, is simply a different arrangement of a three-letter word containing two S's and one F. The number of different arrangements for the general problem was found in Example 6 in Section 10.5 as $C(n, r)$.

11.6 Bernoulli Trials

We assume that the probability of an individual success is known, and we denote it by p. The probability of a failure is then $1 - p$. (Why?) Each complete outcome of interest contains exactly r S's and $(n - r)$ F's. The runs are assumed independent, so the probability of obtaining r S's and $(n - r)$ F's is

$$[P(S)]^r[P(F)]^{n-r} = p^r(1 - p)^{n-r}.$$

There are $C(n, r)$ ways to obtain r successes and $n - r$ failures, and each of these ways has a probability of $p^r(1 - p)^{n-r}$ of happening. It follows from Requirement 4 (Generalized) that

$$\text{The probability of obtaining exactly } r \text{ successes and } n - r \text{ failures} = C(n, r)p^r(1 - p)^{n-r}. \quad (15)$$

Applying this formula to the fire alarm problem, we have $n = 3$ (three alarms), $r = 2$ (two successes), and $p = .8$. Then,

$$C(n, r)p^r(1 - p)^{n-r} = C(3, 2)(.8)^2(1 - .8)^{3-2}$$
$$= 3(.8)^2(.2)$$
$$= .384,$$

as before.

Example 1 A candy distributor knows from experience that 5% of every shipment arrives damaged. He receives 20 shipments and picks 1 candy bar from each. Determine the probability that exactly 3 of these candy bars are damaged.

Solution Since 5% of every shipment is damaged, the probability that any one item is damaged is .05. If we let S denote the event of picking a damaged bar, $P(S) = p = .05$.

Selecting 1 candy bar from each shipment is the same as repeating the process of selecting 1 bar from 1 shipment 20 times with replacement. Since the results of one pick should have no bearing on the results of any other pick, the runs are independent. This is a set of Bernoulli trials with $n = 20$, $r = 3$, and $p = .05$. Using Eq. (15), we calculate the probability of picking exactly three damaged candy bars as

$$C(20, 3)(.05)^3(1 - .05)^{20-3} = (1140)(.05)^3(.95)^{17}$$
$$= .06 \text{ rounded to two decimal places.}$$

Example 2 A manufacturer of ball bearings knows from experience that on the average 90% of the bearings produced meet prescribed tolerances. A quality-control engineer randomly samples 10 bearings from the production line. What is the probability that exactly 9 of these bearings will be within the prescribed tolerances?

Solution On the average, 9 out of 10 bearings meet the prescribed tolerance. If we test 10 bearings, however, we are not guaranteed that 9 of that particular 10 will be satisfactory. There is a chance that all will be satisfactory, and there is a chance that none will be satisfactory.

We let S denote the event of one bearing being satisfactory. Then $P(S) = p = .90$. Since the bearings are picked randomly, the outcome of each test is independent. This is equivalent to a set of Bernoulli trials with $n = 10$ runs, $r = 9$ successes, and $p = .9$. Using Eq. (15), we find that the probability of exactly nine bearings being satisfactory is

$$C(10, 9)(.9)^9(1 - .9)^{10-9} = 10(.9)^9(.1)$$
$$= .39 \text{ rounded to two decimal places.}$$

Example 3 A regular hero is made from three out of four ingredients selected by the customer from ham, salami, provolone, and American cheese. Determine the probability that only two of the next four customers will order provolone for their regular hero.

Solution Each customer is a separate run, and since one customer has no bearing on the next, their orders are independent. Furthermore, we assume that each one of the four ingredients is as likely to be selected as the others by each customer. Letting S denote the event of a customer ordering provolone and F the event that he does not order provolone, we have $p = \frac{3}{4}$. This is obtained by realizing that it is easier to first calculate $1 - p$, the probability of not selecting provolone. Since only one ingredient is omitted from each hero, the chance it will be provolone is $\frac{1}{4}$. Thus $1 - p = \frac{1}{4}$ and $p = \frac{3}{4}$.

Using Eq. (13), with $n = 4$, $r = 2$, and $p = .75$, we find the probability of exactly two customers ordering provolone is $C(4, 2)(.75)^2(.25)^2 = (6)(.5625)(.0625) = .211$ rounded to three decimal places.

Exercises

1. Evaluate formula (15) with
    (a) $n = 8, r = 4, p = .3$
    (b) $n = 7, r = 5, p = .1$
    (c) $n = 15, r = 7, p = .5$.
2. A coin is flipped three times. Determine the probability that two heads will turn up.
3. A card is picked from a regular deck and then replaced, and the deck is shuffled. This process is repeated four times. What is the probability that exactly one of the four cards chosen is a spade?
4. A barrel contains four red balls, two green balls, and two yellow balls. A ball is selected at random and then replaced. This process is repeated five times. What is the probability that exactly two of these five balls picked will not be green?
5. From experience, a manufacturer of baseballs knows that under ordinary circumstances 1% of all baseballs produced are defective. A quality-control engineer

randomly selects 50 baseballs. Determine the probability that exactly 10 of these balls are defective. What conclusions would you draw if 10 were defective?
6. The probability of a licensed driver having an accident within the year is .21. Four licensed drivers are selected at random. Determine the probability that only one of these drivers will be involved in an accident within the year.
7. A vaccine is effective on 97% of people inoculated. A doctor inoculates three people. What is the probability that the vaccine will be effective on only two of these people?
8. An aptitude test given to all applicants for jobs at the Chubby Cat Food Corporation is 90% effective. That is, it accurately predicts the aptitudes of 90% of those tested. Determine the probability that the test will *not* accurately predict the aptitudes of two of the next five people tested.
9. A plumbing manufacturer guarantees the probability of a new faucet working properly when installed correctly is .99. Determine the probability that one of the next 10 faucets installed correctly by a plumber will *not* work properly.

11.7 Expected Value

Although probabilities are a key factor in many decision processes, they are not in general the decisive one. Money is. A case in point involves a corporate director who receives a request for $100,000 from his research department to fund a new project. The director knows that there are only two possible outcomes for the project—either it will succeed or it will fail. It would be useful to know the probabilities associated with each outcome. Even more useful would be a knowledge of the payoffs involved. A 10% chance for success and a 90% chance for failure does not sound appealing. If, however, a success would generate a $1 million profit, the project could be attractive. Even though the probability of success is small, the possible payoff may be large enough to justify the risk.

The notion of large payoffs compensating for small chances of success is familiar to most people. State lotteries and raising capital for oil exploration are but two examples. A risk with little chance of succeeding may be attractive if the payoff is large, whereas a risk with a large chance of succeeding may not be attractive if the payoff is too small. Both factors, probabilities and payoffs, must be balanced against each other.

Once the idea is accepted that both probabilities and payoffs are part of a decision process, one then needs a method of combining these factors into a meaningful quantity. Such a quantity is given mathematically by expectation.

Definition 11.3 Let $A_1, A_2, \ldots, A_n$ be a complete set of numerical outcomes for a process and let $p_1, p_2, \ldots, p_n$ be the respective probabilities associated with each outcome. The *expected value* of the process is

$$E = A_1 p_1 + A_2 p_2 + \cdots + A_n p_n.$$

Expected values are calculated by multiplying each possible outcome by its respective probability and then summing the results. If the numerical outcomes $A_1, A_2, \ldots, A_n$ are in dollars, the expected value is often called the *expected monetary value*. It measures the anticipated *average* return of a given venture. The word "average" is important. An expected monetary value of +$15,000 indicates that on the average a process will return a profit of $15,000. This is not to be interpreted to mean that the process *will* return exactly this amount each time. Sometimes the process will make more than $15,000, sometimes it will make less, and sometimes it may even lose money. On the average, however, the process will make $15,000. If we could run the process 100 times, we could expect a return of 100 times $15,000 or $1.5 million.

To calculate the expected monetary value of the proposed research project considered previously, note that there are only two numerical outcomes: $A_1 = \$1,000,000$ (a $1 million profit if the project succeeds) and $A_2 = -\$100,000$ (a loss of $100,000 if the project fails) where, as always, a minus sign in front of a number indicates a loss. The probabilities associated with each outcome are $p_1 = .10$ for a success and $p_2 = .90$ for a failure. The expected monetary value is

$$E = (1,000,000)(.10) + (-100,000)(.90)$$
$$= 100,000 - 90,000$$
$$= +\$10,000.$$

On the average, the director can expect a profit of $10,000 each time he approves such a project. Obviously, he will not receive $10,000 from each project. Each project can return either $1 million or lose $100,000. But if he approves 50 such projects, he can expect a total return of $(50)(10,000) = \$500,000$.

Obviously there are some shortcomings in this type of analysis. First, the director may not have the opportunity to review 50 projects. This project may indeed be unique. Second, even if the director has such an opportunity, expected monetary values do not take into consideration the practical implications of a sustained string of initial losses.* These shortcomings aside, expected values provide a measure for balancing probabilities against payoffs and are useful in assessing the overall attractiveness of certain ventures.

Example 1 A person selects one card from a regular deck. If the card is an ace, she wins $3. If the card is a picture card, she wins $2. Any other card results in a $1 loss. Is this an attractive game?

Solution There are three numerical outcomes for this game: $3, $2, and $-\$1$, where again the negative sign indicates a loss. The probabilities associated with each outcome are, respectively, $p_1 = \frac{4}{52}$, $p_2 = \frac{12}{52}$, and $p_3 = \frac{36}{52}$. The expected

---

* These shortcomings are partially removed by generalizing expected monetary value to the concept of expected utility value. Such refinements are beyond the scope of this book.

monetary value is $E = (3)(\frac{4}{52}) + (2)(\frac{12}{52}) + (-1)(\frac{36}{52}) = \$0$. The expected monetary value is zero, which indicates that over the long run the person will neither gain nor lose; she will break even. Whether or not this is attractive depends on her reason for playing.

Example 2 A person pays $2 to play the following game. He flips a coin three times and receives $1 each time a head appears. What is the expected monetary value of this game?

Solution The numerical outcomes are $A_1 = -2$ (no heads and the entrance fee is lost), $A_2 = -1$ (one head minus the entrance fee), $A_3 = 0$ (two heads minus the entrance fee), and $A_4 = 1$. Using Bernoulli trials with a success being a head, we calculate

$$p_1 = C(3, 0)(\tfrac{1}{2})^0(1 - \tfrac{1}{2})^{3-0} = 1(1)(\tfrac{1}{2})^3 = \tfrac{1}{8}$$
$$p_2 = C(3, 1)(\tfrac{1}{2})^1(1 - \tfrac{1}{2})^{3-1} = 3(\tfrac{1}{2})(\tfrac{1}{2})^2 = \tfrac{3}{8}$$
$$p_3 = C(3, 2)(\tfrac{1}{2})^2(1 - \tfrac{1}{2})^{3-2} = 3(\tfrac{1}{2})^2(\tfrac{1}{2}) = \tfrac{3}{8}$$

and

$$p_4 = C(3, 3)(\tfrac{1}{2})^3(1 - \tfrac{1}{2})^{3-3} = 1(\tfrac{1}{2})^3(1) = \tfrac{1}{8}.$$

The expected monetary value is

$$E = (-2)(\tfrac{1}{8}) + (-1)(\tfrac{3}{8}) + (0)(\tfrac{3}{8}) + (1)(\tfrac{1}{8}) = -\$0.50.$$

Over the long run, he can expect to lose 50¢ each time he plays the game. One hundred games will result in an approximate loss of $50.

Definition 11.3 is applicable to any process having numerical outcomes; the outcomes need not involve money. A case in point is the following example.

Example 3 A person has six keys of which only one can open a particular lock. Unfortunately, the person does not know which key is the correct one, so he tries them one at a time. How many keys can he expect to try?

Solution The outcomes are $A_1 = 1$, $A_2 = 2$, $A_3 = 3$, $A_4 = 4$, $A_5 = 5$, and $A_6 = 6$. Since each key is equally likely to be the correct one, each outcome is equally likely. Therefore $p_1 = p_2 = p_3 = p_4 = p_5 = p_6 = \tfrac{1}{6}$. The expected value is

$$E = (1)(\tfrac{1}{6}) + (2)(\tfrac{1}{6}) + (3)(\tfrac{1}{6}) + (4)(\tfrac{1}{6}) + (5)(\tfrac{1}{6}) + (6)(\tfrac{1}{6}) = 3\tfrac{1}{2} \text{ keys.}$$

Again, this number must be interpreted as an average. On any one run, the person will try 1, 2, 3, 4, 5, or 6 keys, not $3\tfrac{1}{2}$. However, if this process were repeated many times, the average number of keys for each run would be $3\tfrac{1}{2}$.

Exercises

1. A person buys a parcel of land in New Mexico for $5000 with the intent of selling it at the end of the year for a profit of $10,000. The profit depends on plans for a new dam in the area being approved. If the dam is not approved, the land will be worthless. What is the expected monetary value of this investment if the probability of the dam being approved is $\frac{1}{4}$?

2. A manufacturer is considering a proposal to replace his present process with new machinery at a cost of $50,000. If the new machinery is effective, he will realize an additional profit of $15,000 a year over and above his current profit. If the new machinery is no more effective than his current process, he will have gained nothing and effectively lost $50,000. If the new machinery is worse than his current process, he will sell the new machinery back to the company for $35,000 and rebuild his old process for $10,000, resulting in an effective loss of $25,000. The probabilities of the outcomes are, respectively, .75, .20, and .05. What is the expected monetary value of the new machinery over a 10-year period?

3. Redo Example 2 if the entrance fee is $1.50.

4. A barrel contains five red balls, two white balls, and one green ball. A person selects one ball at random. If it is red, he loses $1, if it is white, he wins $1, and if it is green, he wins $3. Determine the expected monetary value of this game.

5. A roulette wheel has 38 numbers ranging from 1 to 36, with a 0 and a 00. A person selects a number. If that number appears, she wins $35. If that number does not appear, she loses $1. Determine the expected monetary value of the game.

6. A man draws a card from a regular deck. He wins the face value of the card unless it is a picture card, in which case he loses $5. Is this an attractive game?

7. A coin is flipped four times. What is the expected number of times that heads will occur?

8. Three balls are selected from the barrel in Exercise 4 one at a time. Each time a ball is selected, it is returned to the barrel before the next selection. Determine the expected number of red balls that will be picked.

9. One card is selected from each of 10 decks. Determine the expected number of picture cards that will be selected.

APPENDIXES

Appendix

Curve Fitting

Throughout this book, we have used equations to model or represent various business situations without mentioning, for the most part, how the equations were obtained. In practice, such equations are not immediately available to the user but must be derived. The derivations are generally one of two types, theoretical or empirical.

In the theoretical approach, one uses known principles to generate the equations. Indeed, this was the approach used in Chapter 7 to model mathematically the problems presented there. A specific example is given in the beginning of Section 7.3. We used the accepted principle that profit P is the total revenue received from all sales TR minus the total cost TC to generate the equation $P = TR - TC$.

In the empirical approach, one uses past data to generate the equations. Although such methods are no better than the accuracy of the data, they are often the only ones available, especially if there are no theoretical results that apply. As an example, a certain business, knowing that sales volume depends on advertising expenditures, may want to know the exact relationship between these quantities. Does volume increase linearly (as a straight line) with advertising expenditures or perhaps exponentially? Obviously, there is no theoretical principle that can answer this question, since the answer differs from product to product. In this appendix, we present an introduction to curve fitting which is one method of extracting meaningful mathematical equations from a set of data points.

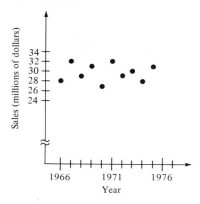

FIGURE A.1

A.1 Constant Curve Fit

The simplest curve fit occurs when the data are relatively constant. For this case, a horizontal straight line (see Section 2.1) may represent a good approximation to the given situation.

Definition A.1 Let a denote the average value of a set of data. The line $y = a$ is the *average-value, straight-line fit* for these data points.

Example 1 The yearly gross sales of a manufacturing firm for the past decade are plotted in Figure A.1. Determine the average-value, straight-line fit for these data.

Solution For convenience, we first tabulate the given data points in Table A.1.

TABLE A.1

Year	Sales (millions)	Year	Sales (millions)
1966	28	1971	32
1967	32	1972	29
1968	29	1973	30
1969	31	1974	28
1970	27	1975	31

Since the data are relatively constant, a straight-line fit is a reasonable approximation. Here the average yearly sales are

$$a = \frac{28 + 32 + 29 + 31 + 27 + 32 + 29 + 30 + 28 + 31}{10} = 29.7$$

The average-value, straight-line fit is $y = 29.7$ which is drawn in Figure A.2.

A.1 Constant Curve Fit   **405**

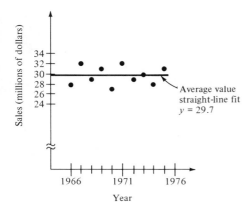

FIGURE A.2

Obviously, if the data are not reasonably constant, the average-value method is a poor approximation which can lead to erroneous conclusions. We would have little confidence in a 1976 projection based on the average-value, straight-line fit for the data given in Figure A.3.

A useful modification to an average-value, straight-line fit is the concept of moving averages. Here several averages are calculated for different time periods, resulting in a set of averages which "move with the data."

Example 2 Determine consecutive 5-year moving averages for the data given in Table A.1.

Solution The first 5-year average a_1 includes data for 1966 through 1970, inclusive.

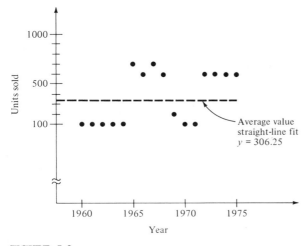

FIGURE A.3

Thus

$$a_1 = \frac{28 + 32 + 29 + 31 + 27}{5} = 29.4.$$

The second 5-year average for the years 1967 through 1971 is

$$a_2 = \frac{32 + 29 + 31 + 27 + 32}{5} = 30.2.$$

Using the data for the years 1968 through 1972, we compute the third 5-year moving average as

$$a_3 = \frac{29 + 31 + 27 + 32 + 29}{5} = 29.6.$$

Continuing in this manner, we also find that $a_4 = 29.8$, $a_5 = 29.2$, and $a_6 = 30.0$.

The arithmetic for computing consecutive moving averages can be simplified if we note that each average differs from its predecessor by the addition and deletion of two pieces of data. In Example 2, a_2 differed from a_1 by the addition of the 1971 data and the deletion of the 1966 data. Similarly, a_3 differed from a_2 by the addition of the 1972 data and the deletion of the 1967 data. In general, each new average can be calculated from the previous average by adding to the previous average the difference between the new data point and the oldest data point divided by the time span under consideration. In Example 2,

$$a_2 = a_1 + \left(\frac{32 - 28}{5}\right)$$

and

$$a_3 = a_2 + \left(\frac{29 - 32}{5}\right).$$

Example 3 Compute and graph consecutive 4-year moving averages for the years 1969 through 1975, inclusive, for the data given in Figure A.3.

Solution The first 4-year moving average encompasses the years 1969 through 1972. Reading the appropriate points from the graph, we find

$$a_1 = \frac{200 + 100 + 100 + 600}{4} = 250.$$

Then

$$a_2 = a_1 + \left(\frac{600 - 200}{4}\right) = 250 + 100 = 350,$$

$$a_3 = a_2 + \left(\frac{600 - 100}{4}\right) = 350 + 125 = 475,$$

and

$$a_4 = a_3 + \left(\frac{600 - 100}{4}\right) = 475 + 125 = 600.$$

These averages are drawn in Figure A.4.

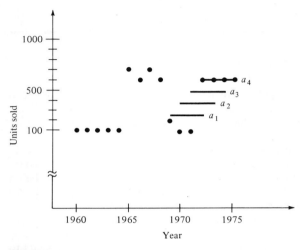

FIGURE A.4

Exercises

1. Find the average-value, straight-line fit for the data given in Table A.2.
2. The gross sales for a small furniture store are given in Table A.3 for a 9-year period. Find the average-value, straight-line fit for these data.
3. Compute consecutive 5-year moving averages for the data given in Table A.2.
4. Compute consecutive 4-year moving averages for the data given in Table A.3.
5. The Spencer Food Company is considering an increase in its advertising budget to bolster sales of its Kitty-Kat high-protein cat food, a product which has particular appeal to upper-income cat owners. As a manager of the company, you are asked for a preliminary opinion on the advisability of such an increase. What is your initial reaction based on the information listed in Table A.4?

TABLE A.2

Price ($)	1	2	3	4	5	6	7	8	9	10
Demand (hundreds)	12	13	9	8	11	10	9	13	8	10

TABLE A.3

Year	1967	1968	1969	1970	1971	1972	1973	1974	1975
Sales (thousands)	28	29	32	27	26	31	28	27	30

TABLE A.4

	1966	1967	1968	1969	1970	1971	1972	1973	1974
Advertising (thousands of $)	140	150	160	170	180	160	160	170	170
Sales (thousands of cases)	28	29	32	26	26	27	28	27	30

A.2 Linear Least-Squares Fit

Empirically obtained data on supply, demand, sales, cost, and other commercial quantities rarely are represented adequately by average-value straight lines. Many times, however, such data can be modeled by the more general straight line $y = mx + b$, with $m \neq 0$.

If the data consist of only two points, we can use the methods developed in Section 2.2 (see Example 5) to fit a straight line through them. If more than two data points are given, one of two situations can occur. First, all the data points can lie on the same straight line. In such a case, which almost never occurs, we simply pick two of the points and construct a straight line through them as before.

The more common situation involves a set of data that do not lie on any straight line but which, nonetheless, seem to be adequately represented by such a curve. A case in point involves the data plotted in Figure A.5. Although a straight line appears to be a reasonable approximation to the data, no one line of the form $y = mx + b$ contains all the given points. Therefore we seek the straight line that best fits the data.

Any straight-line approximation has one y-value on the line for each value of x. This y-value may or may not agree with the given data. Thus, for the values of x at

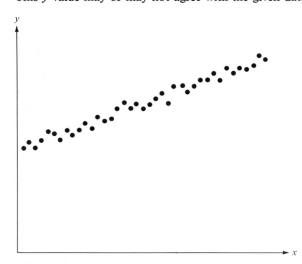

FIGURE A.5

A.2 Linear Least-Squares Fit   409

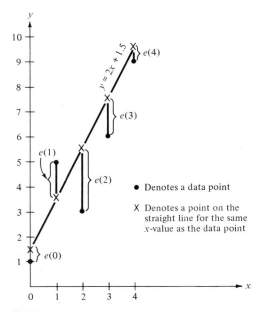

FIGURE A.6

which data are available, we generally have two values of y: one value from the data and a second value from the straight-line approximation to the data. This situation is illustrated in Figure A.6. The error at each x is simply the difference between the y-value of the data point and the y-value obtained from the straight-line approximation. We designate this error as $e(x)$.

Example 1 Calculate the errors made in approximating the data given in Figure A.6 by the line $y = 2x + 1.5$.

Solution The line and the given data points are plotted in Figure A.6. There are errors at $x = 0$, $x = 1$, $x = 2$, $x = 3$, and $x = 4$. Evaluating the equation $y = 2x + 1.5$ at these values of x, we compute Table A.5.

TABLE A.5

Given data		Evaluated from $y = 2x + 1.5$
x	y	y
0	1	1.5
1	5	3.5
2	3	5.5
3	6	7.5
4	9	9.5

It now follows that

$e(0) = 1 - 1.5 = -0.5$
$e(1) = 5 - 3.5 = 1.5$
$e(2) = 3 - 5.5 = -2.5$
$e(3) = 6 - 7.5 = -1.5$

and

$e(4) = 9 - 9.5 = -0.5.$

Note that these errors could have been read directly from the graph.

We can extend this concept of error to the more general situation involving N data points. Let $(x_1, y_1), (x_2, y_2), (x_3, y_3), \ldots, (x_N, y_N)$ be a set of N data points for a particular situation. Any straight-line approximation to this data generates errors $e(x_1), e(x_2), e(x_3), \ldots, e(x_N)$ which individually can be positive, negative, or zero. The latter case occurs when the approximation agrees with the data at a particular point. We define the overall error as follows.

Definition A.2 The *least-squares error* E is the sum of the squares of the individual errors. That is,

$$E = [e(x_1)]^2 + [e(x_2)]^2 + [e(x_3)]^2 + \cdots + [e(x_N)]^2. \qquad (1)$$

The only way the total error E can be zero is for each of the individual errors to be zero. Since each term in Eq. (1) is squared, an equal number of positive and negative individual errors cannot sum to zero.

Example 2 Compute the least-squares error for the approximation used in Example 1.

Solution

$$\begin{aligned} E &= [e(0)]^2 + [e(1)]^2 + [e(2)]^2 + [e(3)]^2 + [e(4)]^2 \\ &= (-0.5)^2 + (1.5)^2 + (-2.5)^2 + (-1.5)^2 + (-0.5)^2 \\ &= 0.25 + 2.25 + 6.25 + 2.25 + 0.25 = 11.25. \end{aligned}$$

Definition A.3 The *least-squares straight line* is the line that minimizes the least-squares error.

It can be shown (although not considered here) that the least-squares straight line is given by the equation $y = mx + b$, where m and b simultaneously satisfy the two equations

$$bN + m \sum_{i=1}^{N} x_i = \sum_{i=1}^{N} y_i \qquad (2)$$

and

$$b \sum_{i=1}^{N} x_i + m \sum_{i=1}^{N} (x_i)^2 = \sum_{i=1}^{N} x_i y_i. \tag{3}$$

Here x_i and y_i denote the values of the ith data point and N is the total number of data points being considered. We also have used the sigma notation introduced in Section 1.11.

Example 3 Find the least-squares straight line for the data listed in Table A.6.

TABLE A.6

x	0	1	2	3	4
y	1	5	3	6	9

Solution A good procedure for calculating the least-squares straight line is to first construct an expanded table similar to Table A.7.

TABLE A.7

	x_i	y_i	$(x_i)^2$	$x_i y_i$
	0	1	0	0
	1	5	1	5
	2	3	4	6
	3	6	9	18
	4	9	16	36
Sum	$\sum_{i=1}^{5} x_i = 10$	$\sum_{i=1}^{5} y_i = 24$	$\sum_{i=1}^{5} (x_i)^2 = 30$	$\sum_{i=1}^{5} x_i y_i = 65$

Then, substituting the appropriate sums into Eqs. (2) and (3) with $N = 5$ since there are 5 data points, we have

$$5b + 10m = 24$$

and

$$10b + 30m = 65.$$

Solving these two equations simultaneously for m and b, we obtain $m = 1.7$ and $b = 1.4$. The equation of the least-squares line is $y = 1.7x + 1.4$, which is graphed with the original data in Figure A.7.

Example 4 Find the least-squares straight line for the data given in Table A.8.

Appendix A: Curve Fitting

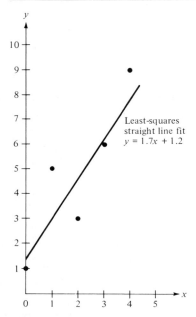

FIGURE A.7

TABLE A.8

x	0	1	2	3	4	5	6
y	2.0	1.0	3.5	5.5	4.5	5.5	6.5

Solution Following the procedure suggested in Example 3, we construct Table A.9.

TABLE A.9

	x_i	y_i	$(x_i)^2$	$x_i y_i$
	0	2	0	0
	1	1	1	1
	2	3.5	4	7
	3	5.5	9	16.5
	4	4.5	16	18
	5	5.5	25	27.5
	6	6.5	36	39
Sum	$\sum_{i=1}^{7} x_i = 21$	$\sum_{i=1}^{7} y_i = 28.5$	$\sum_{i=1}^{7} (x_i)^2 = 91$	$\sum_{i=1}^{7} x_i y_i = 109$

Substituting the appropriate sums into Eqs. (2) and (3) with $N = 7$ since there are 7 data points, we have

$$7b + 21m = 28.5$$

and

$21b + 91m = 109.$

Solving these two equations simultaneously for m and b, we obtain $m = 0.839$ and $b = 1.56$. The equation of the least-squares line is $y = 0.839x + 1.56$, which is drawn with the original data in Figure A.8.

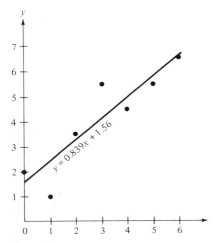

FIGURE A.8

Example 5 Verify that the least-squares line derived in Example 4 yields a smaller least-squares error for the data than the line $y = x + 1$.

Solution The line $y = x + 1$ and the individual errors are drawn in Figure A.9. Reading directly from this graph, we obtain

$$E = [e(0)]^2 + [e(1)]^2 + [e(2)]^2 + [e(3)]^2 + [e(4)]^2 + [e(5)]^2 + [e(6)]^2$$
$$= (1)^2 + (-1)^2 + (0.5)^2 + (1.5)^2 + (-0.5)^2 + (-0.5)^2 + (-0.5)^2$$
$$= 1 + 1 + 0.25 + 2.25 + 0.25 + 0.25 + 0.25$$
$$= 5.25.$$

To compute the error for the least-squares line $y = 0.839x + 1.56$ we could read the individual errors directly from Figure A.8, but the numbers we obtained would be only approximations. More accurately, we first evaluate the equation of the line at the appropriate x-values and then calculate

$$E = (2 - 1.56)^2 + (1 - 2.399)^2 + (3.5 - 3.238)^2 + (5.5 - 4.077)^2$$
$$+ (4.5 - 4.916)^2 + (5.5 - 5.755)^2 + (6.5 - 6.594)^2$$
$$= 0.19 + 1.96 + 0.07 + 2.02 + 0.17 + 0.07 + 0.01$$
$$= 4.49.$$

As expected, the least-squares error is less for the least-squares straight line.

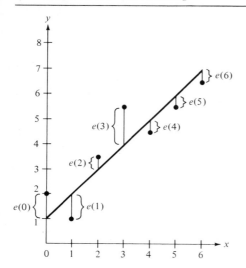

FIGURE A.9

Frequently data are given over equally spaced time intervals, as illustrated in Table A.10. A useful trick when making a least-squares analysis of such data is to code the years as follows. If the total number of years under consideration is odd, pick the middle year in the data and code it year zero. If the total number of years is even, calculate the midpoint of the 2 years closest to the middle and code this midpoint zero. Assign values to all the other years corresponding to the number of years by which they either precede or follow the zero year. For the data presented in Table A.10, 1972 is coded $x = 0$. The year 1971 is coded $x = -1$, and the year 1974 is coded $x = 2$. This coding technique has the practical advantage of reducing the number of calculations required for solving m and b in Eqs. (2) and (3).

TABLE A.10

Year	1970	1971	1972	1973	1974
Sales (millions)	10	13	11	15	14

Example 6 Find the least-squares straight line for the data in Table A.10 and use this line to determine the average yearly increase in sales for 1970 through 1974.

Solution First, coding the years as described above, we compute Table A.11. Substituting the appropriate sums into Eqs. (2) and (3), we find

$$5b + 0m = 63$$

and

$$0b + 10m = 10.$$

TABLE A.11

	x_i	y_i	$(x_i)^2$	$x_i y_i$
1970	−2	10	4	−20
1971	−1	13	1	−13
1972	0	11	0	0
1973	1	15	1	15
1974	2	14	4	28
Sum	$\sum_{i=1}^{5} x_i = 0$	$\sum_{i=1}^{5} y_i = 63$	$\sum_{i=1}^{5} (x_i)^2 = 10$	$\sum_{i=1}^{5} x_i y_i = 10$

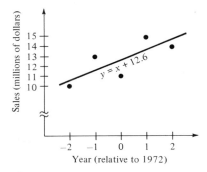

FIGURE A.10

Since $\sum_{i=1}^{5} x_i = 0$, the last two equations are particularly simple to solve. Here $m = 1$ and $b = 12.6$. The equation of the least-squares line becomes $y = x + 12.6$, which is drawn in Figure A.10 with the original data.

The slope of this line is $m = 1$. The average rate of increase in sales with respect to time therefore is $1 million, the dollar figure corresponding to this slope.

Had Table A.10 contained an entry for 1975 too, then there would have been an even number of data points. The 2 years closest to the middle would be 1972 and 1973, so the midpoint between these years corresponds to $x = 0$. Consequently, 1972 is coded as $x = -\frac{1}{2}$, 1973 is coded as $x = \frac{1}{2}$, 1975 becomes $x = 2\frac{1}{2}$, and so on.

Exercises

1. Using the results from Example 6, complete Table A.12 and then compute the least-squares error. Here y_c denotes the y-value obtained from the least-squares straight line by evaluating the equation of the line at the appropriate values of x.
2. Consider the data given in Table A.13.
   (a) Plot the data points and determine whether a straight-line approximation seems reasonable.
   (b) Find the least-squares straight line for the data.
   (c) Calculate the least-squares error for this line.
3. Consider the data given in Table A.14.

416 Appendix A: Curve Fitting

TABLE A.12

x	y	y_c	$e(x)$	$[e(x)]^2$
-2	10			
-1	13			
0	11			
1	15			
2	14			

TABLE A.13

x	0	1	2	3	4	5	6
y	1	5	8	7	12	14	13

TABLE A.14

x	0	1	2	3	4	5	6	7
y	36	49	55	56	67	69	76	85

(a) Plot the data points and determine whether a straight-line approximation seems reasonable.
(b) Find the least-squares straight line for the data.
(c) Calculate the least-squares error for this line.
(d) Draw any other straight line that appears to fit the data reasonably well and compare the least-squares error of this line to the result from part (c).

4. Repeat Exercise 3 for the data given in Table A.15.
5. The annual sales receipts for color television sets for the Village Appliance Shop are given in Table A.16.
   (a) Find the least-squares straight line for these data.
   (b) Determine the sales for 1973 by evaluating the result from part (a) at that time.
   (c) Use the result from part (a) to project the sales for 1978.
6. The number of air conditioners sold each year by the Village Appliance Shop is detailed in Table A.17.
   (a) Find the least-squares straight line for these data.
   (b) Use this line to project the number of units that will be sold in 1978.

TABLE A.15

x	0	1	2	3	4
y	10	13	20	35	52

TABLE A.16

Sales (thousands)	15	18	16	20	18	22	19
Year	1969	1970	1971	1972	1973	1974	1975

TABLE A.17

Year	1970	1971	1972	1973	1974	1975
Units	19	17	21	19	23	20

TABLE A.18

Rain (inches)	2.0	2.2	2.3	3.2	3.8	4.9	5.6
Yield (bushels)	25	25	30	30	40	50	50

7. A farmer in Madison, Iowa, collected the data in Table A.18 relating the number of bushels of wheat obtained from a test plot of land to the amount of rainfall in Madison during the growing season. Find the least-squares straight line for these data and use it to estimate the amount of wheat that could be expected from a rainfall of 3.5 inches.

A.3 Quadratic Least-Squares Fit

Frequently, a set of points is fit appropriately by a quadratic curve of the form (see Section 2.4)

$$y = ax^2 + bx + d. \tag{4}$$

A case in point involves the data plotted in Figure A.11.

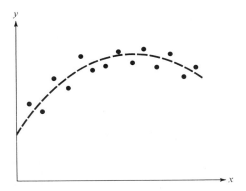

FIGURE A.11

If data consist of exactly three points which do not lie on a straight line, we can find a quadratic curve that contains all the points. In such situations, we generate an exact fit.

Example 1 Find the quadratic curve that exactly fits the data given in Table A.19.

TABLE A.19

x	5	10	15
y	25	50	250

Solution Since the three points are to lie on the same curve, each data point by itself must satisfy the equation of the curve. Substituting the *x*- and *y*-values of the data into Eq. (4), we obtain

$$25a + 5b + d = 25$$
$$100a + 10b + d = 50$$

and

$$225a + 15b + d = 250.$$

These three equations can be solved with the matrix method developed in Chapter 4; the solution is $a = 3.5$, $b = -47.5$, and $d = 175$. The equation of the quadratic curve is

$$y = 3.5x^2 - 47.5x + 175,$$

which is drawn in Figure A.12.

If more than three points are given, almost always there is no quadratic curve that contains all the points. In such cases, for example, the data given in Figure A.11, any quadratic curve generates errors at some (and possibly all) of the data points.

Definition A.4 The *least-squares quadratic curve* is the quadratic curve that minimizes the least-squares error (see Definition A.2).

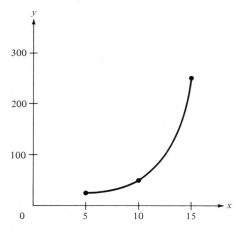

FIGURE A.12

A.3 Quadratic Least-Squares Fit

Being quadratic, the least-squares quadratic curve must satisfy Eq. (4). Since there are three unknowns in Eq. (4), the constants a, b, and c, we expect three equations for these unknowns similar to Eqs. (2) and (3). This is indeed the case. The equations are

$$dN + b \sum_{i=1}^{N} x_i + a \sum_{i=1}^{N} (x_i)^2 = \sum_{i=1}^{N} y_i \tag{5}$$

$$d \sum_{i=1}^{N} x_i + b \sum_{i=1}^{N} (x_i)^2 + a \sum_{i=1}^{N} (x_i)^3 = \sum_{i=1}^{N} x_i y_i \tag{6}$$

and

$$d \sum_{i=1}^{N} (x_i)^2 + b \sum_{i=1}^{N} (x_i)^3 + a \sum_{i=1}^{N} (x_i)^4 = \sum_{i=1}^{N} (x_i)^2 y_i. \tag{7}$$

Here x_i and y_i are the coordinates of the ith data point, and N is the total number of data points under consideration.

Example 2 Find the least-squares quadratic curve for the data given in Table A.20.

TABLE A.20

x	0	1	2	3	4
y	10	14	18	30	50

Solution A good procedure for calculating the least-squares quadratic curve is to first construct an expanded table similar to Table A.21.

Substituting the appropriate sums into Eqs. (5) through (7) with $N = 5$ since there are 5 data points, we obtain

$$5d + 10b + 30a = 122$$
$$10d + 30b + 100a = 340$$

and

$$30d + 100b + 354a = 1156.$$

TABLE A.21

	x_i	y_i	$(x_i)^2$	$(x_i)^3$	$(x_i)^4$	$x_i y_i$	$(x_i)^2 y_i$
	0	10	0	0	0	0	0
	1	14	1	1	1	14	14
	2	18	4	8	16	36	72
	3	30	9	27	81	90	270
	4	50	16	64	256	200	800
Sum	10	122	30	100	354	340	1156

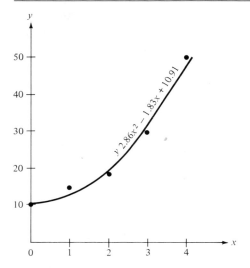

FIGURE A.13

These equations can be solved with the matrix method presented in Chapter 4 (Example 3 in Section 4.6) as $a = 2.86$, $b = -1.83$, and $d = 10.91$. The least-squares quadratic curve is

$$y = 2.86x^2 - 1.83x + 10.91,$$

which is drawn in Figure A.13.

Even if all the data points lie on the same quadratic curve, we can use Eqs. (5) through (7) to obtain this curve. Since the least-squares quadratic curve minimizes the least-squares error, and since the least-squares error is zero for an exact fit, the least-squares quadratic curve will be the exact fit if an exact fit exists.

Example 3 Find the least-squares quadratic curve for the data given in Table A.22.

Solution Following the procedure suggested in Example 2, we first construct Table A.23.

Substituting the appropriate sums into Eqs. (5) through (7) with $N = 6$, we obtain

$$6d + 41b + 327a = 103.5$$
$$41d + 327b + 2879a = 1029.5$$

and

$$327d + 2879b + 26{,}931a = 10{,}195.5.$$

TABLE A.22

x	3	4	6	8	9	11
y	-5.5	-2	8	22	30.5	50.5

TABLE A.23

	x_i	y_i	$(x_i)^2$	$(x_i)^3$	$(x_i)^4$	$x_i y_i$	$(x_i)^2 y_i$
	3	−5.5	9	27	81	−16.5	−49.5
	4	−2	16	64	256	−8	−32
	6	8	36	216	1,296	48	288
	8	22	64	512	4,096	176	1,408
	9	30.5	81	729	6,561	274.5	2,470.5
	11	50.5	121	1331	14,641	555.5	6,110.5
Sum	41	103.5	327	2879	26,931	1029.5	10,195.5

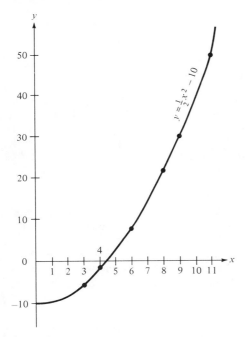

FIGURE A.14

The solution to these simultaneous equations is $a = \frac{1}{2}$, $b = 0$, and $d = -10$. The least-squares quadratic curve is $y = \frac{1}{2}x^2 - 10$, which is drawn in Figure A.14. Note that this curve contains all the data points and therefore is an exact fit.

Exercises

1. Consider the three data points $(-1, 0)$, $(0, 7)$, and $(1, 10)$.
   (a) Using Eq. (4), find the three simultaneous equations required to fit a quadratic curve exactly to these data.
   (b) Using Eqs. (5) through (7), find the three simultaneous equations required to fit a least-squares quadratic curve to these data.

TABLE A.24

x	0	1	2	3	4
y	10	11	15	16	23

TABLE A.25

x	y	y_c	$e(x)$	$[e(x)]^2$
0	10			
1	11			
2	15			
3	16			
4	23			

    (c) Verify that the solutions to both sets of equations found in parts (a) and (b) are $a = -2$, $b = 5$, and $d = 7$.

2. Redo parts (a) and (b) in Exercise 1 for the data points (1, 0), (2, −1), and (3, 0). Verify that the solutions to both sets of equations are $a = 1$, $b = -4$, and $d = 3$.
3. Find the least-squares quadratic curve that best fits the data given in Table A.24.
4. Complete Table A.25 for the data given in Table A.24 and show that the least-squares error is 4.9. Here y_c denotes the y-value obtained from evaluating the least-squares quadratic curve, $y_c = 0.643x^2 + 0.528x + 10.086$, at the appropriate values of x.
5. Find the least-squares straight line for the data given in Table A.24 and then calculate the least-squares error for this line. How does this error compare with the error found in Exercise 4? Draw both the least-squares quadratic curve and the least-squares straight line on the same graph and compare visually.
6. Find the least-squares quadratic curve that best fits the data given in Table A.26. Note that these data have the same y-values as in Table A.24, but that the x-values have been coded according to the procedures presented in Section A.2. What are some of the benefits of such a coding?
7. Consider the data given in Table A.27.
    (a) Find the three equations required to fit a least-squares quadratic curve to these data.
    (b) Verify that a solution to these equations is $a = 0.994$, $b = -0.018$, and $d = -0.990$.

TABLE A.26

x	−2	−1	0	1	2
y	10	11	15	16	23

TABLE A.27

x	−3	−2	−1	0	1	2	3
y	8.1	2.9	0	−1	0	3.1	7.8

(c) Graph the least-squares quadratic curve with the quadratic curve $y = x^2 - 1$ on the same graph.
(d) Determine the least-squares error for both of the curves in part (c) and compare. Note that three of the data points actually lie on the curve $y = x^2 - 1$.

A.4 Exponential Least-Squares Fit

An extremely useful curve for fitting data that appear to be nonlinear (do not follow a straight line) is the exponential curve

$$y = a(b^x), \tag{8}$$

introduced in Section 2.6. Interestingly, all least-squares analysis on such curves can be reduced to the linear least-squares fit developed in Section A.2 if we first transform Eq. (8) into a linear equation.

Taking logarithms of both sides of Eq. (8), we obtain

$$\log y = \log [a(b^x)].$$

Using the elementary properties of logarithms, we can simplify this equation to

$$\log y = \log a + \log b^x$$

or

$$\log y = \log a + x \log b. \tag{9}$$

Finally, setting $Y = \log y$, $A = \log a$, and $B = \log b$, we can rewrite Eq. (9) as

$$Y = Bx + A. \tag{10}$$

Since B and A are constants, Eq. (10) is simply a straight line in the variables Y and x.

Although Eqs. (10) and (9) are different in form from Eq. (8), they do not alter the fundamental relationship between the variables x and y. In fact, no information is either gained or lost by using any one of the three representations. As an example, consider the curve $y = 3(1.5^x)$. Here $a = 3$ and $b = 1.5$, so $A = \log 3 = 0.4771$ and $B = \log 1.5 = 0.1761$. The two curves, $y = 3(1.5^x)$ and $Y = 0.1761x + 0.4771$, corresponding to Eqs. (8) and (10) are plotted in Figures A.15 and A.16, respectively. Suppose now that we are interested in obtaining the value of y corresponding to $x = 4$. Reading directly from Figure A.15, we obtain $y = 15.2$. Using Figure A.16, we find that $Y = 1.18$ corresponds to $x = 4$. Then,

$$y = \text{antilog } Y = \text{antilog } 1.18 = 15.14.$$

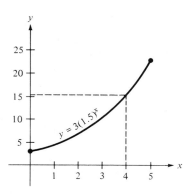

FIGURE A.15

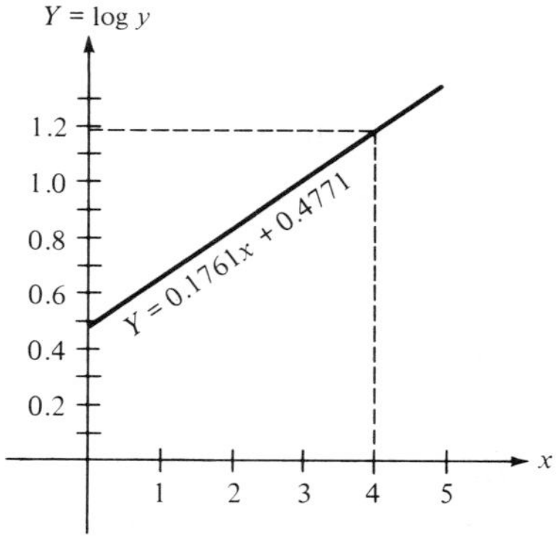

FIGURE A.16

The difference in the two results is due to graphical inaccuracies. Had the graphs been more detailed, both results would have been closer to the actual value $y = 15.1875$.

Since both Eqs. (8) and (10) involve the same relationship between x and y, we can fit an exponential curve to a set of data in two different ways. First, we can find the appropriate constants a and b and substitute them into Eq. (8). Although this is the direct approach, it is algebraically too complex to be of practical use. The second (and recommended) method is first to convert all x- and y-data to x- and Y-data, where $Y = \log y$, and then to find the appropriate constants A and B in Eq. (10). Since Eq. (10) is a straight line, we can use the techniques developed in Section A.2 to compute these constants. The desired numbers a and b can be obtained through antilogs.

Before using the methods developed in Section A.2, we must modify Eqs. (2) and (3), since they were derived for a straight line of the form $y = mx + b$, whereas now we are interested in the line $Y = Bx + A$. The appropriate equations are

$$AN + B \sum_{i=1}^{N} x_i = \sum_{i=1}^{N} Y_i \tag{11}$$

and

$$A \sum_{i=1}^{N} x_i + B \sum_{i=1}^{N} (x_i)^2 = \sum_{i=1}^{N} x_i Y_i. \tag{12}$$

Example 1 Find the least-squares exponential curve that fits the data given in Table A.28.

TABLE A.28

x	0	1	2	3	4	5
y	3.5	5.3	7.4	11.2	14.9	22.7

Solution We first convert the y-data to Y-data and then complete Table A.29 as recommended in Section A.2.

TABLE A.29

	x_i	y_i	$Y_i = \log y_i$	$(x_i)^2$	$x_i Y_i$
	0	3.5	0.5441	0	0.0
	1	5.3	0.7243	1	0.7243
	2	7.4	0.8692	4	1.7384
	3	11.2	1.0492	9	3.1476
	4	14.9	1.1732	16	4.6928
	5	22.7	1.3560	25	6.7800
Sum	15		5.7160	55	17.0831

Substituting the appropriate sums into Eqs. (11) and (12), we have

$$6A + 15B = 5.7160$$

and

$$15A + 55B = 17.0831.$$

The solution to these equations is $A = 0.5537$ and $B = 0.1596$, and the least-squares line has the form $Y = 0.1596x + 0.5537$.

Although the last equation is a valid relationship between x and Y, we typically prefer to have the relationship in the standard form of Eq. (8). Since

$$a = \text{antilog } A = \text{antilog } 0.5537 = 3.58$$

and

$$b = \text{antilog } B = \text{antilog } 0.1596 = 1.44 \text{ rounded to two decimals,}$$

the required exponential curve is $y = 3.58(1.44^x)$, which is drawn in Figure A.17.

Example 2 Profits of a local dress manufacturer for a 5-year period are given in Table A.30. Find the least-squares exponential curve that fits these data and use this curve to project profits for 1978.

TABLE A.30

Year	1972	1973	1974	1975	1976
Profits (thousands of $)	19	20	30	36	44

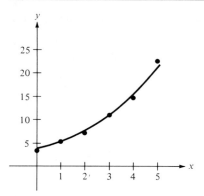

FIGURE A.17

Solution We first code the years according to the procedure given in Section A.2. Then, setting y_i equal to the yearly profits, we complete Table A.31.

Substituting the appropriate sums into Eqs. (11) and (12) we have

$$5A + 0B = 7.2567$$

and

$$0A + 10B = 0.9847.$$

Solving these two equations simultaneously for A and B, we obtain $A = 1.4513$ and $B = 0.09847$. Then,

$$a = \text{antilog } A = \text{antilog } 1.4513 = 28.27$$

and

$$b = \text{antilog } B = \text{antilog } 0.09847 = 1.255,$$

and the required exponential curve is $y = 28.27(1.255^x)$.

TABLE A.31

	x_i	y_i	$Y_i = \log y_i$	$(x_i)^2$	$x_i Y_i$
1972	-2	19	1.2788	4	-2.5576
1973	-1	20	1.3010	1	-1.3010
1974	0	30	1.4771	0	0.0
1975	1	36	1.5563	1	1.5563
1976	2	44	1.6435	4	3.2870
Sum	0		7.2567	10	0.9847

The year 1978 corresponds to $x = 4$. The projected profit is $y = 28.27(1.255^4) = 28.27(2.48) = 70.11 = \$70,110$.

Exercises

1. Find the least-squares exponential curve that fits the data given in Table A.32. Plot both the data and the curve on the same graph.
2. Find the least-squares exponential curve that fits the data given in Table A.33.
3. Redo Exercise 1 by first coding the x-variables so that $x = 2$ corresponds to $x = 0$.
4. Sales of a rapidly growing manufacturing firm for its first 5 years of operation are given in Table A.34. Find the least-squares exponential curve that fits these data and then use this curve to project sales for 1978.
5. Census figures for a midwestern city are given in Table A.35. *Hint:* Code 1950 as $x = 0$ and 1960 as $x = 1$.
   (a) Find the least-squares exponential curve that fits this data.
   (b) Use this curve to project the population in 1980.
   (c) Determine the year when the population was 1000.

TABLE A.32

x	0	1	2	3	4
y	2	17	90	500	2500

TABLE A.33

x	0	1	2	3	4	5
y	17	20	38	60	90	130

TABLE A.34

Year	1973	1974	1975	1976	1977
Sales (thousands)	0.9	2.9	9.5	28.8	100.0

TABLE A.35

Year	1930	1940	1950	1960	1970
Population	4953	7389	11,023	16,445	24,532

A.5 Selecting an Appropriate Curve

In Sections A.1 through A.4 methods were presented which allow the user to develop linear, quadratic, and exponential curves from a given set of empirical data. In practice, one is presented with a set of data and the first problem is to determine which, if any, of the three methods should be employed, that is, which type of curve best fits a given body of data. The starting point is always the same: Plot the given data on graph paper, as in Figure A.5, and take a long, hard look at the resulting plot, called a *scatter diagram*. Typically, a pattern develops. Based on the emergent pattern and a knowledge of various curves, we then select the type of curve that appears to fit the data reasonably well. Having selected the appropriate type, we proceed with a least-squares analysis to compute the actual equation of the curve. In this appendix, we considered only patterns that lend themselves to linear, quadratic, or exponential fits. Fortunately, the majority of business problems of interest assume one of these three forms.

After we have determined a pattern in the scatter diagram, selected an appropriate type of curve, and computed the equation of the least-squares curve, the final step is to test whether or not the resulting curve does in fact model the data adequately. Such tests fall under the areas of statistical inference and hypothesis testing which are both beyond the scope of this book.

Appendix

Mathematics of Finance

Calculations of present and future value of a cash flow rest on summing individual payments over time. Although the tables given in Appendixes C.5 and C.6 frequently suffice in determining these sums, many times they do not. With the advent of modern calculators, however, the formulas used to generate these tables can be used directly.

In this appendix, we derive certain formulas initially presented in Chapter 3, which are suitable for use with a modern calculator. Since the mathematical underpinnings of these formulas is the sum of a finite geometric series, we begin by developing efficient methods for handling such sums. Then we extend our analysis to include the more advanced topics of mortgages, amortization, and installment loans.

B.1 Finite Geometric Series

The sum of a finite number of terms can always be calculated by simple addition. This process can be quite lengthy, however, if one must sum a large number of terms, say 360. Such additions can be simplified considerably if formulas exist for the required sums.

The most common sum in business problems is the *finite geometric series* which is a sum of the form

$$c + cr + cr^2 + cr^3 + \cdots + cr^{n-1}, \tag{1}$$

where c and r are known numbers and n is a given positive integer. When $c = 1$, $r = 2$, and $n = 5$, Eq. (1) becomes

$$1 + 1(2) + 1(2)^2 + 1(2)^3 + 1(2)^4 = 1 + 2 + 4 + 8 + 16.$$

If $c = \frac{1}{2}$, $r = -0.1$, and $n = 6$, Eq. (1) becomes

$$\frac{1}{2} + \frac{1}{2}(-0.1) + \frac{1}{2}(-0.1)^2 + \frac{1}{2}(-0.1)^3 + \frac{1}{2}(-0.1)^4 + \frac{1}{2}(-0.1)^5$$
$$= 0.5 - 0.05 + 0.005 - 0.0005 + 0.00005 - 0.000005.$$

Both the previous two sums are specific examples of a finite geometric series. A useful formula for summing such series is given by the following theorem.

Theorem B.1 The sum of a finite geometric series is

$$c + cr + cr^2 + cr^3 + \cdots + cr^{n-1} = c\left(\frac{1 - r^n}{1 - r}\right). \quad (2)$$

Proof Designate the sum of the first n terms of a geometric series by S, that is,

$$S = c + cr + cr^2 + cr^3 + \cdots + cr^{n-2} + cr^{n-1}. \quad (3)$$

We seek a formula for S. Multiplying S by r, we have

$$rS = cr + cr^2 + cr^3 + cr^4 + \cdots + cr^{n-1} + cr^n. \quad (4)$$

Subtracting Eq. (4) from Eq. (3) and then rearranging terms, we obtain

$$S - rS = (c + cr + cr^2 + cr^3 + \cdots + cr^{n-2} + cr^{n-1})$$
$$\quad - (cr + cr^2 + cr^3 + cr^4 + \cdots + cr^{n-1} + cr^n)$$
$$= c + (cr - cr) + (cr^2 - cr^2) + (cr^3 - cr^3) + \cdots$$
$$\quad + (cr^{n-1} - cr^{n-1}) - cr^n$$
$$= c - cr^n.$$

Finally, if we factor S from the left side and c from the right side of the last equation, we find

$$S(1 - r) = c(1 - r^n),$$

from which

$$S = c\left(\frac{1 - r^n}{1 - r}\right).$$

Example 1 Sum $1 + \frac{1}{2} + \frac{1}{4} + \frac{1}{8} + \cdots + \frac{1}{256} + \frac{1}{512}$.

B.1 Finite Geometric Series

Solution With effort, we could sum the given terms by simple addition. However, if we write the sum as

$$1 + 1(\tfrac{1}{2}) + 1(\tfrac{1}{2})^2 + 1(\tfrac{1}{2})^3 + \cdots + 1(\tfrac{1}{2})^8 + 1(\tfrac{1}{2})^9$$

and recognize it as the first 10 terms of a geometric series with $c = 1$, $r = \tfrac{1}{2}$, and $n = 10$, we can use Eq. (2) to compute

$$S = 1\left[\frac{1 - (\tfrac{1}{2})^{10}}{1 - \tfrac{1}{2}}\right] = \frac{1 - \tfrac{1}{1024}}{\tfrac{1}{2}} = 2\left(\frac{1023}{1024}\right) = \frac{1023}{512} = 1.998.$$

Example 2 Sum $2(4)^5 + 2(4)^6 + \cdots + 2(4)^{10} + 2(4)^{11}$.

Solution This sum can be rewritten as

$$(4)^5[2 + 2(4) + \cdots + 2(4)^5 + 2(4)^6].$$

The terms in the brackets are a geometric series with $c = 2$, $r = 4$, and $n = 7$. Using Eq. (2), we calculate

$$S = (4)^5\left[(2)\frac{1 - (4)^7}{1 - 4}\right] = (1024)(2)\left(\frac{1 - 16{,}384}{1 - 4}\right) = (2048)(5461)$$

$$= 11{,}184{,}128.$$

We can now mathematically derive some of the formulas for future and present values of a cash flow. In Chapter 3, we showed that the net future value of an ordinary annuity at the end of the nth period is

$$\text{Net } FV = R[1 + (1 + i) + \cdots + (1 + i)^{n-3} + (1 + i)^{n-2} + (1 + i)^{n-1}], \tag{5}$$

where i is the interest rate per period. Equation (5) is simply Eq. (11) in Section 3.4 repeated. We agreed to denote the sum of the terms in the square brackets as $s_{\overline{n}|i}$. That is,

$$\text{Net } FV = R s_{\overline{n}|i}, \tag{6}$$

where $s_{\overline{n}|i} = 1 + (1 + i) + \cdots + (1 + i)^{n-3} + (1 + i)^{n-2} + (1 + i)^{n-1}$. We now recognize $s_{\overline{n}|i}$ as a finite geometric series with $c = 1$ and $r = 1 + i$. Using Eq. (2), we have

$$s_{\overline{n}|i} = 1\left[\frac{1 - (1 + i)^n}{1 - (1 + i)}\right]$$

or

$$s_{\overline{n}|i} = \frac{1 - (1 + i)^n}{-i} = \frac{(1 + i)^n - 1}{i} \tag{7}$$

which is precisely the formula given for $s_{\overline{n}|i}$ in Eq. (14) in Chapter 3.

The net present value of an initial investment C_0 which returns R dollars every conversion period was derived as

$$\text{Net } PV = R[(1 + i)^{-1} + (1 + i)^{-2} + (1 + i)^{-3} + \cdots + (1 + i)^{-n}] - C_0, \quad (8)$$

which is Eq. (7) in Section 3.3 repeated. We agreed to denote the sum of the terms in the square brackets as $a_{\overline{n}|i}$. That is,

$$\text{Net } PV = Ra_{\overline{n}|i} - C_0, \quad (9)$$

where $a_{\overline{n}|i} = (1 + i)^{-1} + (1 + i)^{-2} + (1 + i)^{-3} + \cdots + (1 + i)^{-n}$. This too is a finite geometric series, although it requires algebraic manipulation to convert it to the explicit form given by Eq. (1). Using the properties of exponents, specifically, $x^{-n} = (1/x)^n$, we can rewrite $a_{\overline{n}|i}$ as

$$a_{\overline{n}|i} = \frac{1}{1+i} + \left(\frac{1}{1+i}\right)^2 + \left(\frac{1}{1+i}\right)^3 + \cdots + \left(\frac{1}{1+i}\right)^n.$$

Factoring $1/(1 + i)$ from every term but the first, we have

$$a_{\overline{n}|i} = \frac{1}{1+i} + \left(\frac{1}{1+i}\right)\left(\frac{1}{1+i}\right) + \left(\frac{1}{1+i}\right)\left(\frac{1}{1+i}\right)^2 + \cdots + \left(\frac{1}{1+i}\right)\left(\frac{1}{1+i}\right)^{n-1}.$$

$a_{\overline{n}|i}$ is now in the form of Eq. (1) with both c and r equal to $1/(1 + i)$. Using Eq. (2), we find

$$a_{\overline{n}|i} = \left(\frac{1}{1+i}\right)\left[\frac{1 - (1/[1+i])^n}{1 - (1/[1+i])}\right] = \left(\frac{1}{1+i}\right)\left[\frac{1 - (1+i)^{-n}}{1 - (1/[1+i])}\right],$$

which simplifies to

$$a_{\overline{n}|i} = \frac{1 - (1+i)^{-n}}{i}. \quad (10)$$

This is precisely the formula given for $a_{\overline{n}|i}$ in Eq. (10) in Chapter 3.

Exercises

Sum the finite series given in Exercises 1 through 8.
1. $1 + 2 + 4 + 8 + 16 + \cdots + 4096$
2. $5 + \frac{5}{2} + \frac{5}{4} + \frac{5}{8} + \cdots + \frac{5}{1024}$
3. $1 - 3 + 9 - 27 + 81 - 243 + \cdots - 177{,}147 + 531{,}441$
4. $-2 + \frac{2}{3} - \frac{2}{9} + \frac{2}{27} - \frac{2}{81} + \cdots - \frac{5}{59{,}049}$
5. $4^{-2} + 4^{-3} + 4^{-4} + \cdots + 4^{-10}$
6. $2(0.1)^{-4} + 2(0.1)^{-5} + 2(0.1)^{-6} + \cdots + 2(0.1)^{-10}$
7. $(1.2)^{-10} + (1.2)^{-11} + (1.2)^{-12} + \cdots + (1.2)^{-20}$

8. $1 + (1.2)^5 + (1.2)^{10} + \cdots + (1.2)^{30}$
9. Set up and, if available, use a modern calculator to determine $s_{\overline{n}|i}$ for the following values of n and i:
   (a) $n = 30$, $i = 0.0125$
   (b) $n = 80$, $i = 0.0125$
   (c) $n = 10$, $i = 0.05$
   (d) $n = 120$, $i = 0.05$
   (e) $n = 46$, $i = 0.07$
   (f) $n = 46$, $i = 0.09$.

   Compare the results for parts (a), (c), and (e) with those listed in Appendix C.6.
10. Set up and, if available, use a modern calculator to determine $a_{\overline{n}|i}$ for the following values of n and i:
    (a) $n = 41$, $i = 0.01$
    (b) $n = 120$, $i = 0.01$
    (c) $n = 60$, $i = 0.03$
    (d) $n = 80$, $i = 0.03$
    (e) $n = 35$, $i = 0.07$
    (f) $n = 35$, $i = 0.09$.

    Compare the results for parts (a), (c), and (e) with those listed in Appendix C.5.

B.2 Installment Loans and Interest Charges

Consumer credit in the form of automobile loans, vacation loans, home improvement loans, and a host of other cash advances for specific purposes is used daily by millions of people. In the usual case, the loan is repaid in equal monthly installments, so it is an annuity.

The most commonly used procedure for assessing interest is the *add-on method*. The finance charge for each year of the loan is the quoted annual rate applied to the face value of the loan. The monthly installments are obtained by adding the finance charges to the principal and then dividing this result by the total number of months in the life of the loan.

Example 1 A $1000 loan is negotiated for 2 years at 8% interest under the add-on method. Determine the monthly installments.

Solution The interest charge for 1 year is 8% of $1000, or $80. Since the duration of this loan is 2 years, the total finance charge is $2(80) = \$160$. The installment R is obtained by adding this charge to the original principal, $160 + 1000$, and then dividing by 24, the number of months in the life of the loan. Here $R = 1160/24 = \$48.34$.

If the borrower in Example 1 thinks he is paying 8% interest for his loan, he is badly mistaken. In fact, he is paying a good deal more. To determine the actual interest, we assume that the interest is compounded monthly, and we note that all monthly payments are equal. From Section 3.3, we see that Eq. (8) in Chapter 3 or, similarly, Eq. (9) in this chapter is applicable with $C_0 = 0$. (Why?) That is, net $PV = Ra_{\overline{n}|i}$. For installment loans, however, the present value of the annuity is known; it is just the amount of cash being loaned, which under the add-on method is the face value of the loan. Therefore we can solve for

$$a_{\overline{n}|i} = \frac{\text{Net } PV}{R} \tag{11}$$

and use this number with Appendix C.5 to determine i.

Example 2 Determine the true annual interest rate for the loan described in Example 1.

Solution Here $R = 48.34$, Net $PV = 1000$, and $n = 24$. Substituting these values into Eq. (11), we obtain $a_{\overline{24}|i} = 1000/48.34 = 20.69$. It is not feasible to solve Eq. (10) directly for i. An alternative approach is to scan Appendix C.5 for various values of i with $n = 24$. We find $a_{\overline{24}|0.01} = 21.24$ and $a_{\overline{24}|0.0125} = 20.62$, so i, the interest rate per month for this problem, is between 0.01 and 0.0125. For $i = 0.01$ the annual rate is $12(0.01) = 0.12$, and for $i = 0.0125$ the annual rate is $12(0.0125) = 0.15$. Therefore the true interest rate is actually between 12% and 15% and a good deal higher than the quoted rate of 8%.

Example 3 Determine the monthly installment and the true interest rate for a $2200 new car loan for 3 years at $7\frac{1}{2}$% using the add-on method.

Solution The add-on interest per year is $7\frac{1}{2}$% of 2200 or $(0.075)(2200) = \$165$. The total interest charge for 3 years is $3(165) = \$495$. Since $n = 36$ months,

$$R = \frac{2200 + 495}{36} = \$74.87.$$

To find the true interest rate, we use Eq. (11) to obtain $a_{\overline{36}|i} = 2200/74.87 = 29.38$. Scanning Appendix C.5 for various values of i with $n = 36$, we locate $a_{\overline{36}|0.01} = 30.108$ and $a_{\overline{36}|0.0125} = 28.847$. For $i = 0.01$ the annual rate is $12(0.01) = 0.12$, and for $i = 0.0125$ the annual rate is $12(0.0125) = 0.15$, so the actual rate is between 12% and 15%, approximately 14%.

The discrepancy between stated rates and actual rates is due to the borrower not having full use of the loan for its entire duration. Each month he or she repays some of the principal with the installment. The borrower has control of the full amount of the loan for only the first month, although he or she is charged interest as if he or she had full control for the entire life of the loan.

A second widely used procedure for assessing interest is the *discount method*. The total interest charge is calculated exactly as in the add-on method, but this amount is then subtracted or discounted from the face value of the loan, and the difference is the cash received by the borrower. The installment is determined by dividing the face value of the loan by the number of months in the life of the loan.

Example 4 Determine the monthly installment for a $1800 3-year loan discounted at 6%.

Solution With the discount method, $R = 1800/36 = \$50$. The borrower, however, does not receive $1800. Yearly interest charges are 0.06(1800) = $108, which results in a total finance charge of 3(108) = $324. The borrower then receives 1800 − 324 = $1476.

As in the add-on method, the stated rate in the discount method is not the true rate. We can obtain the actual rate as before if we note that the net present value of the loan, Net PV, is the amount the borrower receives, which is the face value of the loan *minus* the finance charges.

Example 5 Determine the actual interest rate for the loan described in Example 4.

Solution Here Net $PV = 1476$, which is the amount received by the borrower. With $R = 50$ and $n = 36$, Eq. (11) becomes $a_{\overline{36}|i} = 1476/50 = 29.52$. Using Appendix C.5, we find $a_{\overline{36}|0.01} = 30.108$ and $a_{\overline{36}|0.0125} = 28.847$. As in Example 3, the actual rate is between 12% and 15%, but this time is closer to 13%.

Exercises

In Exercises 1 through 7, give the actual interest between the two closest rates found in Appendix C.5.

1. Mr. Johnson borrows $3000 from his bank for a new car loan at 8% interest add-on. Determine the monthly installment and the actual interest rate if the loan is to be repaid over (a) 2 years, and (b) 3 years.
2. A bank advertises automobile loans for 11%. If the bank uses the add-on method, determine the actual interest for a $2200 loan for 3 years.
3. The Bakers have decided to apply for a $1000 vacation loan which they plan to repay within a year. (a) How much will they have to repay monthly if the interest rate is 8% add-on? (b) What is the actual rate for this loan?
4. Assume that the loan described in Exercise 3 is discounted rather than an add-on. Determine (a) the monthly installment, (b) the amount of money with which the Bakers can plan a vacation, and (c) the actual interest rate.
5. Mr. Goldberg's application for a $4000 home improvement loan is approved by his bank at 5% discounted for 4 years. Determine (a) his monthly payments, (b) the actual cash he has available for improvements, and (c) the actual interest rate.
6. Ms. Tilson borrowed $2200 from her bank on a new car loan to be repaid over 3 years in monthly installments of $68.90 each. If her bank uses the add-on method, determine (a) the stated interest rate, and (b) the actual interest rate.
7. Mr. Brock receives a personal loan from his bank for $800 at $6\frac{1}{2}$% interest discounted for 18 months. Determine (a) his monthly installments, and (b) the actual interest rate.
8. Show that the actual interest rate using the add-on method depends only on the stated interest rate i and the duration of the loan, not on the amount borrowed.
9. Show that the actual interest rate using the discount method depends only on the stated interest rate i and the duration of the loan, not on the amount borrowed.
10. Prove that the add-on method always results in a higher actual interest rate than the discount method. *Hint:* Use the results of Exercises 8 and 9.

B.3 Mortgages and Amortization

Another common annuity is a mortgage, whereby title is first taken on a piece of property which is then paid for in equal monthly installments. In contrast to the installment loans considered in Section B.2, the monthly interest charge on a mortgage is computed anew each month and then only on the current unpaid balance, a method commonly referred to as the *United States rule*. Therefore the stated interest rate is the actual rate.

The main problem with mortgages is to determine the size of the monthly payment. To formulate the problem mathematically, we set

$P(n)$ = the outstanding balance at the end of the nth month. As such, $P(0)$ represents the initial balance or face value of the mortgage.
$I(n)$ = the amount of interest charged during the nth month.
N = the number of months in the life of the mortgage.

As before, R denotes the monthly payment and i represents the interest rate per conversion period, in this case a month.

The balance at the end of any given month is the balance at the end of the previous month plus the interest charges that have accrued during the month minus the monthly installment. That is,

$$P(n) = P(n - 1) + I(n) - R. \qquad (12)$$

The interest charge for the nth month is the monthly interest rate applied to the outstanding balance which is the balance remaining after all the previous month's transactions have been completed. Therefore

$$I(n) = iP(n - 1). \qquad (13)$$

Solving Eqs. (12) and (13), as we do at the end of this section, we obtain

$$\boxed{R = \frac{P(0)}{a_{N \mid i}}.} \qquad (14)$$

Example 1 Mr. Johnson receives a $25,000 mortgage for 20 years at 6% annual interest. Determine his monthly payments.

Solution Twenty years corresponds to $N = 240$ months, and the monthly interest rate is $i = 0.06/12 = 0.005$. Unfortunately, the table in Appendix C.5 is not extensive enough, so we compute directly using Eq. (10)

$$a_{240 \mid 0.005} = \frac{1 - (1.005)^{-240}}{0.005} = \frac{1 - 0.302096}{0.005} = 139.59.$$

Substituting this value into Eq. (14) with the initial balance $P(0) = 25,000$, we find

$R = 25,000/139.59 = \$179.10$.

Example 2 Mr. Kokowski has agreed to sell a small piece of his property to his neighbor Mr. Brown for $5000. They agree that Mr. Brown will pay this amount in monthly payments over the next 4 years at 7% interest. How much will each payment be?

Solution Here $P(0) = 5000$, $i = 0.07/12 = 0.005833$, and $N = 4(12) = 48$. Using Eq. (10), we calculate $a_{\overline{48}|0.005833} = 41.7605$ hence

$R = 5000/41.7605 = \$119.74$.

Mortgages are representative of many loans in which part of each installment is used both to pay interest and to reduce the principal. As the outstanding balance is reduced, the monthly interest charges also decline. Thus a larger portion of the installment is credited against the principal with each payment. Loans that are repaid in this manner are said to be *amortized*; the procedure itself is known as *amortization*.

An *amortization schedule* is a table that shows the amount of each installment, the portion of the payment that goes toward interest charges, the portion of the payment credited against the principal and, finally, the outstanding balance. Note that the periodic interest is given by Eq. (13). Therefore $R - I(n)$ represents the part of the installment that goes toward reducing the principal.

Example 3 A debt of $1000 is to be amortized over 1 year with monthly payments. Determine the size of the installments and construct an amortization schedule for this loan if the interest rate is 5% under the United States rule.

Solution With $N = 12$, $i = 0.05/12 = 0.004167$, and $P(0) = 1000$, we first compute $a_{\overline{12}|0.004167} = 11.6812$ using Eq. (10), and then we compute

$R = 1000/11.6812 = \$85.61$.

Setting $n = 1$ in Eq. (13), we compute the accrued interest for the first month as

$I(1) = iP(0) = (0.004167)(1000) = \$4.17$.

Since part of the installment must be used to cover this interest, we are left with $85.61 - 4.17 = \$81.44$ for credit against the principal. The outstanding balance at the end of the first month is

$P(1) = 1000 - 81.44 = \$918.56$.

Setting $n = 2$ in Eq. (13), we find that the accrued interest for the second month is

$I(2) = iP(1) = (0.004167)(918.56) = \$3.83$.

TABLE B.1

Amortization Schedule for a $1000 Loan over 12 Months at 5% Interest

Payment Number	Amount of Payment, R ($)	Payment on Interest, $I(n)$ ($)	Payment on Principal, $R - I(n)$ ($)	Outstanding Balance, $P(n)$ ($)
0	—	—	—	1000.00
1	85.61	4.17	81.44	918.56
2	85.61	3.83	81.78	836.78
3	85.61	3.49	82.12	754.66
4	85.61	3.15	82.46	672.20
5	85.61	2.81	82.80	589.40
6	85.61	2.46	83.15	506.25
7	85.61	2.11	83.50	422.75
8	85.61	1.77	83.84	338.91
9	85.61	1.42	84.19	254.72
10	85.61	1.07	84.54	170.18
11	85.61	0.71	84.90	85.28
12	85.64	0.36	85.28	0.00

Therefore we credit $85.61 - 3.83 = \$81.78$ against the outstanding balance. The new balance at the end of the second month is

$$P(2) = 918.56 - 81.78 = \$836.78.$$

Continuing in this manner, we generate the amortization schedule given in Table B.1.

Note that the last payment in Table B.1 is 3¢ more than the other installments. This occurs because we rounded each monthly interest charge up to the nearest penny, effectively rounding the monthly payment on the principal down to the nearest penny. Such rounding in turn increased each outstanding monthly balance by a fraction of a cent, which over a 12-month period amounted to 3¢. Actually, had all calculations been carried to six decimal places, the last payment would have been only $85.58, or 3¢ less than the other installments.

In most amortization schedules, the last payment is less than the other payments. The reason is that the monthly installment is rounded up to the nearest penny, which generally results in an overpayment of a fraction of a cent. Over the lifetime of a mortgage, all the fractional overpayments effect a slight reduction in the last installment. A case in point is the mortgage in Example 1, where the actual payment should be $119.7312.

As we see from Table B.1, interest charges decline with each payment. The effects of interest on early payments are much more dramatic, however, if the mortgage has a longer life than the one in the previous example.

Example 4 Calculate the first two lines of an amortization schedule for the mortgage described in Example 1.

Solution We computed the installment for that mortgage as $179.10. Setting $n = 1$ in Eq. (13), we find that the accrued interest for the first month is

$$I(1) = iP(0) = (0.005)(25{,}000) = \$125.00.$$

Therefore $179.10 - 125.00 = \$54.10$ of the first installment is credited toward the principal, leaving an outstanding balance of $P(1) = \$24{,}945.90$.

Setting $n = 2$ in Eq. (13), we compute the interest charge for the second month as

$$P(2) = iP(1) = (0.005)(24{,}945.90) = \$124.73.$$

We credit $179.10 - 124.73 = \$54.37$ toward the principal, leaving an outstanding balance of $P(2) = \$24{,}891.53$.

We now turn to the derivation of Eq. (14). Substituting Eq. (13) into (12), we obtain

$$P(n) = P(n-1) + iP(n-1) - R$$

or

$$P(n) = (1+i)P(n-1) - R. \tag{15}$$

Evaluating Eq. (15) for successive values of the positive integer n, we find

$n = 1 \quad P(1) = (1+i)P(0) - R$
$n = 2 \quad P(2) = (1+i)P(1) - R$
$n = 3 \quad P(3) = (1+i)P(2) - R$
$n = 4 \quad P(4) = (1+i)P(3) - R$, and so on.

It follows from the last set of equations that

$P(1) = (1+i)P(0) - R$
$P(2) = (1+i)P(1) - R = (1+i)[(1+i)P(0) - R] - R$
$\quad = (1+i)^2 P(0) - R[1 + (1+i)]$
$P(3) = (1+i)P(2) - R$
$\quad = (1+i)\{(1+i)^2 P(0) - R[1 + (1+i)]\} - R$
$\quad = (1+i)^3 P(0) - R[1 + (1+i) + (1+i)^2]$
$P(4) = (1+i)P(3)$
$\quad = (1+i)\{(1+i)^3 P(0) - R[1 + (1+i) + (1+i)^2]\} - R$
$\quad = (1+i)^4 P(0) - R[1 + (1+i) + (1+i)^2 + (1+i)^3]$, and so on.

In general,

$$P(n) = (1+i)^n P(0) - R[1 + (1+i) + (1+i)^2 + \cdots + (1+i)^{n-1}] \tag{16}$$

The terms in the brackets are a geometric series with $c = 1$ and $r = (1 + i)$. Using Eq. (2), we obtain

$$1 + (1 + i) + (1 + i)^2 + \cdots + (1 + i)^{n-1} = \frac{1 - (1 + i)^n}{1 - (1 + i)} = \frac{(1 + i)^n - 1}{i},$$

which can be used to simplify Eq. (16) to

$$P(n) = (1 + i)^n P(0) - R\left[\frac{(1 + i)^n - 1}{i}\right]. \tag{17}$$

To find R, we note that, when $n = N$, $P(N) = 0$; that is, after the last payment is made, the outstanding balance is zero. Setting $n = N$ in Eq. (17), we see that

$$0 = (1 + i)^N P(0) - R\left[\frac{(1 + i)^N - 1}{i}\right],$$

from which

$$R = \frac{(1 + i)^N}{\{[(1 + i)^N - 1]/i\}} P(0).$$

Algebraically simplifying this last equation, we obtain Eq. (14).

Exercises

1. Determine the monthly installments for a $36,000 mortgage over 30 years at $7\frac{1}{2}\%$.
2. Determine the monthly installments for a $30,000 mortgage over 25 years at 8%.
3. Mr. O'Toole agrees to sell his business to Mr. Johnson and also to act as the mortgager. The mortgage will be for $8000 over 5 years at 5%.
   (a) Determine the monthly payments.
   (b) Complete the first three lines of an amortization schedule for this transaction.
4. Ms. Tilson agrees to sell some property to a friend and also to act as the mortgager. The mortgage will be for $4500 over 4 years at 4%.
   (a) Determine the monthly installment.
   (b) Complete the first three lines of an amortization schedule for this transaction.
   (c) Using Eq. (17), calculate the outstanding principal after 2 years.
   (d) After 2 years, Ms. Tilson's friend wishes to refinance the outstanding balance over the next 5 years at 6%. What are the new monthly installments?
5. With the aid of a computer, calculate the amortization schedule for the mortgage in Exercise 1.
6. With the aid of a computer, calculate the amortization schedule for the mortgage in Exercise 2.
7. Using Eq. (17), find the outstanding balance after 10 years for the mortgage in Exercise 1.
8. Complete an amortization schedule for a $800 loan to be amortized over 1 year with monthly payments at 6% interest under the United States rule.

Appendix C

Tables

TABLE C.1
Common logarithms

x	0	1	2	3	4	5	6	7	8	9
1.0	0000	0043	0086	0128	0170	0212	0253	0294	0334	0374
1.1	0414	0453	0492	0531	0569	0607	0645	0682	0719	0755
1.2	0792	0828	0864	0899	0934	0969	1004	1038	1072	1106
1.3	1139	1173	1206	1239	1271	1303	1335	1367	1399	1430
1.4	1461	1492	1523	1553	1584	1614	1644	1673	1703	1732
1.5	1761	1790	1818	1847	1875	1903	1931	1959	1987	2014
1.6	2041	2068	2095	2122	2148	2175	2201	2227	2253	2279
1.7	2304	2330	2355	2380	2405	2430	2455	2480	2504	2529
1.8	2553	2577	2601	2625	2648	2672	2695	2718	2742	2765
1.9	2788	2810	2833	2856	2878	2900	2923	2945	2967	2989
2.0	3010	3032	3054	3075	3096	3118	3139	3160	3181	3201
2.1	3222	3243	3263	3284	3304	3324	3345	3365	3385	3404
2.2	3424	3444	3464	3483	3502	3522	3541	3560	3579	3598
2.3	3617	3636	3655	3674	3692	3711	3729	3747	3766	3784
2.4	3802	3820	3838	3856	3874	3892	3909	3927	3945	3962
2.5	3979	3997	4014	4031	4048	4065	4082	4099	4116	4133
2.6	4150	4166	4183	4200	4216	4232	4249	4265	4281	4298
2.7	4314	4330	4346	4362	4378	4393	4409	4425	4440	4456
2.8	4472	4487	4502	4518	4533	4548	4564	4579	4594	4609
2.9	4624	4639	4654	4669	4683	4698	4713	4728	4742	4757
3.0	4771	4786	4800	4814	4829	4843	4857	4871	4886	4900
3.1	4914	4928	4942	4955	4969	4983	4997	5011	5024	5038
3.2	5051	5065	5079	5092	5105	5119	5132	5145	5159	5172
3.3	5185	5198	5211	5224	5237	5250	5263	5276	5289	5302
3.4	5315	5328	5340	5353	5366	5378	5391	5403	5416	5428
3.5	5441	5453	5465	5478	5490	5502	5514	5527	5539	5551
3.6	5563	5575	5587	5599	5611	5623	5635	5647	5658	5670
3.7	5682	5694	5705	5717	5729	5740	5752	5763	5775	5786
3.8	5798	5809	5821	5832	5843	5855	5866	5877	5888	5899
3.9	5911	5922	5933	5944	5955	5966	5977	5988	5999	6010
4.0	6021	6031	6042	6053	6064	6075	6085	6096	6107	6117
4.1	6128	6138	6149	6160	6170	6180	6191	6201	6212	6222
4.2	6232	6243	6253	6263	6274	6284	6294	6304	6314	6325
4.3	6335	6345	6355	6365	6375	6385	6395	6405	6415	6425
4.4	6435	6444	6454	6464	6474	6484	6493	6503	6513	6522
4.5	6532	6542	6551	6561	6571	6580	6590	6599	6609	6618
4.6	6628	6637	6646	6656	6665	6675	6684	6693	6702	6712
4.7	6721	6730	6739	6749	6758	6767	6776	6785	6794	6803
4.8	6812	6821	6830	6839	6848	6857	6866	6875	6884	6893
4.9	6902	6911	6920	6928	6937	6946	6955	6964	6972	6981
5.0	6990	6998	7007	7016	7024	7033	7042	7050	7059	7067
5.1	7076	7084	7093	7101	7110	7118	7126	7135	7143	7152
5.2	7160	7168	7177	7185	7193	7202	7210	7218	7226	7235
5.3	7243	7251	7259	7267	7275	7284	7292	7300	7308	7316
5.4	7324	7332	7340	7348	7356	7364	7372	7380	7388	7396
x	0	1	2	3	4	5	6	7	8	9

TABLE C.1 (continued)

x	0	1	2	3	4	5	6	7	8	9
5.5	7404	7412	7419	7427	7435	7443	7451	7459	7466	7474
5.6	7482	7490	7497	7505	7513	7520	7528	7536	7543	7551
5.7	7559	7566	7574	7582	7589	7597	7604	7612	7619	7627
5.8	7634	7642	7649	7657	7664	7672	7679	7686	7694	7701
5.9	7709	7716	7723	7731	7738	7745	7752	7760	7767	7774
6.0	7782	7789	7796	7803	7810	7818	7825	7832	7839	7846
6.1	7853	7860	7868	7875	7882	7889	7896	7903	7910	7917
6.2	7924	7931	7938	7945	7952	7959	7966	7973	7980	7987
6.3	7993	8000	8007	8014	8021	8028	8035	8041	8048	8055
6.4	8062	8069	8075	8082	8089	8096	8102	8109	8116	8122
6.5	8129	8136	8142	8149	8156	8162	8169	8176	8182	8189
6.6	8195	8202	8209	8215	8222	8228	8235	8241	8248	8254
6.7	8261	8267	8274	8280	8287	8293	8299	8306	8312	8319
6.8	8325	8331	8338	8344	8351	8357	8363	8370	8376	8382
6.9	8388	8395	8401	8407	8414	8420	8426	8432	8439	8445
7.0	8451	8457	8463	8470	8476	8482	8488	8494	8500	8506
7.1	8513	8519	8525	8531	8537	8543	8549	8555	8561	8567
7.2	8573	8579	8585	8591	8597	8603	8609	8615	8621	8627
7.3	8633	8639	8645	8651	8657	8663	8669	8675	8681	8686
7.4	8692	8698	8704	8710	8716	8722	8727	8733	8739	8745
7.5	8751	8756	8762	8768	8774	8779	8785	8791	8797	8802
7.6	8808	8814	8820	8825	8831	8837	8842	8848	8854	8859
7.7	8865	8871	8876	8882	8887	8893	8899	8904	8910	8915
7.8	8921	8927	8932	8938	8943	8949	8954	8960	8965	8971
7.9	8976	8982	8987	8993	8998	9004	9009	9015	9020	9025
8.0	9031	9036	9042	9047	9053	9058	9063	9069	9074	9079
8.1	9085	9090	9096	9101	9106	9112	9117	9122	9128	9133
8.2	9138	9143	9149	9154	9159	9165	9170	9175	9180	9186
8.3	9191	9196	9201	9206	9212	9217	9222	9227	9232	9238
8.4	9243	9248	9253	9258	9263	9269	9274	9279	9284	9289
8.5	9294	9299	9304	9309	9315	9320	9325	9330	9335	9340
8.6	9345	9350	9355	9360	9365	9370	9375	9380	9385	9390
8.7	9395	9400	9405	9410	9415	9420	9425	9430	9435	9440
8.8	9445	9450	9455	9460	9465	9469	9474	9479	9484	9489
8.9	9494	9499	9504	9509	9513	9518	9523	9528	9533	9538
9.0	9542	9547	9552	9557	9562	9566	9571	9576	9581	9586
9.1	9590	9595	9600	9605	9609	9614	9619	9624	9628	9633
9.2	9638	9643	9647	9652	9657	9661	9666	9671	9675	9680
9.3	9685	9689	9694	9699	9703	9708	9713	9717	9722	9727
9.4	9731	9736	9741	9745	9750	9754	9759	9763	9768	9773
9.5	9777	9782	9786	9791	9795	9800	9805	9809	9814	9818
9.6	9823	9827	9832	9836	9841	9845	9850	9854	9859	9863
9.7	9868	9872	9877	9881	9886	9890	9894	9899	9903	9908
9.8	9912	9917	9921	9926	9930	9934	9939	9943	9948	9952
9.9	9956	9961	9965	9969	9974	9978	9983	9987	9991	9996
x	0	1	2	3	4	5	6	7	8	9

TABLE C.2
Exponential functions

x	e^x	e^{-x}	x	e^x	e^{-x}
0.00	1.00000	1.00000	1.50	4.48169	0.22313
0.01	1.01005	0.99005	1.60	4.95303	0.20190
0.02	1.02020	0.98020	1.70	5.47395	0.18268
0.03	1.03045	0.97045	1.80	6.04965	0.16530
0.04	1.04081	0.96079	1.90	6.68589	0.14957
0.05	1.05127	0.95123	2.00	7.38906	0.13534
0.06	1.06184	0.94176	2.10	8.16617	0.12246
0.07	1.07251	0.93239	2.20	9.02501	0.11080
0.08	1.08329	0.92312	2.30	9.97418	0.10026
0.09	1.09417	0.91393	2.40	11.02318	0.09072
0.10	1.10517	0.90484	2.50	12.18249	0.08208
0.11	1.11628	0.89583	2.60	13.46374	0.07427
0.12	1.12750	0.88692	2.70	14.87973	0.06721
0.13	1.13883	0.87810	2.80	16.44465	0.06081
0.14	1.15027	0.86936	2.90	18.17415	0.05502
0.15	1.16183	0.86071	3.00	20.08554	0.04979
0.16	1.17351	0.85214	3.10	22.19795	0.04505
0.17	1.18530	0.84366	3.20	24.53253	0.04076
0.18	1.19722	0.83527	3.30	27.11264	0.03688
0.19	1.20925	0.82696	3.40	29.96410	0.03337
0.20	1.22140	0.81873	3.50	33.11545	0.03020
0.21	1.23368	0.81058	3.60	36.59823	0.02732
0.22	1.24608	0.80252	3.70	40.44730	0.02472
0.23	1.25860	0.79453	3.80	44.70118	0.02237
0.24	1.27125	0.78663	3.90	49.40245	0.02024
0.25	1.28403	0.77880	4.00	54.59815	0.01832
0.30	1.34986	0.74082	4.10	60.34029	0.01657
0.35	1.41907	0.70469	4.20	66.68633	0.01500
0.40	1.49182	0.67032	4.30	73.69979	0.01357
0.45	1.56831	0.63763	4.40	81.45087	0.01228
0.50	1.64872	0.60653	4.50	90.01713	0.01111
0.55	1.73325	0.57695	4.60	99.48432	0.01005
0.60	1.82212	0.54881	4.70	109.94717	0.00910
0.65	1.91554	0.52205	4.80	121.51042	0.00823
0.70	2.01375	0.49659	4.90	134.28978	0.00745
0.75	2.11700	0.47237	5.00	148.41316	0.00674
0.80	2.22554	0.44933	5.50	244.69193	0.00409
0.85	2.33965	0.42741	6.00	403.42879	0.00248
0.90	2.45960	0.40657	6.50	665.14163	0.00150
0.95	2.58571	0.38674	7.00	1096.63316	0.00091
1.00	2.71828	0.36788	7.50	1808.04241	0.00055
1.10	3.00417	0.33287	8.00	2980.95799	0.00034
1.20	3.32012	0.30119	8.50	4914.76884	0.00020
1.30	3.66930	0.27253	9.00	8103.08393	0.00012
1.40	4.05520	0.24660	10.00	22026.46579	0.00005

TABLE C.3
Values of $(1 + i)^n$

N	I=.0025	I=.0050	I=.0075	I=.0100	I=.0125
1	1.002500	1.005000	1.007500	1.010000	1.012500
2	1.005006	1.010025	1.015056	1.020100	1.025156
3	1.007519	1.015075	1.022669	1.030301	1.037971
4	1.010038	1.020151	1.030339	1.040604	1.050945
5	1.012563	1.025251	1.038067	1.051010	1.064082
6	1.015094	1.030378	1.045852	1.061520	1.077383
7	1.017632	1.035529	1.053696	1.072135	1.090850
8	1.020176	1.040707	1.061599	1.082857	1.104486
9	1.022726	1.045911	1.069561	1.093685	1.118292
10	1.025283	1.051140	1.077583	1.104622	1.132271
11	1.027846	1.056396	1.085664	1.115668	1.146424
12	1.030416	1.061678	1.093807	1.126825	1.160755
13	1.032992	1.066986	1.102010	1.138093	1.175264
14	1.035574	1.072321	1.110276	1.149474	1.189955
15	1.038163	1.077683	1.118603	1.160969	1.204829
16	1.040759	1.083071	1.126992	1.172579	1.219890
17	1.043361	1.088487	1.135445	1.184304	1.235138
18	1.045969	1.093929	1.143960	1.196147	1.250577
19	1.048584	1.099399	1.152540	1.208109	1.266210
20	1.051206	1.104896	1.161184	1.220190	1.282037
21	1.053834	1.110420	1.169893	1.232392	1.298063
22	1.056468	1.115972	1.178667	1.244716	1.314288
23	1.059109	1.121552	1.187507	1.257163	1.330717
24	1.061757	1.127160	1.196414	1.269735	1.347351
25	1.064411	1.132796	1.205387	1.282432	1.364193
26	1.067072	1.138460	1.214427	1.295256	1.381245
27	1.069740	1.144152	1.223535	1.308209	1.398511
28	1.072414	1.149873	1.232712	1.321291	1.415992
29	1.075096	1.155622	1.241957	1.334504	1.433692
30	1.077783	1.161400	1.251272	1.347849	1.451613
31	1.080478	1.167207	1.260656	1.361327	1.469759
32	1.083179	1.173043	1.270111	1.374941	1.488131
33	1.085887	1.178908	1.279637	1.388690	1.506732
34	1.088602	1.184803	1.289234	1.402577	1.525566
35	1.091323	1.190727	1.298904	1.416603	1.544636
36	1.094051	1.196681	1.308645	1.430769	1.563944
37	1.096787	1.202664	1.318460	1.445076	1.583493
38	1.099528	1.208677	1.328349	1.459527	1.603287
39	1.102277	1.214721	1.338311	1.474123	1.623328
40	1.105033	1.220794	1.348349	1.488864	1.643619
41	1.107796	1.226898	1.358461	1.503752	1.664165
42	1.110565	1.233033	1.368650	1.518790	1.684967
43	1.113341	1.239198	1.378915	1.533978	1.706029
44	1.116125	1.245394	1.389256	1.549318	1.727354
45	1.118915	1.251621	1.399676	1.564811	1.748946
46	1.121712	1.257879	1.410173	1.580459	1.770808
47	1.124517	1.264168	1.420750	1.596263	1.792943
48	1.127328	1.270489	1.431405	1.612226	1.815355
49	1.130146	1.276842	1.442141	1.628348	1.838047
50	1.132972	1.283226	1.452957	1.644632	1.861022
51	1.135804	1.289642	1.463854	1.661078	1.884285
52	1.138644	1.296090	1.474833	1.677689	1.907839
53	1.141490	1.302571	1.485894	1.694466	1.931687
54	1.144344	1.309083	1.497038	1.711410	1.955833
55	1.147205	1.315629	1.508266	1.728525	1.980281
56	1.150073	1.322207	1.519578	1.745810	2.005034
57	1.152948	1.328818	1.530975	1.763268	2.030097
58	1.155830	1.335462	1.542457	1.780901	2.055473
59	1.158720	1.342139	1.554026	1.798710	2.081167
60	1.161617	1.348850	1.565681	1.816697	2.107181

TABLE C.3 (continued)

N	I=.0150	I=.0175	I=.0200	I=.0250	I=.0300
1	1.015000	1.017500	1.020000	1.025000	1.030000
2	1.030225	1.035306	1.040400	1.050625	1.060900
3	1.045678	1.053424	1.061208	1.076891	1.092727
4	1.061364	1.071859	1.082432	1.103813	1.125509
5	1.077284	1.090617	1.104081	1.131408	1.159274
6	1.093443	1.109702	1.126162	1.159693	1.194052
7	1.109845	1.129122	1.148686	1.188686	1.229874
8	1.126493	1.148882	1.171659	1.218403	1.266770
9	1.143390	1.168987	1.195093	1.248863	1.304773
10	1.160541	1.189444	1.218994	1.280085	1.343916
11	1.177949	1.210260	1.243374	1.312087	1.384234
12	1.195618	1.231439	1.268242	1.344889	1.425761
13	1.213552	1.252990	1.293607	1.378511	1.468534
14	1.231756	1.274917	1.319479	1.412974	1.512590
15	1.250232	1.297228	1.345868	1.448298	1.557967
16	1.268986	1.319929	1.372786	1.484506	1.604706
17	1.288020	1.343028	1.400241	1.521618	1.652848
18	1.307341	1.366531	1.428246	1.559659	1.702433
19	1.326951	1.390445	1.456811	1.598650	1.753506
20	1.346855	1.414778	1.485947	1.638616	1.806111
21	1.367058	1.439537	1.515666	1.679582	1.860295
22	1.387564	1.464729	1.545980	1.721571	1.916103
23	1.408377	1.490361	1.576899	1.764611	1.973587
24	1.429503	1.516443	1.608437	1.808726	2.032794
25	1.450945	1.542981	1.640606	1.853944	2.093778
26	1.472710	1.569983	1.673418	1.900293	2.156591
27	1.494800	1.597457	1.706886	1.947800	2.221289
28	1.517222	1.625413	1.741024	1.996495	2.287928
29	1.539981	1.653858	1.775845	2.046407	2.356566
30	1.563080	1.682800	1.811362	2.097568	2.427262
31	1.586526	1.712249	1.847589	2.150007	2.500080
32	1.610324	1.742213	1.884541	2.203757	2.575083
33	1.634479	1.772702	1.922231	2.258851	2.652335
34	1.658996	1.803725	1.960676	2.315322	2.731905
35	1.683881	1.835290	1.999890	2.373205	2.813862
36	1.709140	1.867407	2.039887	2.432535	2.898278
37	1.734777	1.900087	2.080685	2.493349	2.985227
38	1.760798	1.933338	2.122299	2.555682	3.074783
39	1.787210	1.967172	2.164745	2.619574	3.167027
40	1.814018	2.001597	2.208040	2.685064	3.262038
41	1.841229	2.036625	2.252200	2.752190	3.359899
42	1.868847	2.072266	2.297244	2.820995	3.460696
43	1.896880	2.108531	2.343189	2.891520	3.564517
44	1.925333	2.145430	2.390053	2.963808	3.671452
45	1.954213	2.182975	2.437854	3.037903	3.781596
46	1.983526	2.221177	2.486611	3.113851	3.895044
47	2.013279	2.260048	2.536344	3.191697	4.011895
48	2.043478	2.299599	2.587070	3.271490	4.132252
49	2.074130	2.339842	2.638812	3.353277	4.256219
50	2.105242	2.380789	2.691588	3.437109	4.383906
51	2.136821	2.422453	2.745420	3.523036	4.515423
52	2.168873	2.464846	2.800328	3.611112	4.650886
53	2.201406	2.507980	2.856335	3.701390	4.790412
54	2.234428	2.551870	2.913461	3.793925	4.934125
55	2.267944	2.596528	2.971731	3.888773	5.082149
56	2.301963	2.641967	3.031165	3.985992	5.234613
57	2.336493	2.688202	3.091789	4.085642	5.391651
58	2.371540	2.735245	3.153624	4.187783	5.553401
59	2.407113	2.783112	3.216697	4.292478	5.720003
60	2.443220	2.831816	3.281031	4.399790	5.891603

TABLE C.3 (continued)

N	I=.0350	I=.0400	I=.0450	I=.0500	I=.0550
1	1.035000	1.040000	1.045000	1.050000	1.055000
2	1.071225	1.081600	1.092025	1.102500	1.113025
3	1.108718	1.124864	1.141166	1.157625	1.174241
4	1.147523	1.169859	1.192519	1.215506	1.238825
5	1.187686	1.216653	1.246182	1.276282	1.306960
6	1.229255	1.265319	1.302260	1.340096	1.378843
7	1.272279	1.315932	1.360862	1.407100	1.454679
8	1.316809	1.368569	1.422101	1.477455	1.534687
9	1.362897	1.423312	1.486095	1.551328	1.619094
10	1.410599	1.480244	1.552969	1.628895	1.708144
11	1.459970	1.539454	1.622853	1.710339	1.802092
12	1.511069	1.601032	1.695881	1.795856	1.901207
13	1.563956	1.665074	1.772196	1.885649	2.005774
14	1.618695	1.731676	1.851945	1.979932	2.116091
15	1.675349	1.800944	1.935282	2.078928	2.232476
16	1.733986	1.872981	2.022370	2.182875	2.355263
17	1.794676	1.947900	2.113377	2.292018	2.484802
18	1.857489	2.025817	2.208479	2.406619	2.621466
19	1.922501	2.106849	2.307860	2.526950	2.765647
20	1.989789	2.191123	2.411714	2.653298	2.917757
21	2.059431	2.278768	2.520241	2.785963	3.078234
22	2.131512	2.369919	2.633652	2.925261	3.247537
23	2.206114	2.464716	2.752166	3.071524	3.426152
24	2.283328	2.563304	2.876014	3.225100	3.614590
25	2.363245	2.665836	3.005434	3.386355	3.813392
26	2.445959	2.772470	3.140679	3.555673	4.023129
27	2.531567	2.883369	3.282010	3.733456	4.244401
28	2.620172	2.998703	3.429700	3.920129	4.477843
29	2.711878	3.118651	3.584036	4.116136	4.724124
30	2.806794	3.243398	3.745318	4.321942	4.983951
31	2.905031	3.373133	3.913857	4.538039	5.258069
32	3.006708	3.508059	4.089981	4.764941	5.547262
33	3.111942	3.648381	4.274030	5.003189	5.852362
34	3.220860	3.794316	4.466362	5.253348	6.174242
35	3.333590	3.946089	4.667348	5.516015	6.513825
36	3.450266	4.103933	4.877378	5.791816	6.872085
37	3.571025	4.268090	5.096860	6.081407	7.250050
38	3.696011	4.438813	5.326219	6.385477	7.648803
39	3.825372	4.616366	5.565899	6.704751	8.069487
40	3.959260	4.801021	5.816365	7.039989	8.513309
41	4.097834	4.993061	6.078101	7.391988	8.981541
42	4.241258	5.192784	6.351615	7.761588	9.475525
43	4.389702	5.400495	6.637438	8.149667	9.996679
44	4.543342	5.616515	6.936123	8.557150	10.546497
45	4.702359	5.841176	7.248248	8.985008	11.126554
46	4.866941	6.074823	7.574420	9.434258	11.738515
47	5.037284	6.317816	7.915268	9.905971	12.384133
48	5.213589	6.570528	8.271456	10.401270	13.065260
49	5.396065	6.833349	8.643671	10.921333	13.783849
50	5.584927	7.106683	9.032636	11.467400	14.541961
51	5.780399	7.390951	9.439105	12.040770	15.341769
52	5.982713	7.686589	9.863865	12.642808	16.185566
53	6.192108	7.994052	10.307739	13.274949	17.075773
54	6.408832	8.313814	10.771587	13.938696	18.014940
55	6.633141	8.646367	11.256308	14.635631	19.005762
56	6.865301	8.992222	11.762842	15.367412	20.051079
57	7.105587	9.351910	12.292170	16.135783	21.153888
58	7.354282	9.725987	12.845318	16.942572	22.317352
59	7.611682	10.115026	13.423357	17.789701	23.544806
60	7.878091	10.519627	14.027408	18.679186	24.839770

448 Appendix C: Tables

TABLE C.3 (continued)

N	I=.0600	I=.0650	I=.0700	I=.0750	I=.0800
1	1.060000	1.065000	1.070000	1.075000	1.080000
2	1.123600	1.134225	1.144900	1.155625	1.166400
3	1.191016	1.207950	1.225043	1.242297	1.259712
4	1.262477	1.286466	1.310796	1.335469	1.360489
5	1.338226	1.370087	1.402552	1.435629	1.469328
6	1.418519	1.459142	1.500730	1.543302	1.586874
7	1.503630	1.553987	1.605781	1.659049	1.713824
8	1.593848	1.654996	1.718186	1.783478	1.850930
9	1.689479	1.762570	1.838459	1.917239	1.999005
10	1.790848	1.877137	1.967151	2.061032	2.158925
11	1.898299	1.999151	2.104852	2.215609	2.331639
12	2.012196	2.129096	2.252192	2.381780	2.518170
13	2.132928	2.267487	2.409845	2.560413	2.719624
14	2.260904	2.414874	2.578534	2.752444	2.937194
15	2.396558	2.571841	2.759032	2.958877	3.172169
16	2.540352	2.739011	2.952164	3.180793	3.425943
17	2.692773	2.917046	3.158815	3.419353	3.700018
18	2.854339	3.106654	3.379932	3.675804	3.996019
19	3.025600	3.308587	3.616528	3.951489	4.315701
20	3.207135	3.523645	3.869684	4.247851	4.660957
21	3.399564	3.752682	4.140562	4.566440	5.033834
22	3.603537	3.996606	4.430402	4.908923	5.436540
23	3.819750	4.256386	4.740530	5.277092	5.871464
24	4.048935	4.533051	5.072367	5.672874	6.341181
25	4.291871	4.827699	5.427433	6.098340	6.848475
26	4.549383	5.141500	5.807353	6.555715	7.396353
27	4.822346	5.475697	6.213868	7.047394	7.988061
28	5.111687	5.831617	6.648838	7.575948	8.627106
29	5.418388	6.210672	7.114257	8.144144	9.317275
30	5.743491	6.614366	7.612255	8.754955	10.062657
31	6.088101	7.044300	8.145113	9.411577	10.867669
32	6.453387	7.502179	8.715271	10.117445	11.737083
33	6.840590	7.989821	9.325340	10.876253	12.676050
34	7.251025	8.509159	9.978114	11.691972	13.690134
35	7.686087	9.062255	10.676581	12.568870	14.785344
36	8.147252	9.651301	11.423942	13.511536	15.968172
37	8.636087	10.278636	12.223618	14.524901	17.245626
38	9.154252	10.946747	13.079271	15.614268	18.625276
39	9.703507	11.658286	13.994820	16.785339	20.115298
40	10.285718	12.416075	14.974458	18.044239	21.724521
41	10.902861	13.223119	16.022670	19.397557	23.462483
42	11.557033	14.082622	17.144257	20.852374	25.339482
43	12.250455	14.997993	18.344355	22.416302	27.366640
44	12.985482	15.972862	19.628460	24.097524	29.555972
45	13.764611	17.011098	21.002452	25.904839	31.920449
46	14.590487	18.116820	22.472623	27.847702	34.474085
47	15.465917	19.294413	24.045707	29.936279	37.232012
48	16.393872	20.548550	25.728907	32.181500	40.210573
49	17.377504	21.884205	27.529930	34.595113	43.427419
50	18.420154	23.306679	29.457025	37.189746	46.901613
51	19.525364	24.821613	31.519017	39.978977	50.653742
52	20.696885	26.435018	33.725348	42.977400	54.706041
53	21.938698	28.153294	36.086122	46.200705	59.082524
54	23.255020	29.983258	38.612151	49.665758	63.809126
55	24.650322	31.932170	41.315001	53.390690	68.913856
56	26.129341	34.007761	44.207052	57.394992	74.426965
57	27.697101	36.218265	47.301545	61.699616	80.381122
58	29.358927	38.572452	50.612653	66.327087	86.811612
59	31.120463	41.079662	54.155539	71.301619	93.756540
60	32.987691	43.749840	57.946427	76.649240	101.257064

TABLE C.4
Values of $(1 + i)^{-n}$

N	I=.0025	I=.0050	I=.0075	I=.0100	I=.0125
1	0.997506	0.995025	0.992556	0.990099	0.987654
2	0.995019	0.990075	0.985167	0.980296	0.975461
3	0.992537	0.985149	0.977833	0.970590	0.963418
4	0.990062	0.980248	0.970554	0.960980	0.951524
5	0.987593	0.975371	0.963329	0.951466	0.939777
6	0.985130	0.970518	0.956158	0.942045	0.928175
7	0.982674	0.965690	0.949040	0.932718	0.916716
8	0.980223	0.960885	0.941975	0.923483	0.905398
9	0.977779	0.956105	0.934963	0.914340	0.894221
10	0.975340	0.951348	0.928003	0.905287	0.883181
11	0.972908	0.946615	0.921095	0.896324	0.872277
12	0.970482	0.941905	0.914238	0.887449	0.861509
13	0.968062	0.937219	0.907432	0.878663	0.850873
14	0.965648	0.932556	0.900677	0.869963	0.840368
15	0.963239	0.927917	0.893973	0.861349	0.829993
16	0.960837	0.923300	0.887318	0.852821	0.819746
17	0.958441	0.918707	0.880712	0.844377	0.809626
18	0.956051	0.914136	0.874156	0.836017	0.799631
19	0.953667	0.909588	0.867649	0.827740	0.789759
20	0.951289	0.905063	0.861190	0.819544	0.780009
21	0.948916	0.900560	0.854779	0.811430	0.770379
22	0.946550	0.896080	0.848416	0.803396	0.760868
23	0.944190	0.891622	0.842100	0.795442	0.751475
24	0.941835	0.887186	0.835831	0.787566	0.742197
25	0.939486	0.882772	0.829609	0.779768	0.733034
26	0.937143	0.878380	0.823434	0.772048	0.723984
27	0.934806	0.874010	0.817304	0.764404	0.715046
28	0.932475	0.869662	0.811220	0.756836	0.706219
29	0.930150	0.865335	0.805181	0.749342	0.697500
30	0.927830	0.861030	0.799187	0.741923	0.688889
31	0.925517	0.856746	0.793238	0.734577	0.680384
32	0.923209	0.852484	0.787333	0.727304	0.671984
33	0.920906	0.848242	0.781472	0.720103	0.663688
34	0.918610	0.844022	0.775654	0.712973	0.655494
35	0.916319	0.839823	0.769880	0.705914	0.647402
36	0.914034	0.835645	0.764149	0.698925	0.639409
37	0.911754	0.831487	0.758461	0.692005	0.631515
38	0.909481	0.827351	0.752814	0.685153	0.623719
39	0.907213	0.823235	0.747210	0.678370	0.616019
40	0.904950	0.819139	0.741648	0.671653	0.608413
41	0.902694	0.815064	0.736127	0.665003	0.600902
42	0.900443	0.811009	0.730647	0.658419	0.593484
43	0.898197	0.806974	0.725208	0.651900	0.586157
44	0.895957	0.802959	0.719810	0.645445	0.578920
45	0.893723	0.798964	0.714451	0.639055	0.571773
46	0.891494	0.794989	0.709133	0.632728	0.564714
47	0.889271	0.791034	0.703854	0.626463	0.557742
48	0.887053	0.787098	0.698614	0.620260	0.550856
49	0.884841	0.783182	0.693414	0.614119	0.544056
50	0.882635	0.779286	0.688252	0.608039	0.537339
51	0.880433	0.775409	0.683128	0.602019	0.530705
52	0.878238	0.771551	0.678043	0.596058	0.524153
53	0.876048	0.767713	0.672995	0.590156	0.517682
54	0.873863	0.763893	0.667986	0.584313	0.511291
55	0.871684	0.760093	0.663013	0.578528	0.504979
56	0.869510	0.756311	0.658077	0.572800	0.498745
57	0.867342	0.752548	0.653178	0.567129	0.492587
58	0.865179	0.748804	0.648316	0.561514	0.486506
59	0.863021	0.745079	0.643490	0.555954	0.480500
60	0.860869	0.741372	0.638700	0.550450	0.474568

TABLE C.4 (continued)

N	I=.0150	I=.0175	I=.0200	I=.0250	I=.0300
1	0.985222	0.982801	0.980392	0.975610	0.970874
2	0.970662	0.965898	0.961169	0.951814	0.942596
3	0.956317	0.949285	0.942322	0.928599	0.915142
4	0.942184	0.932959	0.923845	0.905951	0.888487
5	0.928260	0.916913	0.905731	0.883854	0.862609
6	0.914542	0.901143	0.887971	0.862297	0.837484
7	0.901027	0.885644	0.870560	0.841265	0.813092
8	0.887711	0.870412	0.853490	0.820747	0.789409
9	0.874592	0.855441	0.836755	0.800728	0.766417
10	0.861667	0.840729	0.820348	0.781198	0.744094
11	0.848933	0.826269	0.804263	0.762145	0.722421
12	0.836387	0.812058	0.788493	0.743556	0.701380
13	0.824027	0.798091	0.773033	0.725420	0.680951
14	0.811849	0.784365	0.757875	0.707727	0.661118
15	0.799852	0.770875	0.743015	0.690466	0.641862
16	0.788031	0.757616	0.728446	0.673625	0.623167
17	0.776385	0.744586	0.714163	0.657195	0.605016
18	0.764912	0.731780	0.700159	0.641166	0.587395
19	0.753607	0.719194	0.686431	0.625528	0.570286
20	0.742470	0.706825	0.672971	0.610271	0.553676
21	0.731498	0.694668	0.659776	0.595386	0.537549
22	0.720688	0.682720	0.646839	0.580865	0.521893
23	0.710037	0.670978	0.634156	0.566697	0.506692
24	0.699544	0.659438	0.621721	0.552875	0.491934
25	0.689206	0.648096	0.609531	0.539391	0.477606
26	0.679021	0.636950	0.597579	0.526235	0.463695
27	0.668986	0.625995	0.585862	0.513400	0.450189
28	0.659099	0.615228	0.574375	0.500878	0.437077
29	0.649359	0.604647	0.563112	0.488661	0.424346
30	0.639762	0.594248	0.552071	0.476743	0.411987
31	0.630308	0.584027	0.541246	0.465115	0.399987
32	0.620993	0.573982	0.530633	0.453771	0.388337
33	0.611816	0.564111	0.520229	0.442703	0.377026
34	0.602774	0.554408	0.510028	0.431905	0.366045
35	0.593866	0.544873	0.500028	0.421371	0.355383
36	0.585090	0.535502	0.490223	0.411094	0.345032
37	0.576443	0.526292	0.480611	0.401067	0.334983
38	0.567924	0.517240	0.471187	0.391285	0.325226
39	0.559531	0.508344	0.461948	0.381741	0.315754
40	0.551262	0.499601	0.452890	0.372431	0.306557
41	0.543116	0.491008	0.444010	0.363347	0.297628
42	0.535089	0.482563	0.435304	0.354485	0.288959
43	0.527182	0.474264	0.426769	0.345839	0.280543
44	0.519391	0.466107	0.418401	0.337404	0.272372
45	0.511715	0.458090	0.410197	0.329174	0.264439
46	0.504153	0.450212	0.402154	0.321146	0.256737
47	0.496702	0.442469	0.394268	0.313313	0.249259
48	0.489362	0.434858	0.386538	0.305671	0.241999
49	0.482130	0.427379	0.378958	0.298216	0.234950
50	0.475005	0.420029	0.371528	0.290942	0.228107
51	0.467985	0.412805	0.364243	0.283846	0.221463
52	0.461069	0.405705	0.357101	0.276923	0.215013
53	0.454255	0.398727	0.350099	0.270169	0.208750
54	0.447542	0.391869	0.343234	0.263579	0.202670
55	0.440928	0.385130	0.336504	0.257151	0.196767
56	0.434412	0.378506	0.329906	0.250879	0.191036
57	0.427992	0.371996	0.323437	0.244760	0.185472
58	0.421667	0.365598	0.317095	0.238790	0.180070
59	0.415435	0.359310	0.310878	0.232966	0.174825
60	0.409296	0.353130	0.304782	0.227284	0.169733

TABLE C.4 (continued)

Table C.4 Values of $(1+i)^{-n}$

N	I=.0350	I=.0400	I=.0450	I=.0500	I=.0550
1	0.966184	0.961538	0.956938	0.952381	0.947867
2	0.933511	0.924556	0.915730	0.907029	0.898452
3	0.901943	0.888996	0.876297	0.863838	0.851614
4	0.871442	0.854804	0.838561	0.822702	0.807217
5	0.841973	0.821927	0.802451	0.783526	0.765134
6	0.813501	0.790315	0.767896	0.746215	0.725246
7	0.785991	0.759918	0.734828	0.710681	0.687437
8	0.759412	0.730690	0.703185	0.676839	0.651599
9	0.733731	0.702587	0.672904	0.644609	0.617629
10	0.708919	0.675564	0.643928	0.613913	0.585431
11	0.684946	0.649581	0.616199	0.584679	0.554911
12	0.661783	0.624597	0.589664	0.556837	0.525982
13	0.639404	0.600574	0.564272	0.530321	0.498561
14	0.617782	0.577475	0.539973	0.505068	0.472569
15	0.596891	0.555265	0.516720	0.481017	0.447933
16	0.576706	0.533908	0.494469	0.458112	0.424581
17	0.557204	0.513373	0.473176	0.436297	0.402447
18	0.538361	0.493628	0.452800	0.415521	0.381466
19	0.520156	0.474642	0.433302	0.395734	0.361579
20	0.502566	0.456387	0.414643	0.376889	0.342729
21	0.485571	0.438834	0.396787	0.358942	0.324862
22	0.469151	0.421955	0.379701	0.341850	0.307926
23	0.453286	0.405726	0.363350	0.325571	0.291873
24	0.437957	0.390121	0.347703	0.310068	0.276657
25	0.423147	0.375117	0.332731	0.295303	0.262234
26	0.408838	0.360689	0.318402	0.281241	0.248563
27	0.395012	0.346817	0.304691	0.267848	0.235605
28	0.381654	0.333477	0.291571	0.255094	0.223322
29	0.368748	0.320651	0.279015	0.242946	0.211679
30	0.356278	0.308319	0.267000	0.231377	0.200644
31	0.344230	0.296460	0.255502	0.220359	0.190184
32	0.332590	0.285058	0.244500	0.209866	0.180269
33	0.321343	0.274094	0.233971	0.199873	0.170871
34	0.310476	0.263552	0.223896	0.190355	0.161963
35	0.299977	0.253415	0.214254	0.181290	0.153520
36	0.289833	0.243669	0.205028	0.172657	0.145516
37	0.280032	0.234297	0.196199	0.164436	0.137930
38	0.270562	0.225285	0.187750	0.156605	0.130739
39	0.261413	0.216621	0.179665	0.149148	0.123924
40	0.252572	0.208289	0.171929	0.142046	0.117463
41	0.244031	0.200278	0.164525	0.135282	0.111339
42	0.235779	0.192575	0.157440	0.128840	0.105535
43	0.227806	0.185168	0.150661	0.122704	0.100033
44	0.220102	0.178046	0.144173	0.116861	0.094818
45	0.212659	0.171198	0.137964	0.111297	0.089875
46	0.205468	0.164614	0.132023	0.105997	0.085190
47	0.198520	0.158283	0.126338	0.100949	0.080748
48	0.191806	0.152195	0.120898	0.096142	0.076539
49	0.185320	0.146341	0.115692	0.091564	0.072549
50	0.179053	0.140713	0.110710	0.087204	0.068767
51	0.172998	0.135301	0.105942	0.083051	0.065182
52	0.167148	0.130097	0.101380	0.079096	0.061783
53	0.161496	0.125093	0.097014	0.075330	0.058563
54	0.156035	0.120282	0.092837	0.071743	0.055509
55	0.150758	0.115656	0.088839	0.068326	0.052616
56	0.145660	0.111207	0.085013	0.065073	0.049873
57	0.140734	0.106930	0.081353	0.061974	0.047273
58	0.135975	0.102817	0.077849	0.059023	0.044808
59	0.131377	0.098863	0.074497	0.056212	0.042472
60	0.126934	0.095060	0.071289	0.053536	0.040258

TABLE C.4 (continued)

N	I=.0600	I=.0650	I=.0700	I=.0750	I=.0800
1	0.943396	0.938967	0.934579	0.930233	0.925926
2	0.889996	0.881659	0.873439	0.865333	0.857339
3	0.839619	0.827849	0.816298	0.804961	0.793832
4	0.792094	0.777323	0.762895	0.748801	0.735030
5	0.747258	0.729881	0.712986	0.696559	0.680583
6	0.704961	0.685334	0.666342	0.647962	0.630170
7	0.665057	0.643506	0.622750	0.602755	0.583490
8	0.627412	0.604231	0.582009	0.560702	0.540269
9	0.591898	0.567353	0.543934	0.521583	0.500249
10	0.558395	0.532726	0.508349	0.485194	0.463193
11	0.526788	0.500212	0.475093	0.451343	0.428883
12	0.496969	0.469683	0.444012	0.419854	0.397114
13	0.468839	0.441017	0.414964	0.390562	0.367698
14	0.442301	0.414100	0.387817	0.363313	0.340461
15	0.417265	0.388827	0.362446	0.337966	0.315242
16	0.393646	0.365095	0.338735	0.314387	0.291890
17	0.371364	0.342813	0.316574	0.292453	0.270269
18	0.350344	0.321890	0.295864	0.272049	0.250249
19	0.330513	0.302244	0.276508	0.253069	0.231712
20	0.311805	0.283797	0.258419	0.235413	0.214548
21	0.294155	0.266476	0.241513	0.218989	0.198656
22	0.277505	0.250212	0.225713	0.203711	0.183941
23	0.261797	0.234941	0.210947	0.189498	0.170315
24	0.246979	0.220602	0.197147	0.176277	0.157699
25	0.232999	0.207138	0.184249	0.163979	0.146018
26	0.219810	0.194496	0.172195	0.152539	0.135202
27	0.207368	0.182625	0.160930	0.141896	0.125187
28	0.195630	0.171479	0.150402	0.131997	0.115914
29	0.184557	0.161013	0.140563	0.122788	0.107328
30	0.174110	0.151186	0.131367	0.114221	0.099377
31	0.164255	0.141959	0.122773	0.106252	0.092016
32	0.154957	0.133295	0.114741	0.098839	0.085200
33	0.146186	0.125159	0.107235	0.091943	0.078889
34	0.137912	0.117520	0.100219	0.085529	0.073045
35	0.130105	0.110348	0.093663	0.079562	0.067635
36	0.122741	0.103613	0.087535	0.074011	0.062625
37	0.115793	0.097289	0.081809	0.068847	0.057986
38	0.109239	0.091351	0.076457	0.064044	0.053690
39	0.103056	0.085776	0.071455	0.059576	0.049713
40	0.097222	0.080541	0.066780	0.055419	0.046031
41	0.091719	0.075625	0.062412	0.051553	0.042621
42	0.086527	0.071010	0.058329	0.047956	0.039464
43	0.081630	0.066676	0.054513	0.044610	0.036541
44	0.077009	0.062606	0.050946	0.041498	0.033834
45	0.072650	0.058785	0.047613	0.038603	0.031328
46	0.068538	0.055197	0.044499	0.035910	0.029007
47	0.064658	0.051828	0.041587	0.033404	0.026859
48	0.060998	0.048665	0.038867	0.031074	0.024869
49	0.057546	0.045695	0.036324	0.028906	0.023027
50	0.054288	0.042906	0.033948	0.026889	0.021321
51	0.051215	0.040287	0.031727	0.025013	0.019742
52	0.048316	0.037829	0.029651	0.023268	0.018280
53	0.045582	0.035520	0.027711	0.021645	0.016925
54	0.043001	0.033352	0.025899	0.020135	0.015672
55	0.040567	0.031316	0.024204	0.018730	0.014511
56	0.038271	0.029405	0.022621	0.017423	0.013436
57	0.036105	0.027610	0.021141	0.016208	0.012441
58	0.034061	0.025925	0.019758	0.015077	0.011519
59	0.032133	0.024343	0.018465	0.014025	0.010666
60	0.030314	0.022857	0.017257	0.013046	0.009876

TABLE C.5
Values of $a_{\overline{n}|i}$

N	I=.0025	I=.0050	I=.0075	I=.0100	I=.0125
1	0.997506	0.995025	0.992556	0.990099	0.987654
2	1.992525	1.985099	1.977723	1.970395	1.963115
3	2.985062	2.970248	2.955556	2.940985	2.926534
4	3.975124	3.950496	3.926110	3.901966	3.878058
5	4.962718	4.925866	4.889440	4.853431	4.817835
6	5.947848	5.896384	5.845598	5.795476	5.746010
7	6.930522	6.862074	6.794638	6.728195	6.662726
8	7.910745	7.822959	7.736613	7.651678	7.568124
9	8.888524	8.779064	8.671576	8.566018	8.462345
10	9.863864	9.730412	9.599580	9.471305	9.345526
11	10.836772	10.677027	10.520675	10.367628	10.217803
12	11.807254	11.618932	11.434913	11.255077	11.079312
13	12.775316	12.556151	12.342345	12.133740	11.930185
14	13.740963	13.488708	13.243022	13.003703	12.770553
15	14.704203	14.416625	14.136995	13.865053	13.600546
16	15.665040	15.339925	15.024313	14.717874	14.420292
17	16.623481	16.258632	15.905025	15.562251	15.229918
18	17.579533	17.172768	16.779181	16.398269	16.029549
19	18.533200	18.082356	17.646830	17.226008	16.819308
20	19.484488	18.987419	18.508020	18.045553	17.599316
21	20.433405	19.887979	19.362799	18.856983	18.369695
22	21.379955	20.784059	20.211215	19.660379	19.130563
23	22.324145	21.675681	21.053315	20.455821	19.882037
24	23.265980	22.562866	21.889146	21.243387	20.624235
25	24.205466	23.445638	22.718755	22.023156	21.357269
26	25.142609	24.324018	23.542189	22.795204	22.081253
27	26.077416	25.198028	24.359493	23.559608	22.796299
28	27.009891	26.067689	25.170713	24.316443	23.502518
29	27.940041	26.933024	25.975893	25.065785	24.200018
30	28.867871	27.794054	26.775080	25.807708	24.888906
31	29.793388	28.650800	27.568318	26.542285	25.569290
32	30.716596	29.503284	28.355650	27.269589	26.241274
33	31.637503	30.351526	29.137122	27.989693	26.904962
34	32.556112	31.195548	29.912776	28.702666	27.560456
35	33.472431	32.035371	30.682656	29.408580	28.207858
36	34.386465	32.871016	31.446805	30.107505	28.847267
37	35.298220	33.702504	32.205266	30.799510	29.478783
38	36.207700	34.529854	32.958080	31.484663	30.102501
39	37.114913	35.353089	33.705290	32.163033	30.718520
40	38.019863	36.172228	34.446938	32.834686	31.326933
41	38.922557	36.987291	35.183065	33.499689	31.927835
42	39.822999	37.798300	35.913713	34.158108	32.521319
43	40.721196	38.605274	36.638921	34.810008	33.107475
44	41.617154	39.408232	37.358730	35.455454	33.686395
45	42.510876	40.207196	38.073181	36.094508	34.258168
46	43.402370	41.002185	38.782314	36.727236	34.822882
47	44.291641	41.793219	39.486168	37.353699	35.380624
48	45.178695	42.580318	40.184782	37.973959	35.931481
49	46.063536	43.363500	40.878195	38.588079	36.475537
50	46.946170	44.142786	41.566447	39.196118	37.012876
51	47.826604	44.918195	42.249575	39.798136	37.543581
52	48.704842	45.689747	42.927618	40.394194	38.067734
53	49.580890	46.457459	43.600614	40.984351	38.585417
54	50.454753	47.221353	44.268599	41.568664	39.096708
55	51.326437	47.981445	44.931612	42.147192	39.601687
56	52.195947	48.737757	45.589689	42.719992	40.100431
57	53.063288	49.490305	46.242868	43.287121	40.593019
58	53.928467	50.239109	46.891184	43.848635	41.079524
59	54.791489	50.984189	47.534674	44.404589	41.560024
60	55.652358	51.725561	48.173374	44.955038	42.034592

TABLE C.5 (continued)

N	I=.0150	I=.0175	I=.0200	I=.0250	I=.0300
1	0.985222	0.982801	0.980392	0.975610	0.970874
2	1.955883	1.948699	1.941561	1.927424	1.913470
3	2.912200	2.897984	2.883883	2.856024	2.828611
4	3.854385	3.830943	3.807729	3.761974	3.717098
5	4.782645	4.747855	4.713460	4.645828	4.579707
6	5.697187	5.648998	5.601431	5.508125	5.417191
7	6.598214	6.534641	6.471991	6.349391	6.230283
8	7.485925	7.405053	7.325481	7.170137	7.019692
9	8.360517	8.260494	8.162237	7.970866	7.786109
10	9.222185	9.101223	8.982585	8.752064	8.530203
11	10.071118	9.927492	9.786848	9.514209	9.252624
12	10.907505	10.739550	10.575341	10.257765	9.954004
13	11.731532	11.537641	11.348374	10.983185	10.634955
14	12.543382	12.322006	12.106249	11.690912	11.296073
15	13.343233	13.092880	12.849264	12.381378	11.937935
16	14.131264	13.850497	13.577709	13.055003	12.561102
17	14.907649	14.595083	14.291872	13.712198	13.166118
18	15.672561	15.326863	14.992031	14.353364	13.753513
19	16.426168	16.046057	15.678462	14.978891	14.323799
20	17.168639	16.752881	16.351433	15.589162	14.877475
21	17.900137	17.447549	17.011209	16.184549	15.415024
22	18.620824	18.130269	17.658048	16.765413	15.936917
23	19.330861	18.801248	18.292204	17.332110	16.443608
24	20.030405	19.460686	18.913926	17.884986	16.935542
25	20.719611	20.108782	19.523456	18.424376	17.413148
26	21.398632	20.745732	20.121036	18.950611	17.876842
27	22.067617	21.371726	20.706898	19.464011	18.327031
28	22.726717	21.986955	21.281272	19.964889	18.764108
29	23.376076	22.591602	21.844385	20.453550	19.188455
30	24.015838	23.185849	22.396456	20.930293	19.600441
31	24.646146	23.769877	22.937702	21.395407	20.000428
32	25.267139	24.343859	23.468335	21.849178	20.388766
33	25.878954	24.907970	23.988564	22.291881	20.765792
34	26.481728	25.462378	24.498592	22.723786	21.131837
35	27.075595	26.007251	24.998619	23.145157	21.487220
36	27.660684	26.542753	25.488842	23.556251	21.832252
37	28.237127	27.069045	25.969453	23.957318	22.167235
38	28.805052	27.586285	26.440641	24.348603	22.492462
39	29.364583	28.094629	26.902589	24.730344	22.808215
40	29.915845	28.594230	27.355479	25.102775	23.114772
41	30.458961	29.085238	27.799489	25.466122	23.412400
42	30.994050	29.567801	28.234794	25.820607	23.701359
43	31.521232	30.042065	28.661562	26.166446	23.981902
44	32.040622	30.508172	29.079963	26.503849	24.254274
45	32.552337	30.966263	29.490160	26.833024	24.518713
46	33.056490	31.416474	29.892314	27.154170	24.775449
47	33.553192	31.858943	30.286582	27.467483	25.024708
48	34.042554	32.293801	30.673120	27.773154	25.266707
49	34.524683	32.721181	31.052078	28.071369	25.501657
50	34.999688	33.141209	31.423606	28.362312	25.729764
51	35.467673	33.554014	31.787849	28.646158	25.951227
52	35.928742	33.959719	32.144950	28.923081	26.166240
53	36.382997	34.358446	32.495049	29.193249	26.374990
54	36.830539	34.750316	32.838283	29.456829	26.577660
55	37.271467	35.135445	33.174788	29.713979	26.774428
56	37.705879	35.513951	33.504694	29.964858	26.965464
57	38.133871	35.885947	33.828131	30.209617	27.150936
58	38.555538	36.251545	34.145226	30.448407	27.331005
59	38.970973	36.610855	34.456104	30.681373	27.505831
60	39.380269	36.963986	34.760887	30.908656	27.675564

Table C.5 Values of $a_{\overline{n}|i}$

TABLE C.5 (continued)

N	I=.0350	I=.0400	I=.0450	I=.0500	I=.0550
1	0.966184	0.961538	0.956938	0.952381	0.947867
2	1.899694	1.886095	1.872668	1.859410	1.846320
3	2.801637	2.775091	2.748964	2.723248	2.697933
4	3.673079	3.629895	3.587526	3.545951	3.505150
5	4.515052	4.451822	4.389977	4.329477	4.270284
6	5.328553	5.242137	5.157872	5.075692	4.995530
7	6.114544	6.002055	5.892701	5.786373	5.682967
8	6.873956	6.732745	6.595886	6.463213	6.334566
9	7.607687	7.435332	7.268790	7.107822	6.952195
10	8.316605	8.110896	7.912718	7.721735	7.537626
11	9.001551	8.760477	8.528917	8.306414	8.092536
12	9.663334	9.385074	9.118581	8.863252	8.618518
13	10.302738	9.985648	9.682852	9.393573	9.117079
14	10.920520	10.563123	10.222825	9.898641	9.589648
15	11.517411	11.118387	10.739546	10.379658	10.037581
16	12.094117	11.652296	11.234015	10.837770	10.462162
17	12.651321	12.165669	11.707191	11.274066	10.864609
18	13.189682	12.659297	12.159992	11.689587	11.246074
19	13.709837	13.133939	12.593294	12.085321	11.607654
20	14.212403	13.590326	13.007936	12.462210	11.950382
21	14.697974	14.029160	13.404724	12.821153	12.275244
22	15.167125	14.451115	13.784425	13.163003	12.583170
23	15.620410	14.856842	14.147775	13.488574	12.875042
24	16.058368	15.246963	14.495478	13.798642	13.151699
25	16.481515	15.622080	14.828209	14.093945	13.413933
26	16.890352	15.982769	15.146611	14.375185	13.662495
27	17.285365	16.329586	15.451303	14.643034	13.898100
28	17.667019	16.663063	15.742874	14.898127	14.121422
29	18.035767	16.983715	16.021889	15.141074	14.333101
30	18.392045	17.292033	16.288889	15.372451	14.533745
31	18.736276	17.588494	16.544391	15.592811	14.723929
32	19.068865	17.873551	16.788891	15.802677	14.904198
33	19.390208	18.147646	17.022862	16.002549	15.075069
34	19.700684	18.411198	17.246758	16.192904	15.237033
35	20.000661	18.664613	17.461012	16.374194	15.390552
36	20.290494	18.908282	17.666041	16.546852	15.536068
37	20.570525	19.142579	17.862240	16.711287	15.673999
38	20.841087	19.367864	18.049990	16.867893	15.804738
39	21.102500	19.584485	18.229656	17.017041	15.928662
40	21.355072	19.792774	18.401584	17.159086	16.046125
41	21.599104	19.993052	18.566109	17.294368	16.157464
42	21.834883	20.185627	18.723550	17.423208	16.262999
43	22.062689	20.370795	18.874210	17.545912	16.363032
44	22.282791	20.548841	19.018383	17.662773	16.457851
45	22.495450	20.720040	19.156347	17.774070	16.547726
46	22.700918	20.884654	19.288371	17.880066	16.632915
47	22.899438	21.042936	19.414709	17.981016	16.713664
48	23.091244	21.195131	19.535607	18.077158	16.790203
49	23.276564	21.341472	19.651298	18.168722	16.862751
50	23.455618	21.482185	19.762008	18.255925	16.931518
51	23.628616	21.617485	19.867950	18.338977	16.996699
52	23.795765	21.747582	19.969330	18.418073	17.058483
53	23.957260	21.872675	20.066345	18.493403	17.117045
54	24.113295	21.992957	20.159181	18.565146	17.172555
55	24.264053	22.108612	20.248021	18.633472	17.225170
56	24.409713	22.219819	20.333034	18.698545	17.275043
57	24.550448	22.326749	20.414387	18.760519	17.322316
58	24.686423	22.429567	20.492236	18.819542	17.367124
59	24.817800	22.528430	20.566733	18.875754	17.409596
60	24.944734	22.623490	20.638022	18.929290	17.449854

TABLE C.5 (continued)

N	I=.0600	I=.0650	I=.0700	I=.0750	I=.0800
1	0.943396	0.938967	0.934579	0.930233	0.925926
2	1.833393	1.820626	1.808018	1.795565	1.783265
3	2.673012	2.648476	2.624316	2.600526	2.577097
4	3.465106	3.425799	3.387211	3.349326	3.312127
5	4.212364	4.155679	4.100197	4.045885	3.992710
6	4.917324	4.841014	4.766540	4.693846	4.622880
7	5.582381	5.484520	5.389289	5.296601	5.206370
8	6.209794	6.088751	5.971299	5.857304	5.746639
9	6.801692	6.656104	6.515232	6.378887	6.246888
10	7.360087	7.188830	7.023582	6.864081	6.710081
11	7.886875	7.689042	7.498674	7.315424	7.138964
12	8.383844	8.158725	7.942686	7.735278	7.536078
13	8.852683	8.599742	8.357651	8.125840	7.903776
14	9.294984	9.013842	8.745468	8.489154	8.244237
15	9.712249	9.402669	9.107914	8.827120	8.559479
16	10.105895	9.767764	9.446649	9.141507	8.851369
17	10.477260	10.110577	9.763223	9.433960	9.121638
18	10.827603	10.432466	10.059087	9.706009	9.371887
19	11.158116	10.734710	10.335595	9.959078	9.603599
20	11.469921	11.018507	10.594014	10.194491	9.818147
21	11.764077	11.284983	10.835527	10.413480	10.016803
22	12.041582	11.535196	11.061240	10.617191	10.200744
23	12.303379	11.770137	11.272187	10.806689	10.371059
24	12.550358	11.990739	11.469334	10.982967	10.528758
25	12.783356	12.197877	11.653583	11.146946	10.674776
26	13.003166	12.392373	11.825779	11.299485	10.809978
27	13.210534	12.574998	11.986709	11.441381	10.935165
28	13.406164	12.746477	12.137111	11.573378	11.051078
29	13.590721	12.907490	12.277674	11.696165	11.158406
30	13.764831	13.058676	12.409041	11.810386	11.257783
31	13.929086	13.200635	12.531814	11.916638	11.349799
32	14.084043	13.333929	12.646555	12.015478	11.434999
33	14.230230	13.459088	12.753790	12.107421	11.513888
34	14.368141	13.576609	12.854009	12.192950	11.586934
35	14.498246	13.686957	12.947672	12.272511	11.654568
36	14.620987	13.790570	13.035208	12.346522	11.717193
37	14.736780	13.887859	13.117017	12.415370	11.775179
38	14.846019	13.979210	13.193473	12.479414	11.828869
39	14.949075	14.064986	13.264928	12.538593	11.878582
40	15.046297	14.145527	13.331709	12.594409	11.924613
41	15.138016	14.221152	13.394120	12.645962	11.967235
42	15.224543	14.292161	13.452449	12.693918	12.006699
43	15.306173	14.358837	13.506962	12.738528	12.043240
44	15.383182	14.421443	13.557908	12.780026	12.077074
45	15.455832	14.480228	13.605522	12.818629	12.108402
46	15.524370	14.535426	13.650020	12.854539	12.137409
47	15.589028	14.587254	13.691608	12.887943	12.164267
48	15.650027	14.635919	13.730474	12.919017	12.189136
49	15.707572	14.681615	13.766799	12.947922	12.212163
50	15.761861	14.724521	13.800746	12.974812	12.233485
51	15.813076	14.764808	13.832473	12.999825	12.253227
52	15.861393	14.802637	13.862124	13.023093	12.271506
53	15.906974	14.838157	13.889836	13.044737	12.288432
54	15.949976	14.871509	13.915735	13.064872	12.304103
55	15.990543	14.902825	13.939939	13.083602	12.318614
56	16.028814	14.932230	13.962560	13.101025	12.332050
57	16.064919	14.959840	13.983701	13.117233	12.344491
58	16.098980	14.985766	14.003458	13.132309	12.356010
59	16.131113	15.010109	14.021924	13.146334	12.366676
60	16.161428	15.032966	14.039181	13.159381	12.376552

TABLE C.6
Values of $s_{\overline{n}|i}$

N.	I=.0025	I=.0050	I=.0075	I=.0100	I=.0125
1	1.000000	1.000000	1.000000	1.000000	1.000000
2	2.002500	2.005000	2.007500	2.010000	2.012500
3	3.007506	3.015025	3.022556	3.030100	3.037656
4	4.015025	4.030100	4.045225	4.060401	4.075627
5	5.025063	5.050251	5.075565	5.101005	5.126572
6	6.037625	6.075502	6.113631	6.152015	6.190654
7	7.052719	7.105879	7.159484	7.213535	7.268038
8	8.070351	8.141409	8.213180	8.285671	8.358888
9	9.090527	9.182116	9.274779	9.368527	9.463374
10	10.113253	10.228026	10.344339	10.462213	10.581666
11	11.138536	11.279167	11.421922	11.566835	11.713937
12	12.166383	12.335562	12.507586	12.682503	12.860361
13	13.196799	13.397240	13.601393	13.809328	14.021116
14	14.229791	14.464226	14.703404	14.947421	15.196380
15	15.265365	15.536548	15.813695	16.096896	16.386335
16	16.303529	16.614230	16.932282	17.257864	17.591164
17	17.344287	17.697301	18.059274	18.430443	18.811053
18	18.387648	18.785788	19.194718	19.614748	20.046192
19	19.433617	19.879717	20.338679	20.810895	21.296769
20	20.482201	20.979115	21.491219	22.019004	22.562979
21	21.533407	22.084011	22.652403	23.239194	23.845016
22	22.587240	23.194431	23.822296	24.471586	25.143078
23	23.643708	24.310403	25.000963	25.716302	26.457367
24	24.702818	25.431955	26.188471	26.973465	27.788084
25	25.764575	26.559115	27.384884	28.243200	29.135435
26	26.828986	27.691911	28.590271	29.525631	30.499628
27	27.896059	28.830370	29.804698	30.820888	31.880873
28	28.965799	29.974522	31.028233	32.129097	33.279384
29	30.038213	31.124395	32.260945	33.450388	34.695377
30	31.113309	32.280017	33.502902	34.784892	36.129069
31	32.191092	33.441417	34.754174	36.132740	37.580682
32	33.271570	34.608624	36.014830	37.494068	39.050441
33	34.354749	35.781667	37.284941	38.869009	40.538571
34	35.440636	36.960575	38.564578	40.257699	42.045303
35	36.529237	38.145378	39.853813	41.660276	43.570870
36	37.620560	39.336105	41.152716	43.076878	45.115505
37	38.714612	40.532785	42.461361	44.507647	46.679449
38	39.811398	41.735449	43.779822	45.952724	48.262942
39	40.910927	42.944127	45.108170	47.412251	49.866229
40	42.013204	44.158847	46.446482	48.886373	51.489557
41	43.118237	45.379642	47.794830	50.375237	53.133177
42	44.226033	46.606540	49.153291	51.878989	54.797341
43	45.336598	47.839572	50.521941	53.397779	56.482308
44	46.449939	49.078770	51.900856	54.931757	58.188337
45	47.566064	50.324164	53.290112	56.481075	59.915691
46	48.684979	51.575785	54.689788	58.045885	61.664637
47	49.806692	52.833664	56.099961	59.626344	63.435445
48	50.931208	54.097832	57.520711	61.222608	65.228388
49	52.058536	55.368321	58.952116	62.834834	67.043743
50	53.188683	56.645163	60.394257	64.463182	68.881790
51	54.321654	57.928389	61.847214	66.107814	70.742812
52	55.457459	59.218031	63.311068	67.768892	72.627097
53	56.596102	60.514121	64.785901	69.446581	74.534936
54	57.737593	61.816692	66.271796	71.141047	76.466623
55	58.881937	63.125775	67.768834	72.852457	78.422456
56	60.029141	64.441404	69.277100	74.580982	80.402736
57	61.179214	65.763611	70.796679	76.326792	82.407771
58	62.332162	67.092429	72.327654	78.090060	84.437868
59	63.487993	68.427891	73.870111	79.870960	86.493341
60	64.646713	69.770031	75.424137	81.669670	88.574508

TABLE C.6 (continued)

N	I=.0150	I=.0175	I=.0200	I=.0250	I=.0300
1	1.000000	1.000000	1.000000	1.000000	1.000000
2	2.015000	2.017500	2.020000	2.025000	2.030000
3	3.045225	3.052806	3.060400	3.075625	3.090900
4	4.090903	4.106230	4.121608	4.152516	4.183627
5	5.152267	5.178089	5.204040	5.256329	5.309136
6	6.229551	6.268706	6.308121	6.387737	6.468410
7	7.322994	7.378408	7.434283	7.547430	7.662462
8	8.432839	8.507530	8.582969	8.736116	8.892336
9	9.559332	9.656412	9.754628	9.954519	10.159106
10	10.702722	10.825399	10.949721	11.203382	11.463879
11	11.863262	12.014844	12.168715	12.483466	12.807796
12	13.041211	13.225104	13.412090	13.795553	14.192030
13	14.236830	14.456543	14.680332	15.140442	15.617790
14	15.450382	15.709533	15.973938	16.518953	17.086324
15	16.682138	16.984449	17.293417	17.931927	18.598914
16	17.932370	18.281677	18.639285	19.380225	20.156881
17	19.201355	19.601607	20.012071	20.864730	21.761588
18	20.489376	20.944635	21.412312	22.386349	23.414435
19	21.796716	22.311166	22.840559	23.946007	25.116868
20	23.123667	23.701611	24.297370	25.544658	26.870374
21	24.470522	25.116389	25.783317	27.183274	28.676486
22	25.837580	26.555926	27.298984	28.862856	30.536780
23	27.225144	28.020655	28.844963	30.584427	32.452884
24	28.633521	29.511016	30.421862	32.349038	34.426470
25	30.063024	31.027459	32.030300	34.157764	36.459264
26	31.513969	32.570440	33.670906	36.011708	38.553042
27	32.986678	34.140422	35.344324	37.912001	40.709634
28	34.481479	35.737880	37.051210	39.859801	42.930923
29	35.998701	37.363293	38.792235	41.856296	45.218850
30	37.538681	39.017150	40.568079	43.902703	47.575416
31	39.101762	40.699950	42.379441	46.000271	50.002678
32	40.688288	42.412200	44.227030	48.150278	52.502759
33	42.298612	44.154413	46.111570	50.354034	55.077841
34	43.933092	45.927115	48.033802	52.612885	57.730177
35	45.592088	47.730840	49.994478	54.928207	60.462082
36	47.275969	49.566129	51.994367	57.301413	63.275944
37	48.985109	51.433537	54.034255	59.733948	66.174223
38	50.719885	53.333624	56.114940	62.227297	69.159449
39	52.480684	55.266962	58.237238	64.782979	72.234233
40	54.267894	57.234134	60.401983	67.402554	75.401260
41	56.081912	59.235731	62.610023	70.087617	78.663298
42	57.923141	61.272357	64.862223	72.839808	82.023196
43	59.791988	63.344623	67.159468	75.660803	85.483892
44	61.688868	65.453154	69.502657	78.552323	89.048409
45	63.614201	67.598584	71.892710	81.516131	92.719861
46	65.568414	69.781559	74.330564	84.554034	96.501457
47	67.551940	72.002736	76.817176	87.667885	100.396501
48	69.565219	74.262784	79.353519	90.859582	104.408396
49	71.608698	76.562383	81.940590	94.131072	108.540648
50	73.682828	78.902225	84.579401	97.484349	112.796867
51	75.788070	81.283014	87.270989	100.921458	117.180773
52	77.924892	83.705466	90.016409	104.444494	121.696197
53	80.093765	86.170312	92.816737	108.055606	126.347082
54	82.295171	88.678292	95.673072	111.756996	131.137495
55	84.529599	91.230163	98.586534	115.550921	136.071620
56	86.797543	93.826690	101.558264	119.439694	141.153768
57	89.099506	96.468658	104.589430	123.425687	146.388381
58	91.435999	99.156859	107.681218	127.511329	151.780033
59	93.807539	101.892104	110.834843	131.699112	157.333434
60	96.214652	104.675216	114.051539	135.991590	163.053437

Table C.6 Values of $s_{\overline{n}|i}$

TABLE C.6 (continued)

N	I=.0350	I=.0400	I=.0450	I=.0500	I=.0550
1	1.000000	1.000000	1.000000	1.000000	1.000000
2	2.035000	2.040000	2.045000	2.050000	2.055000
3	3.106225	3.121600	3.137025	3.152500	3.168025
4	4.214943	4.246464	4.278191	4.310125	4.342266
5	5.362466	5.416323	5.470710	5.525631	5.581091
6	6.550152	6.632975	6.716892	6.801913	6.888051
7	7.779408	7.898294	8.019152	8.142008	8.266894
8	9.051687	9.214226	9.380014	9.549109	9.721573
9	10.368496	10.582795	10.802114	11.026564	11.256260
10	11.731393	12.006107	12.288209	12.577893	12.875354
11	13.141992	13.486351	13.841179	14.206787	14.583498
12	14.601962	15.025805	15.464032	15.917127	16.385591
13	16.113030	16.626838	17.159913	17.712983	18.286798
14	17.676986	18.291911	18.932109	19.598632	20.292572
15	19.295681	20.023588	20.784054	21.578564	22.408663
16	20.971030	21.824531	22.719337	23.657492	24.641140
17	22.705016	23.697512	24.741707	25.840366	26.996403
18	24.499691	25.645413	26.855084	28.132385	29.481205
19	26.357180	27.671229	29.063562	30.539004	32.102671
20	28.279682	29.778079	31.371423	33.065954	34.868318
21	30.269471	31.969202	33.783137	35.719252	37.786076
22	32.328902	34.247970	36.303378	38.505214	40.864310
23	34.460414	36.617889	38.937030	41.430475	44.111847
24	36.666528	39.082604	41.689196	44.501999	47.537998
25	38.949857	41.645908	44.565210	47.727099	51.152588
26	41.313102	44.311745	47.570645	51.113454	54.965981
27	43.759060	47.084214	50.711324	54.669126	58.989109
28	46.290627	49.967583	53.993333	58.402583	63.233510
29	48.910799	52.966286	57.423033	62.322712	67.711354
30	51.622677	56.084938	61.007070	66.438848	72.435478
31	54.429471	59.328335	64.752388	70.760790	77.419429
32	57.334502	62.701469	68.666245	75.298829	82.677498
33	60.341210	66.209527	72.756226	80.063771	88.224760
34	63.453152	69.857909	77.030256	85.066959	94.077122
35	66.674013	73.652225	81.496618	90.320307	100.251364
36	70.007603	77.598314	86.163966	95.836323	106.765189
37	73.457869	81.702246	91.041344	101.628139	113.637274
38	77.028895	85.970336	96.138205	107.709546	120.887324
39	80.724906	90.409150	101.464424	114.095023	128.536127
40	84.550278	95.025516	107.030323	120.799774	136.605614
41	88.509537	99.826536	112.846688	127.839763	145.118923
42	92.607371	104.819598	118.924789	135.231751	154.100464
43	96.848629	110.012382	125.276404	142.993339	163.575989
44	101.238331	115.412877	131.913842	151.143006	173.572669
45	105.781673	121.029392	138.849965	159.700156	184.119165
46	110.484031	126.870568	146.098214	168.685164	195.245719
47	115.350973	132.945390	153.672633	178.119422	206.984234
48	120.388257	139.263206	161.587902	188.025393	219.368367
49	125.601846	145.833734	169.859357	198.426663	232.433627
50	130.997910	152.667084	178.503028	209.347996	246.217476
51	136.582837	159.773767	187.535665	220.815396	260.759438
52	142.363236	167.164718	196.974797	232.856165	276.101207
53	148.345950	174.851306	206.838634	245.498974	292.286773
54	154.538058	182.845359	217.146373	258.773922	309.362546
55	160.946890	191.159173	227.917959	272.712618	327.377486
56	167.580031	199.805540	239.174268	287.348249	346.383247
57	174.445332	208.797762	250.937110	302.715662	366.434326
58	181.550919	218.149672	263.229280	318.851445	387.588214
59	188.905201	227.875659	276.074597	335.794017	409.905566
60	196.516883	237.990685	289.497954	353.583718	433.450372

TABLE C.6 (continued)

N	I=.0600	I=.0650	I=.0700	I=.0750	I=.0800
1	1.000000	1.000000	1.000000	1.000000	1.000000
2	2.060000	2.065000	2.070000	2.075000	2.080000
3	3.183600	3.199225	3.214900	3.230625	3.246400
4	4.374616	4.407175	4.439943	4.472922	4.506112
5	5.637093	5.693641	5.750739	5.808391	5.866601
6	6.975319	7.063728	7.153291	7.244020	7.335929
7	8.393838	8.522870	8.654021	8.787322	8.922803
8	9.897468	10.076856	10.259803	10.446371	10.636628
9	11.491316	11.731852	11.977989	12.229849	12.487558
10	13.180795	13.494423	13.816448	14.147087	14.486562
11	14.971643	15.371560	15.783599	16.208119	16.645487
12	16.869941	17.370711	17.888451	18.423728	18.977126
13	18.882138	19.499808	20.140643	20.805508	21.495297
14	21.015066	21.767295	22.550488	23.365921	24.214920
15	23.275970	24.182169	25.129022	26.118365	27.152114
16	25.672528	26.754010	27.888054	29.077242	30.324283
17	28.212880	29.493021	30.840217	32.258035	33.750226
18	30.905653	32.410067	33.999033	35.677388	37.450244
19	33.759992	35.516722	37.378965	39.353192	41.446263
20	36.785591	38.825309	40.995492	43.304681	45.761964
21	39.992727	42.348954	44.865177	47.552532	50.422921
22	43.392290	46.101636	49.005739	52.118972	55.456755
23	46.995828	50.098242	53.436141	57.027895	60.893296
24	50.815577	54.354628	58.176671	62.304987	66.764759
25	54.864512	58.887679	63.249038	67.977862	73.105940
26	59.156383	63.715378	68.676470	74.076201	79.954415
27	63.705766	68.856877	74.483823	80.631916	87.350768
28	68.528112	74.332574	80.697691	87.679310	95.338830
29	73.639798	80.164192	87.346529	95.255258	103.965936
30	79.058186	86.374864	94.460786	103.399403	113.283211
31	84.801677	92.989230	102.073041	112.154358	123.345868
32	90.889778	100.033530	110.218154	121.565935	134.213537
33	97.343165	107.535710	118.933425	131.683380	145.950620
34	104.183755	115.525531	128.258765	142.559633	158.626670
35	111.434780	124.034690	138.236878	154.251606	172.316804
36	119.120867	133.096945	148.913460	166.820476	187.102148
37	127.268119	142.748247	160.337402	180.332012	203.070320
38	135.904206	153.026883	172.561020	194.856913	220.315945
39	145.058458	163.973630	185.640292	210.471181	238.941221
40	154.761966	175.631916	199.635112	227.256520	259.056519
41	165.047684	188.047990	214.609570	245.300759	280.781040
42	175.950545	201.271110	230.632240	264.698315	304.243523
43	187.507577	215.353732	247.776496	285.550689	329.583005
44	199.758032	230.351725	266.120851	307.966991	356.949646
45	212.743514	246.324587	285.749311	332.064515	386.505617
46	226.508125	263.335685	306.751763	357.969354	418.426067
47	241.098612	281.452504	329.224386	385.817055	452.900152
48	256.564529	300.746917	353.270093	415.753334	490.132164
49	272.958401	321.295467	378.999000	447.934835	530.342737
50	290.335905	343.179672	406.528929	482.529947	573.770156
51	308.756059	366.486351	435.985955	519.719693	620.671769
52	328.281422	391.307963	467.504971	559.698670	671.325510
53	348.978308	417.742981	501.230319	602.676070	726.031551
54	370.917006	445.896275	537.316442	648.876776	785.114075
55	394.172027	475.879533	575.928593	698.542534	848.923201
56	418.822348	507.811702	617.243594	751.933224	917.837058
57	444.951689	541.819463	661.450646	809.328216	992.264022
58	472.648790	578.037728	708.752191	871.027832	1072.645144
59	502.007718	616.610180	759.364844	937.354919	1159.456755
60	533.128181	657.689842	813.520383	1008.656538	1253.213296

Solutions to Selected Odd-Numbered Problems

Chapter 1

Section 1.1

1. -3 3. 1.6 5. $-9\frac{1}{2}$ 7. 162 9. -162 11. $-\frac{2}{3}$ 13. -14.03 15. 4
17. -4 19. $-\frac{4}{5}$ 21. -0.55 23. -4 25. 4 27. -4 29. -7.7
31. 4 33. 2 35. -6.62 37. 30 39. $\frac{11}{112}$

Section 1.2

1. Yes   3. No   5. No   7. Yes   9. $x = 5$ 11. $x = -2$ 13. $p = -4$
15. $t = 7$ 17. $x = -3$ 19. $p = -\frac{23}{6}$ 21. $a = \frac{13}{3}$ 23. $y = \frac{1}{3}$
25. $t = -\frac{172}{7}$

Section 1.3

1. $0.67, 0.667$ 3. $0.24, 0.235$ 5. $2.87, 2.871$

Section 1.4

1. $\frac{1}{3}$ 3. π^{15} 5. $(1.7)^{5.1}$ 7. y^{10} 9. $(3.1)^{24}$ 11. $2^{-5} = \frac{1}{32}$
13. $10^{-3} = \frac{1}{1000}$ 15. $27^{1/3} = 3$ 17. $2^{-9} = \frac{1}{512}$ 19. $\frac{2}{3}$ 21. $\frac{6}{5}$

Section 1.5

1. $x = 2$ 3. $y = \pm 3$ 5. $b = \pm 2$ 7. $p = 1.3^{1/5}$ 9. $t = (9.3)^{1/9.3}$
11. $x = 3, x = 2$ 13. $p = -\frac{3}{2} + \sqrt{\frac{17}{2}}, p = -\frac{3}{2} - \sqrt{\frac{17}{2}}$ 15. $x = -3$
17. $N = \frac{1}{3}, N = -1$ 19. $t = (1 \pm \sqrt{21})/10$ 21. $x = 0, x = 2$

Section 1.6

1. 5.46×10^2 3. 1.021×10^1 5. 1.04×10^{-3} 7. 1.0004×10^1 9. 3356
11. 3300 13. 0.0863

Section 1.7

1. 0.1004 3. 2.1004 5. 0.6561 7. -0.3439 9. 1.8710 11. 1.1804
13. -0.9948 15. 5.1640 17. 483 19. 18.3 21. 0.183 23. 61.7
25. 0.0903 27. 6.943 29. 0.007743 31. 3745
33. (a) 2337 (b) 33.16 (c) 587,900

Section 1.8

1. 0.987 3. 0.9447 5. 0.1074 7. 0.6668 9. 4.39 11. 4.401 13. 0.02434

Section 1.9

1. (a) $A: (3, 2)$ $B: (9, 6)$ $C: (10, 0)$ $D: (4, -6)$ $E: (8, -4)$ $F: (-6, 5)$
   $G: (-2, 1)$ $H: (0, 5)$ $I: (-5, -3)$ $J: (-1, -4)$ $K: (0, -7)$
   (b) A, B, C

3. a.
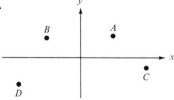

b.
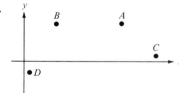

c.
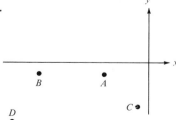

5.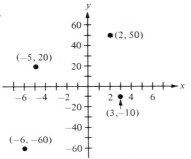

7. 0, 0   9. Same *y*-coordinate for each point on the line

Section 1.10

1. 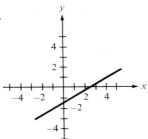   3.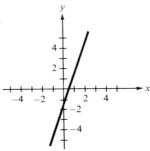

5. 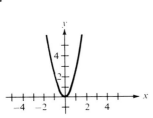   7. 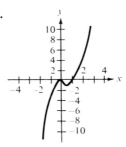   9.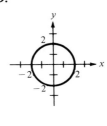

11. (a)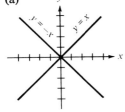

(b) Opposite slopes

13. All three points lie on the graph.

15. (a)

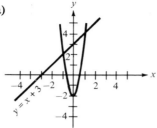

(b) Points of intersection: (1.105, 4.105), (−0.905, 2.095)

Section 1.11

1. (a) $x_1^2 + x_2^2 + x_3^2$
   (b) $2x_3 + 2x_4 + 2x_5 + 2x_6 + 2x_7 + 2x_8 + 2x_9 + 2x_{10} + 2x_{11}$
   (c) $(x_2 + y_2) + (x_3 + y_3) + (x_4 + y_4) + (x_5 + y_5) + (x_6 + y_6) + (x_7 + y_7)$
   (d) $(3m_{99} + 4) + (3m_{100} + 4) + (3m_{101} + 4) + (3m_{102} + 4)$
   $+ (3m_{103} + 4) + (3m_{104} + 4) + (3m_{105} + 4)$

3. (a) $\sum_{i=2}^{29} 3i^2$ (b) $\sum_{i=2}^{29} i(3)^2$ (c) $\sum_{i=2}^{29} 2(3^i)$ (d) $\sum_{i=2}^{29} (-1)^i(3i^2)$

5. (a) 15 (b) 31 (c) 151 (d) 3 (e) 26 (f) 465 (g) They are not equal.

9. Average: $1/n \sum_{i=1}^{n} G_i$

Chapter 2

Section 2.1

1. (a) Yes (b) No (c) No (d) Yes (e) Yes (f) Yes (g) No
   (h) Yes (i) No (j) Yes
3. $V = 5000 - 1000t$ is a linear equation.
5. (a)

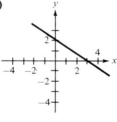

(b)

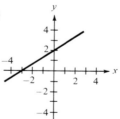

(c)

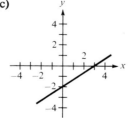

(d)

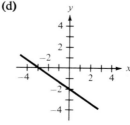

(e) 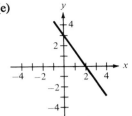   (f)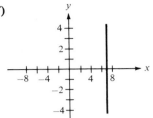

(g) 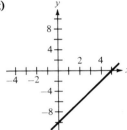   (h)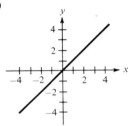

7. $3.25 million   9. $1 ($P = 20$ nickels)

Section 2.2

1. (a) $-\frac{2}{3}$ (b) $\frac{2}{3}$ (c) $\frac{2}{3}$ (d) $-\frac{2}{3}$ (e) $-\frac{3}{2}$ (f) ∞ (g) 2 (h) 1
3. $A = \frac{2}{11}t + 1$ 5. $P = 0.05S + 2000$

Section 2.3

1. (a) $C = 10x + 20,000$ (b) $R = 12x$ (c) $14,000 loss   (d) 10,000 units

3.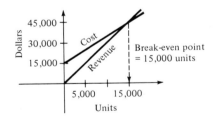

5. (a) 15,000 units (b) 4286 units
   (c) Yes—revenue increases faster for each unit sold at 50¢.   7. $8

Section 2.4

1. (a) Yes (b) No (c) No (d) Yes (e) No (f) Yes (g) Yes (h) Yes
3. (a) $x = 3, -2$ (b) $x = 5/3, -1$ (c) $x = -\sqrt{7}/2, +\sqrt{7}/2$
   (d) $x = (3 \pm \sqrt{21})/2$ (e) $x = (1 \pm \sqrt{17})/2$
5. (a) $1000   (b) $4000   (c) $t = 2.485$ years   (d) $t = 2.85$ years
   (e) $t = 2$ years

Section 2.5

1. (a) Yes; fifth degree in x (b) No (c) Yes; fourth degree in x

Section 2.6

1. Yes 3. No 5. Yes 7. No 9. Yes

Chapter 3

Section 3.1

1. $2431.01 3. $3149.28 5. $3362.22 7. $2145.94 9. $1077.01
11. $P(1460) = \$400(1 + 0.05/365)^{1460}$ 13. $P(1095) = \$1000(1 + 0.06/365)^{1095}$

Section 3.2

1. $18,937.16 3. $20,892.29 5. $11,881.41 7. $6892.50 9. $10,859.82
11. (a) $1532.79 (b) $6719.58 13. No 15. First offer 17. Third offer
19. Venture A is preferable. 23. 10.5% 27. 6.88 years 29. 13.87 years

Section 3.3

1. Not a profitable investment; $NPV = -\$52.23$ 3. $3084.47
5. PV(bank) = $2000; PV(lend) = $2042.44; lending is more profitable.
7. PV of first opportunity = $5734.96; PV of second opportunity = $5582.38; first opportunity is more profitable. 9. $5189.83 11. $8696.88

Section 3.4

1. (a) $12,006.11 (b) $12,486.35 3. (a) $1389.40 (b) $1406.77
5. $10\{[(1 + 0.05/52)^{52} - 1]/(0.05/52)\} = 10(53.2956) = \$532.96$
7. $790.66 9. $7995.44

Section 3.5

1. 6.1364% 3. 6.1837% 5. 4.0811% 7. 10.3813%

Chapter 4

Section 4.2

1. $A: 2 \times 2$ $B: 1 \times 5$ $C: 2 \times 3$ $D: 4 \times 3$ $E: 3 \times 1$

3. (a) $\begin{bmatrix} 3 & 2 \\ 0 & 0 \end{bmatrix}$; $\begin{bmatrix} -1 & 4 \\ 0 & -2 \end{bmatrix}$ (b) $\begin{bmatrix} 1 \\ 3 \\ 1 \end{bmatrix}$; $\begin{bmatrix} 1 \\ 1 \\ -3 \end{bmatrix}$

(c) $A + B$ and $A - B$ do not exist. (d) $\begin{bmatrix} 10 & 10 & 10 \\ 10 & 10 & 10 \\ 10 & 10 & 10 \end{bmatrix}$; $\begin{bmatrix} -8 & -6 & -4 \\ -2 & 0 & 2 \\ 4 & 6 & 8 \end{bmatrix}$

(e) $A + B$ and $A - B$ do not exist. (f) $\begin{bmatrix} 9 & -4 \\ 2 & 9 \\ 0 & 4 \\ 1 & -1 \\ 4 & 1 \end{bmatrix}$; $\begin{bmatrix} -1 & 2 \\ 2 & 5 \\ -2 & -4 \\ 5 & -3 \\ -2 & 1 \end{bmatrix}$

5. $\begin{bmatrix} -5 & 11 \\ -6.2 & 0.1 \\ -8.6 & 0 \end{bmatrix}$ 7. No 9. $x = \frac{3}{2}; y = \frac{10}{3}, z = 2$

11. (a) $\begin{bmatrix} 1 & 0 \\ 3 & -1 \end{bmatrix}$; $\begin{bmatrix} 2 & 0 \\ -1 & 1 \end{bmatrix}$ (b) $[1 \quad 2 \quad -1]$; $[0 \quad 1 \quad 2]$

(c) $\begin{bmatrix} 1 & 2 \\ 0 & 1 \\ 3 & 1 \end{bmatrix}$; $\begin{bmatrix} 2 & 4 \\ 1 & -1 \end{bmatrix}$ (d) $\begin{bmatrix} 1 & 4 & 7 \\ 2 & 5 & 8 \\ 3 & 6 & 9 \end{bmatrix}$; $\begin{bmatrix} 9 & 6 & 3 \\ 8 & 5 & 2 \\ 7 & 4 & 1 \end{bmatrix}$

(e) $[1 \quad 3 \quad 5 \quad 7]$; $\begin{bmatrix} 2 \\ 4 \\ 6 \\ 8 \end{bmatrix}$

(f) $\begin{bmatrix} 4 & 2 & -1 & 3 & 1 \\ -1 & 7 & 0 & -2 & 1 \end{bmatrix}$; $\begin{bmatrix} 5 & 0 & 1 & -2 & 3 \\ -3 & 2 & 4 & 1 & 0 \end{bmatrix}$

13. (a) No (b) No
15. (a) 6000 gallons premium, 10,000 gallons regular, 2000 gallons lead-free
    (b) 4000 gallons premium, 7500 gallons regular, 1000 gallons lead-free
    (c) 17,500 gallons premium, 25,800 gallons regular, 1000 gallons lead-free
    (d) More sales of lead-free reported than on hand

Section 4.3

1. (a) 3×1 (b) Does not exist (c) 1×4 (d) 3×3 (e) 4×4
   (f) 4×3 (g) 1×4 (h) Does not exist (i) 4×3

3. $AB = \begin{bmatrix} 17 & 11 \\ 4 & 32 \end{bmatrix}$; $BA = \begin{bmatrix} 15 & 30 & 20 & -5 \\ 7 & 6 & 2 & 7 \\ 7 & 18 & 13 & -7 \\ 11 & 6 & 0 & 15 \end{bmatrix}$

5. $XZ = \begin{bmatrix} 4 & 2 & 12 \\ 2 & 1 & 4 \\ 6 & 3 & 12 \end{bmatrix}$; $ZX = 17$; XY does not exist.

9. Two such matrices are $A = \begin{bmatrix} 1 & 2 \\ 2 & 1 \end{bmatrix}$ and $B = \begin{bmatrix} 2 & 3 \\ 3 & 2 \end{bmatrix}$.

11. (a) The entries of HP represent the total value of Dolin's stock at the close of each day.
    (b) PH does not exist.

13. $AI = \begin{bmatrix} 35 & -17 \\ -9 & 23 \end{bmatrix} = IA$

15. $AI = IA$ if A is a square matrix and I is an identity matrix having the same order as A.

Section 4.4

1. (a) Linear (b) Linear (c) Nonlinear (d) Linear (e) Nonlinear
   (f) Nonlinear (g) Linear
3. (a) No (b) No (c) Yes
5. (a) $\begin{bmatrix} 1 & -2 & 3 & -1 & | & 4 \\ 2 & 0 & -1 & 4 & | & 8 \end{bmatrix}$ (b) $\begin{bmatrix} 2 & -3 & 4 & | & 2 \\ 1 & 1 & -2 & | & 0 \\ -1 & 2 & 5 & | & 1 \end{bmatrix}$

   (d) $\begin{bmatrix} 8 & 1 & 1 & | & 3 \\ 4 & 2 & -1 & | & 2 \end{bmatrix}$ (g) $\begin{bmatrix} 1 & -2 & | & 1 \\ 1 & 2 & | & -1 \\ 2 & -1 & | & 4 \\ 1 & -2 & | & 1 \end{bmatrix}$

7. (a) $r + s - 3t = 0$ (b) $r + + 2t = 4$
       $2r + 4s + 5t = 6$ $s + 2t = 3$
                                    $2r + + t = 1$

9. (a) $C = 80{,}000 + 5B$ (b) $C - S = 0$
       $S = 8B$ $C - 5B = 80{,}000$
                                    $S - 8B = 0$

Section 4.5

1. $\begin{bmatrix} 1 & 3 & | & -1 \\ 0 & 1 & | & -6 \\ 0 & 0 & | & -21 \end{bmatrix}$ 3. $\begin{bmatrix} 1 & 3 & 4 & 1 & | & 2 \\ 0 & 1 & 0 & 2 & | & 1 \\ 0 & 0 & 1 & 1 & | & -2 \end{bmatrix}$ 5. $\begin{bmatrix} 1 & \frac{1}{2} & | & 2 \\ 0 & 1 & | & \frac{5}{3} \end{bmatrix}$

7. $\begin{bmatrix} 1 & \frac{1}{2} & 0 & | & \frac{7}{2} \\ 0 & 1 & \frac{1}{3} & | & \frac{1}{3} \\ 0 & 0 & 1 & | & -8 \end{bmatrix}$

Section 4.6

1. (a) $x = 1, y = 1$ (b) $a = \frac{11}{26}, b = -\frac{9}{26}$ (c) $s = 1.156, t = -0.0845$
   (d) $x = \frac{23}{8}, y = -2, z = -\frac{17}{8}$ (e) $a = -\frac{8}{11}, b = \frac{37}{11}, c = \frac{31}{11}$
   (f) $u = \frac{26}{17}, v = \frac{1}{17}, w = \frac{11}{17}$
   (g) $a = -2.0488, b = 0.8293, c = -5.4146, d = -3.0244$
   (h) $p = 1.61, q = -0.72$
3. (a) $x = 4 - 2\frac{1}{2}z, y = -1 - \frac{1}{2}z, z$ arbitrary
   (b) $a = 0.7145(1 + c), b = -0.143(1 + c), c$ arbitrary, $d = 0$
   (c) $u = -w, v = w, w$ arbitrary
   (d) $u = v - 2w, v$ arbitrary, w arbitrary

Section 4.7

1. $\begin{bmatrix} 2 & -3 \\ -1 & 2 \end{bmatrix}$ 3. $\begin{bmatrix} -2 & 1 \\ \frac{3}{2} & -\frac{1}{2} \end{bmatrix}$ 5. $\begin{bmatrix} -\frac{3}{2} & -1 & \frac{7}{2} \\ 1 & 1 & -2 \\ \frac{1}{2} & 0 & -\frac{1}{2} \end{bmatrix}$

7. $\begin{bmatrix} 7 & -3 & -\frac{17}{2} & \frac{43}{2} \\ -2 & 1 & \frac{5}{2} & -\frac{13}{2} \\ 0 & 0 & -\frac{1}{2} & \frac{3}{2} \\ 0 & 0 & \frac{1}{2} & -\frac{1}{2} \end{bmatrix}$

9. The first part of the augmented matrix cannot be reduced to the identity matrix.
11. $x = 1, y = 1$ 13. $a = -0.7273, b = 3.3636, c = 2.8182$

Chapter 5

Section 5.1

1. (a) True   (b) True   (c) True   (d) True   (e) False   (f) False
   (g) True   (h) True   (i) False

3.

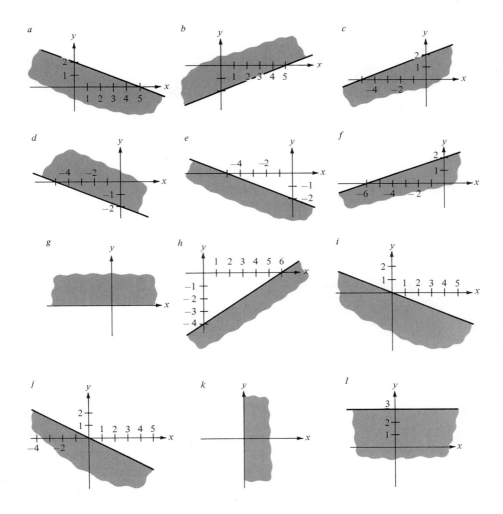

Section 5.2

1.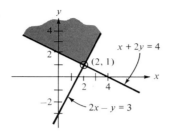

3.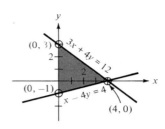

5.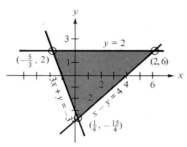

7.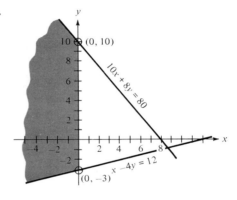

9.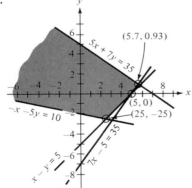

11.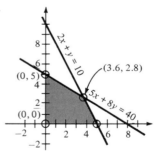

Section 5.3

1.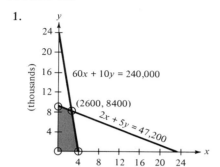

   Maximum = 24,600 at $x = 2600, y = 8400$

3.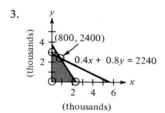

   Maximum = 84,000 at $x = 800, y = 2400$, and $x = 0, y = 2800$

5.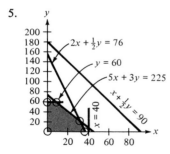

   Minimum = 8370 at $x = 9, y = 60$
   $x + \frac{1}{2}y \leq 90$ and $x \leq 40$ superfluous

7.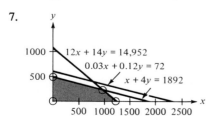

   Maximum = 2644 at $x = 980, y = 228$

Section 5.4

1. Minimum cost = $9,575,000, plant A operation = 31 days, plant B operation = $13\frac{1}{2}$ days
3. (a) Maximum profit = $246 obtained by producing 2600 type I and 8400 type II
   (b) Maximum profit = $377.60 obtained by producing 0 type I and 9440 type II
5. Maximum profit = $84,000 at 0 Sleeper and 2800 Johnny Firm, 800 Sleeper and 2400 Johnny Firm, or any point on the line connecting the two points.
7. Minimum cost = $318.18 using 90.91 gallons of juice and 909.09 gallons of drink
9. Minimum cost = $8370 with 9 sofas and 60 chairs
11. Maximize $P = 7x + 22y + 2z$
    subject to $\quad 3x + 10y + \quad z \leq 30{,}000$
    $\quad\quad\quad\quad\quad 5x + 8y + \quad z \leq 40{,}000$
    $\quad\quad\quad\quad\quad 0.1x + 0.6y + 0.1z \leq 120$
    $\quad\quad\quad\quad\quad x \geq 0, y \geq 0, z \geq 0$
13. Minimize $C = 0.02w + 0.02x + 0.01y + 0.03z$
    subject to $\quad 10w + x + y + 5z \geq 10$
    $\quad\quad\quad\quad\quad 2w + 40x + y + 10z \geq 20$
    $\quad\quad\quad\quad\quad 0.5w + 3x + 6y + 3z \geq 6$
    $\quad\quad\quad\quad\quad w \geq 0, x \geq 0, y \geq 0, z \geq 0$

Section 5.5

1. Maximize $3x + 2y + 0s_1 + 0s_2$
   subject to $\quad 60x + 10y + s_1 \quad\quad\quad = 240{,}000$
   $\quad\quad\quad\quad\quad 2x + 5y \quad\quad + s_2 = 47{,}200$
   assuming $x, y, s_1, s_2 \geq 0$
3. Minimize $2a + 7b + 0s_1 + 0s_2$
   subject to $\quad a - 3b - s_1 \quad\quad = 2$
   $\quad\quad\quad\quad\quad 2a + 4b \quad\quad - s_2 = 1$
   assuming $a, b, s_1, s_2 \geq 0$
5. Maximize $2x + 3y + 4z + 6w + 0s_1 + 0s_2 + 0s_3$
   subject to $\quad x + y + z + w + s_1 \quad\quad\quad\quad\quad = 15$
   $\quad\quad\quad\quad\quad 7x + 5y + 3z + 2w \quad\quad + s_2 \quad\quad = 120$
   $\quad\quad\quad\quad\quad 3x + 5y + 10z + 15w \quad\quad\quad\quad + s_3 = 100$
   assuming $x, y, z, w, s_1, s_2, s_3 \geq 0$

Section 5.6

1. Maximum = 24,600 at $x = 2600, y = 8400$
3. Maximum = 98,000 at $x = 500, y = 1300$
5. Maximum = 22.2857 at $x = 0, y = 9.43, z = 0.86$
7. Maximum = 40 at $x = 0, y = 4, z = 0$
9. Maximum = 19.7143 at $a = 3.86, b = 0, c = 0, d = 0.43$

Section 5.7

1. Maximize $2x + y$
   subject to $\quad x + 2y \leq 2$
   $\quad\quad\quad\quad\quad -3x + 4y \leq 7$
   $\quad\quad\quad\quad\quad x \geq 0, y \geq 0$

3. Maximize $100x + 100y + 20z$
   subject to $\quad 0.6x + 0.4y \quad\quad\quad\quad \leq 1.0$
   $\quad\quad\quad\quad\quad 0.05x + 0.1y + 0.08z \leq 0.25$
   $\quad\quad\quad\quad\quad x \geq 0, y \geq 0$
5. Maximum = 98,000 at $x = 1, y = 0, z = 2.5$
7. Minimum = 84,000 at $r = 3.75, s = 0$
9. Minimum = 52.5 at $x = 1.5, y = 0, z = 0.3$

Chapter 6

Section 6.1

1. Yes   3. No   5. Yes   7. No   9. Yes   11. Yes   13. Yes
15. (a) Yes   (b) Yes   (c) No

Section 6.2

1. (a) 0, 1, 2, 3, 4, 5, 6, 7, 8, 9, 10   (b) Integer numbers
   (c) 225, 196, 169, 144, 121, 100, 81, 64, 49, 36, 25
3. Yes; inverse is also a function.
5. Yes; inverse is not a function.
7. (a) 4   (b) 34   (c) -6 (d) $(a + b)^2 + 3(a + b) - 6$
   (e) $x^2 + 3x - 6 + (2x + 3)\Delta x + \Delta x^2$
9. (a) 18   (b) $d + 2d^2 + d^3$ (c) $(x + y) + 2(x + y)^2 + (x + y)^3$
   (d) $2a + 8a^2 + 8a^3$

Section 6.3

1. (a) 7   (b) 5   (c) -5 (d) -8 3. (a) 3   (b) 12   (c) 31   (d) -18
5. $-1/(x_1 x_2)$

Section 6.4

1. (a)   (b) 0   (c) 4

[Graph of $f(x)$ with Slope = -4, showing a parabola]

3. (a)   (b)   (c) 5

[Graph of $f(x)$, showing a V-shaped curve]

Slope at $x = 2$ is 5.

	y_2	$\dfrac{y_2 - y_1}{x_2 - x_1}$
	20	7
	12	6
	8.75	5.5
	7.04	5.2
	6.51	5.1

5. $-3, -3, -3$ 7. $2x - 2, 0, -10$ 9. $6x^2 - 4x + 3, 5, 115$
11. $2/(x + 2)^2, \frac{2}{9}, \frac{1}{2}$ 13. $-2/x^3, -2, \frac{1}{32}$
15. From the left instantaneous rate of change equals zero; from the right instantaneous rate of change equals zero.
17. (a) $6.5 million per year   (b) $10 million per year   (c) $80 million
    (d) $75.5 million

Section 6.5

1. $f'(x) = 5x^4 - 21x^2 + 4$ 3. $f'(x) = 7x^6 + 18x^2 - 8x$ 5. $f'(x) = x^4 - x^2 - x$
7. $f'(x) = 7x^6 + 20x^4 + 6x$ 9. $f'(x) = 10e^x$ 11. $f'(x) = 4x^3 - 14x + 7e^x$
13. $f'(x) = 3x^2 + 4x + 8e^x$ 15. $f'(x) = 1$ 17. $f'(x) = 7x^6 + 30x^4 + 3$
19. $f'(x) = 5x^4 - x^3 + 7e^x$ 21. $f'(x) = x^3 + x^2 - x$ 23. $f'(x) = 3x^2 - 6x + 2$
25. (a) $ds/dx = 6000 - 100x$ (b) Yes; yes
27. (a) 2 miles per 4 minutes = 30 miles per hour

(b)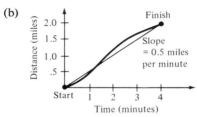

(c) 0.65 miles per minute = 39 miles per hour

(d)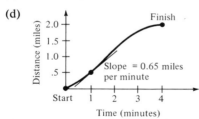

29. (a) No   (b) Yes   (c) No   (d) Yes

Section 6.6

1. $9x^8 + 24x^7 + 7x^6 + 18x^5 + 6x^2 + 12x$ 3. $-5x^{-6} - 3x^{-4} - 2x^{-3}$
5. $-\frac{5}{2}x^{-7/2} - \frac{1}{2}x^{-3/2}$ 7. $5(x^7 + 6x)^4(7x^6 + 6)$ 9. $7e^{7x}(x^7 + x^6)$
11. $(x^{10} + 15x^6 - 35x^4 - 21)/(x^{10} + 6x^6 + 9x^2)$ 13. $(10x^3 + 3x^2)e^{10x}$
15. $5x^4 - 15x^{-6}$ 17. $-5/x^6 - 4/x^5$ 19. $\frac{3}{2}x^{1/2} + \frac{35}{2}x^{3/2}$ 21. $4x^3 + 6x$
23. $4x/(3x^4 + 13x^2 + 13)$

Section 6.7

1. $5x^4 + 8x + 3, 20x^3 + 8, 28, 168$ 3. $4x^3 + 6x + 4, 12x^2 + 6, 18, 54$
5. $(10x^3 + 3x^2)e^{10x}, (100x^3 + 60x^2 + 6x)e^{10x}, 66e^{10}, 1052e^{20}$

7. (a) $-0.15t^2 + 0.5t + 0.3$ (b) 0.65 mile per minute = 39 miles per hour
   (c) $-0.3t + 0.5$ (d) 0.2 mile per minute^2, -0.4 mile per minute^2; at $t = 3$ the car is decelerating or slowing down

Chapter 7

Section 7.1

1.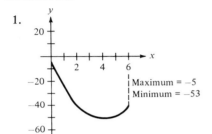

3.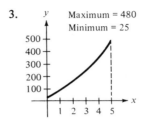

5.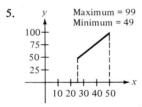

7. Maximum = 23; minimum = -9 9. Maximum = 7; minimum = -9
11. Maximum = 2908; Minimum = 2692   13. Maximum = 12; minimum = $-\frac{220}{3}$
15. Maximum = 75; minimum = -25 17. (a) 22,000   (b) $232,000
19. (a) 100 units   (b) $5000   (c) $25,000   21. 45 units
23. Maximum = 6; minimum = 1   25. 5, 3, 2, 3.5   27. No; No
29. Relative minimum at $x = 10$, relative maximum at $x = 4$

Section 7.3

1. (a) $P = -\frac{1}{2}x^2 + 180x - 5000$ (b) $0 \le x \le 300$ (c) $11,200
3. (a) $TR = 10x$; $TC = x^2/4000 - 5x + 50,000$
   (b) $P = -x^2/4000 + 15x - 50,000$ (c) $0 \le x \le 50,000$ (d) $175,000
5. 10,000 units   7. (a) $TR = 500x - 2x^2$ (b) $P = -2x^2 + 200x - 5000$
7. (a) $TR = 500x - 2x^2$ (b) $P = -2x^2 + 200x - 5000$ (c) $0 \le x \le 40$
   (d) 40 units   9. (a) 15   (b) $200 - x/2$ (c) $2.25 - (5 \times 10^{-5})x$
11. (a) $x/2500 + 8$ (b) 50   (c) 0.25

Section 7.4

1. 60   3. 50   5. (a) 100   (b) 10

7. (a) 800 (b) 3 (c) 2400 9.

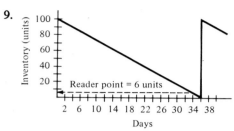

13. $TC = (bQ)/x + (k + 2MQ)x/2$; $EOQ = \sqrt{2bQ/(k + 2MQ)}$

Section 7.5

1. $62.5 million   3. $62.5 million   5. $2383.36 increase   7. $100 million

Chapter 8

Section 8.1

1. 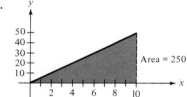   3.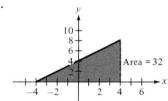

5. 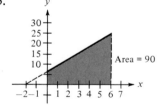   7.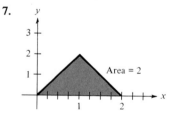

9.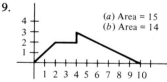

11. 212   13. 110.5   15. 45.60764   17. 1734.27   19. 1094

Section 8.2

1. $x^3/3 + C$ 3. $x^2/2 + C$ 5. $x^4/4 + x^3/3 + C$ 7. $2x^3 + 3x^2/2 + C$
9. $e^x + C$ 11. $2x^3 + e^{5x}/5 + C$ 13. $x^2 - 3e^s + C$ 15. $-e^{-x} + C$
17. $(e^x - e^{-x})/3 + C$ 19. $8x^{3/2}/3 + 9x^{4/3}/4 + C$

Section 8.3

1. $\frac{992}{3}$ 3. $\frac{63}{2}$ 5. $29\frac{1}{4}$ 7. 32   9. 402.42879   11. $130 + (e^{20} - e^{-5})/5$
13. -417.23948 15. 20.03575   17. 7342.16   19. 210.98   23. 111
25. 10   27. 0.63212

Section 8.4

1. $1004   3. $671.67, $4671.67   5. $356.96   7. $7695, $92,340
9. 33,128 cases   11. $1.84 million, $6.075 million

Section 8.5

1. 18.14   3. 324   5. $z^4/4 \,|_1^{1.2239} = 0.31$ 7. $z^2/2 \,|_{6.1}^{15.15} = 96.16$
9. $z^2/6 \,|_{-50}^{4} = -414$ 11. 1.71828   13. 9.85883

Chapter 9

Section 9.1

1. (b), (c), (e), and (i) are true; (a), (d), (f), (g), (h), (j), (k), and (l) are false.
3. (b), (c), (d), (e), (f), (g), (i), (j), (k), and (l) are true; (a) and (h) are false.
5. {Ford, Chevrolet, Pontiac}, {Ford, Chevrolet}, {Ford, Pontiac}, {Chevrolet, Pontiac}, {Ford}, {Chevrolet}, {Pontiac}, ∅.
7. No. $A = \{1, 2\}$ and $B = \{1, 3\}$ satisfy this definition, but A is not a subset of B.

Section 9.2

1. (a) {1, 2, 5, 6}   (b) {1, 2}   (c) {5, 6}   (d) {3, 4, 5, 6}   (e) {5}   (f) {6}
   (g) {5, 6}   (h) {1, 2}   (i) {3, 4, 5, 6}   (j) {1, 2, 3, 6}.
3. (a) {Buffalo}   (b) {Buffalo, Nashville, Tampa}   (c) ∅
   (d) {Buffalo, Cleveland}   (e) {Nashville, Tampa}   (f) {Cleveland}
   (g) cannot be calculated without a knowledge of U.
5. (a) {a, b, c, x, y, z}   (b) {a, b, c, r, s, t, u, v}   (c) does not exist.
7. (a) $R \cap S$ (b) $T \cap S'$ (c) $T \cap R'$ (d) T' (e) $(T \cap R) \cap S'$.
9. (a) {GM, GE, IBM, CBS}   (b) {GE}   (c) {GE}   (d) {CBS, Exxon, GE}
   (e) {IBM, GE, RCA, NBC}.

Section 9.3

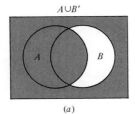

$A \cup B'$

(a)

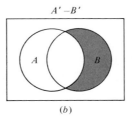

$A' - B'$

(b)

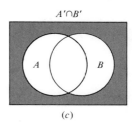

$A' \cap B'$

(c)

3.

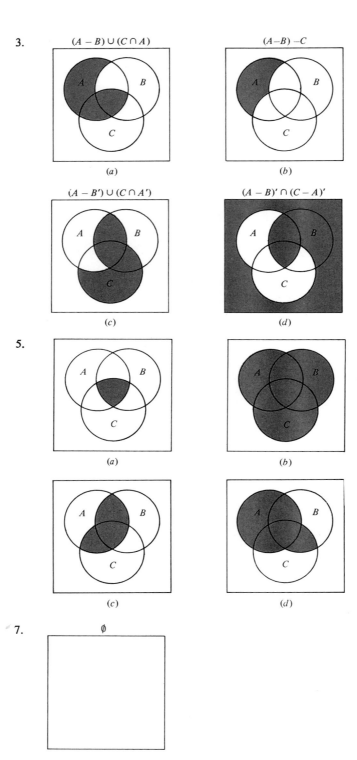

5.

7. ∅

Section 9.4

1. {green, red, white} 3. {−3, −2, −1, 1, 2, 3}
5. {2, 3, 4, 5, 6, 7, 8, 9, 10, jack, queen, king, ace}
7. Designate the architects as A, B, and C. Then
   U = {ABC, ACB, BAC, BCA, CAB, CBA} where, in particular, ABC denotes A first, B second, and C third.
9. {the individual passes and will be a good manager, the individual fails but would have been a good manager, the individual passes and is unsuited for management, the individual fails and would have been unsuited for management}
11. Designate each applicant by his or her first initial. Then
    U = {JM, JP, JT, MJ, MP, MT, PJ, PM, PT, TJ, TM, TP}.
13. Only one way—all the applicants are assigned jobs.

Chapter 10

Section 10.1

1.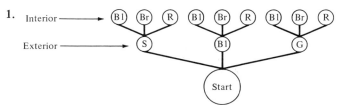

A total of nine combinations

3.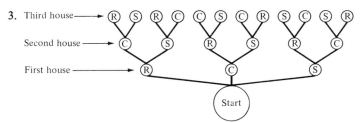

A total of 12 ways.

5.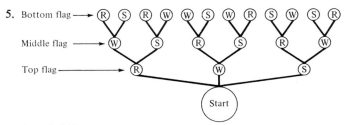

A total of 12 messages.

480 Solutions

7. Designate real by R and artificial by A.

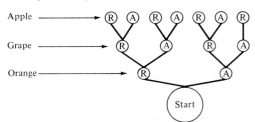

A total of seven different combinations

9.

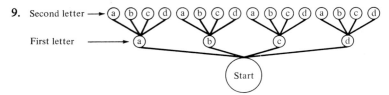

A total of 16 combinations

11.

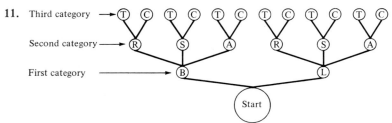

A total of 12 sets.

13. Designate the four types of lighting as 1, 2, 3, and 4 respectively

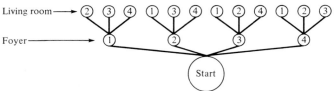

A Total of 12 combinations

15.

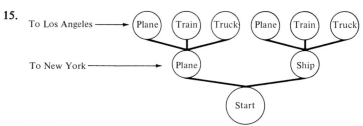

A total of six ways

Section 10.2

1. $(15)(20) = 300$ 3. $(10)(10)(26)(26)(26) = 1,757,600$ 5. $(6)(10)(7) = 420$
7. $(7)(8)(4)(3)(8)(12) = 64,512$ 9. $(14)(13)(12)(11) = 24,024$
11. $(3)(5)(4)(3)(6)(5)(4)(3)(2) = 129,600$ 13. $(9)(8)(7)(6)(5) = 15,120$
15. Yes; there are $(2)(3) = 6$ ways.

Section 10.3

1. $P(31, 6) = 530,122,320$ 3. $P(9, 5) = 15,120$ 5. $P(4, 3) = 24$
7. $P(4, 2) = 12$ 9. $141!$ 11. $P(95, 80)$ 13. $5! = 120$
15. $P(n, r) = (n)(n - 1)(n - 2) \cdots (n - r + 1)$

$$= (n)(n - 1)(n - 2) \cdots (n - r + 1) \left[\frac{(n - r)(n - r - 1) \cdots (3)(2)(1)}{(n - r)(n - r - 1) \cdots (3)(2)(1)} \right]$$

$$= \frac{(n)(n - 1)(n - 2) \cdots (n - r + 1)(n - r)(n - r - 1) \cdots (3)(2)(1)}{(n - r)(n - r - 1) \cdots (3)(2)(1)}$$

$$= \frac{n!}{(n - r)!}$$

Section 10.4

1. (a) 15   (b) 210   (c) 171   (d) 455   (e) 21   (f) 1   (g) 1
   (h) is not defined.
3. $C(30, 3) = 4060$ 5. $C(14, 5) = 2002$ 7. $C(1000, 50)$ 9. $C(645, 17)$
11. From Exercise 15 in Section 10.3, we have $P(n, r) = \dfrac{n!}{(n - r)!}$; therefore

$$C(n, r) = \frac{P(n, r)}{r!} = \frac{n!/(n - r)!}{r!} = \frac{n!}{(n - r)!\, r!}.$$

13. $C(n - 1, r - 1) + C(n - 1, r)$

$$= \frac{(n - 1)!}{[(n - 1) - (r - 1)]!\,(r - 1)!} + \frac{(n - 1)!}{(n - 1 - r)!\, r!}$$

$$= \frac{(n - 1)!}{(n - r)!\,(r - 1)!} + \frac{(n - 1)!}{(n - r - 1)!\, r!}$$

$$= \frac{(n - 1)!}{(n - r)(n - r - 1)!\,(r - 1)!} + \frac{(n - 1)!}{(n - r - 1)!\, r(r - 1)!}$$

$$= \frac{(n - 1)!}{(n - r - 1)!\,(r - 1)!} \left[\frac{1}{n - r} + \frac{1}{r} \right]$$

$$= \frac{(n - 1)!}{(n - r - 1)!\,(r - 1)!} \left[\frac{n}{(n - r)r} \right]$$

$$= \frac{n(n - 1)!}{(n - r)(n - r - 1)!\, r(r - 1)!}$$

$$= \frac{n!}{(n - r)!\, r!}$$

$$= C(n, r)$$

Section 10.5

1. $C(5, 1)C(10, 1)C(5, 2)C(6, 1) = 3000$ 3. $C(10, 2)C(8, 3) = 2520$
5. $C(52, 20)C(7, 3)C(31, 5)C(11, 3)$ 7. $C(3, 2)C(20, 4)C(22, 4) = 106{,}323{,}525$
9. $\dfrac{11!}{4!\,4!\,2!} = 34{,}650$ 11. $\dfrac{3!\,6!}{3!\,4!\,2!} = 15$ 13. $\dfrac{10!}{6!\,4!} = 210$ 15. $\dfrac{12!}{6!\,6!} = 924$
17. $C(4, 2)C(4, 3) = 24$ 19. $C(4, 2)C(4, 2)C(44, 1) = 1584$
21. $C(13, 4)C(13, 4)C(13, 4)C(13, 1) = 4{,}751{,}836{,}375$ 23. 6
25. (a) 10!   (b) 4! 2! 3! 1! 4!   (c) 4! 3! 4! 2! 1!

Chapter 11

Section 11.1

1. (a) and (c) have equally likely outcomes; (b), (d), and (e) do not.
3. (a) Rolling a 2   (b) Rolling a 2 or a 12   (c) Rolling a 2, 3, or a 4 consisting of two 2s   (d) Rolling at least one 1   (e) Rolling a 10, 11, or 12
5. (a) {HHH, HHT, HTH, HTT, THH, THT, TTH, TTT}
   (b) {HHT, HTH, THH}   (c) {HTT, THT, TTH, TTT}   (d) $\frac{3}{8}$ (e) $\frac{4}{8} = \frac{1}{2}$
7. Label each dress by color and number so that G1 designates the first green dress and R1 designates the first red dress, etc. Then
   (a) {G1-G2, G1-G3, G1-R1, G1-R2, G1-W1, G1-W2, G2-G3, G2-R1, G2-R2, G2-W1, G2-W2, G3-R1, G3-R2, G3-W1, G3-W2, R1-R2, R1-W1, R1-W2, R2-W1, R2-W2, W1-W2}
   (b) {G1-R1, G1-R2, G1-W1, G1-W2, G2-R1, G2-R2, G2-W1, G2-W2, G3-R1, G3-R2, G3-W1, G3-W2, R1-W1, R1-W2, R2-W1, R2-W2}
   (c) $\frac{16}{21}$
9. $\frac{4}{6} = \frac{2}{3}$ 11. (a) $\frac{73}{100}$ (b) $\frac{45}{100}$ (c) $\frac{31}{100}$ 13. $4/C(5, 2) = \frac{4}{10} = .4$
15. The probability of sliced beef is $\frac{1}{3}$, while the probability of both sliced beef and ice cream is $\frac{1}{12}$.

Section 11.2

1. $C(65, 12)/C(80, 12)$ 3. $(10)(5)(4)(4)(3)(3)(2)(2)(1)(1)/10! = .0079$
5. (a) $C(12, 6)/C(48, 6)$ (b) $C(36, 6)/C(48, 6)$
   (c) $C(12, 2)C(12, 2)C(24, 2)/C(48, 6)$
7. A production run of 3000 radios will result in 6 defective ones.
   (a) $C(2994, 10)/C(3000, 10)$
   (b) 0, since there are only 6 defective ones in the entire lot
   (c) $C(6, 2)C(2994, 8)/C(3000, 10)$
9. Four of the 80 cars on the lot will experience difficulty. The probability of no difficulties in the next 10 cars is $C(76, 10)/C(80, 10)$. The probability of 1 difficulty is $C(4, 1)C(76, 9)/C(80, 10)$
11. $(52)(3)(48)(44)(40)/C(52, 5)$ 13. $C(20, 2)C(80, 5)/C(100, 7)$
15. $P(365, 22)/(365)^{22}$
17. $C(6, 1)C(3, 1)C(1, 1)/C(10, 3) = (6)(3)(1)/120 = .15$
19. $C(9, 2)C(11, 1)/C(20, 3) = (36)(11)/1140 = .35$

Section 11.3

1. 38%
3. The three events are mutually exclusive, and their union is U. The sum of the probabilities should equal $P(U) = 1$ but does not; the sum is 1.23.
5. (a) The probability that the customer is not female
   (b) The probability that the customer is either female or makes a purchase
   (c) The probability that the customer is female and makes a purchase
   (d) The probability that the customer is female but does not make a purchase
   (e) The probability that the customer is male and makes a purchase
   (f) The probability that the customer is either male or does not make a purchase
7. .11
9. 1.01, which is absurd since probabilities must be less than or equal to 1. At least one of the given probabilities must be incorrect.
11. 19%
13. Let O_i designate the event of the ith outcome occurring and let $P(O_i) = x$ since each event is equally likely. Then $1 = P(U) = P(O_1 \cup O_2 \cup \cdots \cup O_{14}) = P(O_1) + P(O_2) + \cdots + P(O_{14}) = x + x + \cdots + x = 14x$. Hence $1 = 14x$ or $x = \frac{1}{14}$.
15. Denote $P(\text{tail}) = x$; then $P(\text{head}) = 2x$ and $P(\text{tail}) + P(\text{head}) = x + 2x = 3x$. But $P(\text{tail}) + P(\text{head}) = P(\text{tail or head}) = P(U) = 1$. Therefore $3x = 1$ or $x = \frac{1}{3}$.
17. (a) $P(U) = n(U)/n(U) = 1$ (b) $P(\emptyset) = n(\emptyset)/n(U) = 0/n(U) = 0$
    (c) Since E is a subset of U, $n(E) \leq n(U)$, hence $n(E)/n(U) \leq 1$. Furthermore, $n(E) \geq 0$ and $n(U) \geq 0$, since sets cannot have a negative number of elements, so $n(E)/n(U) \geq 0$.
    (d) If $E_1, E_2, \ldots, E_n$ are mutually exclusive, then $n(E_1 \cup E_2 \cup \cdots \cup E_n) = n(E_1) + n(E_2) + \cdots + n(E_n)$. Therefore

$$P(E_1 \cup E_2 \cup \cdots \cup E_n) = \frac{n(E_1 \cup E_2 \cup \cdots \cup E_n)}{n(U)}$$

$$= \frac{n(E_1) + n(E_2) + \cdots + n(E_n)}{n(U)}$$

$$= \frac{n(E_1)}{n(U)} + \frac{n(E_2)}{n(U)} + \cdots + \frac{n(E_n)}{n(U)}$$

$$= P(E_1) + P(E_2) + \cdots + P(E_n).$$

19. $1 = P(U) = P(E \cup E') = P(E) + P(E')$. Since $1 = P(E) + P(E')$, it follows that $P(E') = 1 - P(E)$.
21. $P(E \cup F) = P[(E - F) \cup (F - E) \cup (E \cap F)]$
    $= P(E - F) + P(F - E) + P(E \cap F)$
    $= [P(E) - P(E \cap F)] + [P(F) - P(F \cap E)] + P(E \cap F)$
    $= P(E) + P(F) - P(F \cap E)$
    $= P(E) + P(F) - P(E \cap F)$

Section 11.4

1. $\frac{3}{36} = .083$; $\frac{2}{30} = .067$
3. (a) $11{,}000/25{,}000 = .44$ (b) $15{,}000/25{,}000 = .60$ (c) $8000/11{,}000 = .73$

(d) $3000/15,000 = .20$ (e) $12,000/25,000 = .48$
5. (a) $.05/.45 = .11$ (b) $\dfrac{.05}{1 - .73} = \dfrac{.05}{.27} = .19$
7. $(.70)(.20) = .14$ 9. $(.11)/(.19) = .58$

Section 11.5

1. (a) $\frac{1}{4}$ (b) $.01$ (c) $.72$ (d) $.48$
3. $P(E) = .4/.3 = 1.33$, which is absurd since probabilities must be less than or equal to 1. At least one of the given probabilities is wrong.
5. $U = \{HHH, HHT, HTH, HTT, THH, THT, TTH, TTT\}$
   $E = \{HHH, HHT, HTH, HTT, THH, THT, TTH\}$
   $F = \{HHT, HTH, HTT, THH, THT, TTH\}$
   and $E \cap F = F$. Then $P(E \cap F) = \frac{6}{8}$, $P(E) = \frac{7}{8}$, $P(F) = \frac{6}{8}$, and
   $P(E \cap F) \neq P(E)\, P(F)$,
   so the events are *not* independent.
7. Since the first card is replaced and the deck reshuffled, $P(E \mid F) = \frac{4}{52} = P(E)$.
9. (a) $(\frac{3}{7})(\frac{3}{7}) = .18$ (b) $(\frac{3}{7})(\frac{2}{7}) = .12$ 11. $(.97)^3 = .91$
13. $(1 - .21)^4 = (.79)^4 = .39$
15. There are six female lawyers, so the probability is $(\frac{6}{30})(\frac{5}{29}) = .03$.
17. (a) $(.99)^2 = .98$
    (b) The same assuming the manufacturer produces a large number of faucets

Section 11.6

1. (a) $C(8, 4)(.3)^4(1 - .3)^{8-4} = 70(.0081)(.2401) = .14$
   (b) $C(7, 5)(.1)^5(1 - .1)^{7-5} = 21(.00001)(.81) = .00017$
   (c) $C(15, 7)(.5)^7(1 - .5)^{15-7}(6,435)(\frac{1}{128})(\frac{1}{256}) = .20$
3. $C(4, 1)(\frac{1}{4})^1(1 - \frac{1}{4})^{4-1} = .42$ 5. $C(50, 10)(.01)^{10}(.99)^{40}$
7. $C(3, 2)(.97)^2(.03)^1 = .08$
9. $C(10, 1)(.01)(.99)^9 = .091$

Section 11.7

1. $-\$1250$ 3. 0 5. $-\$0.05$
7. $\displaystyle\sum_{j=0}^{4} jC(4, j)(\tfrac{1}{2})^j(\tfrac{1}{2})^{4-j} = 2$ 9. $\displaystyle\sum_{j=0}^{10} jC(10, j)(\tfrac{12}{52})^j(\tfrac{40}{52})^{10-j}$

Appendix A

Section A.1

1. $y = 10.3$ 3. $a_1 = 10.6; a_2 = 10.2; a_3 = 9.4; a_4 = 10.2; a_5 = 10.2; a_6 = 10$
5. An increase does not seem advisable.

Appendix A

Section A.2

1.

x	y	y_c	$e(x)$	$[e(x)]^2$
-2	10	10.6	-0.6	0.36
-1	13	11.6	1.4	1.96
0	11	12.6	-1.6	2.56
1	15	13.6	1.4	1.96
2	14	14.6	-0.6	0.36

The least-squares error is 7.2.
3. (b) $y_c = 6.32x + 39.5$ (c) $E = 49.54$
5. (a) $y = 0.79x + 18.29$ for x coded around 1972   (b) 19.08   (c) 23.03
7. $y_c = 9.08x + 4.58$. For $x = 3.5$, $y_c = 36.36$.

Section A.3

1. (a) $0 = a - b + d$ (b) $3d \quad + 2a = 17$
   $\quad\quad 7 \quad\quad + d \quad\quad\quad 2b \quad\quad = 10$
   $\quad\quad 10 = a + b + d \quad\quad 2d \quad + 2a = 10$
3. $y = 0.643x^2 + 0.528x + 10.086$
5. Least-squares straight line is $y = 3.1x + 8.8$ with $E = 9.9$.

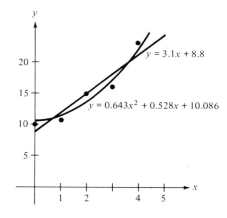

7. $\quad 7d \quad\quad + 28a = 20.9$
   $\quad\quad\quad 28b \quad\quad\quad = -0.5$
   $\quad 28d \quad + 196a = 167.1$
   $E = 0.057$ for the least-squares curve, while $E = 0.070$ for $y = x^2 - 1$.

Section A.4

1. $y = (2.42)(5.84)^x$

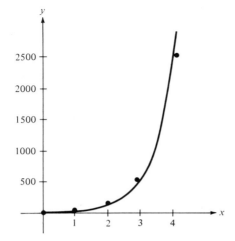

3. $y = (82.51)(5.84)^x$ in coded x.
5. (a) $y = 11{,}023(1.49)^x$ in coded x
   (b) 1980 corresponds to $x = 3$, hence $y = 11{,}023(1.49)^3 = 36{,}464$
   (c) When $y = 1000$, $x = -6.02$ which is approximately 1890.

Appendix B

Section B.1

1. 8191   3. 398,581   5. 0.083   7. $(1.2)^{-10}(5.19)$
9. (a) 36.129069   (b) 136.118795   (c) 12.577892   (d) 6958.239572
   (e) 306.751770   (f) 574.186015

Section B.2

1. (a) \$145; $12\% < i < 15\%$ (b) \$103.33; $12\% < i < 15\%$
3. (a) \$90   (b) $12\% < i < 15\%$
5. (a) \$83.33   (b) \$3200   (c) $9\% < i < 12\%$
7. (a) \$44.44   (b) $12\% < i < 15\%$

Section B.3

1. \$251.72
3. (a) \$150.97

(b)

Payment number	Amount of payment	Interest	Payment on principal	Outstanding balance
0	—	—	—	\$8000.00
1	\$150.97	\$33.34	\$117.63	\$7882.37
2	\$150.97	\$32.85	\$118.12	\$7764.25

7. \$31,245.70

INDEX

A

$a_{\overline{n}|i}$, 102, 111, 432
Addition
 of matrices, 121
 of signed numbers, 3
Add-on installment loan, 433
Amortization, 437
 schedule, 437
Annuity, 107
 due, 107
 ordinary, 107, 110
 simple, 107
Antiderivative, 288
 complete, 289
Antilog, 28
Approximate year, 92
Approximating areas, 282
Area, 279
 by integration, 296
Associative law, 6
Augmented matrix, 135, 139
Average rate of change, 218
 as a slope, 222
Average value, straight line fit, 404

B

Bernoulli trials, 393
Bounded region, 169
Break-even
 analysis, 72
 point, 73

C

$C(n, r)$, 351
Cartesian coordinate system, 33
Cash flow, 98. *See also* Annuity
 net present value of, 99, 102
 present value of, 98
Chain rule, 245
Coefficient matrix, 135
Column vector, 118
Combinations, 348, 351
Common logarithms. *See* Logarithms
Commutative law, 6
Complement, 319
Complete antiderivative, 289
Compound interest, 87
Conditional probability, 381
Continuous conversion rate, 115
Continuous variable, 260
Conversion period, 90
Coordinates, 35
Cornerpoints, 164
Cost, 72, 155, 261, 301
 marginal, 267, 301
Curve fitting, 403

D

Definite integral, 294
DeMorgan's laws, 329
Dependent variable, 215
Derivative
 applications of, 249

488 Index

Derivative (*continued*)
 definition, 236
 rules of computation for, 236, 242
 second, 246
Difference of sets, 317
Differentiation. *See* Derivative
Discount installment loan, 434
Discrete variable, 260
Disjoint sets, 317
Distributive law, 7
Division of signed numbers, 5
Domain, 209, 216
Dual, 199

E

e, 115, 238
Econometrics, 274
Economic order quantity, 272
Effective interest, 114
Element, 119, 313
Elementary row operations, 138, 140
Empty set, 315
End points, 254
Equal probability, 362, 364, 367
Equality of matrices, 120
Equality of sets, 314
Equations
 exponential, 84, 115
 linear, 54, 133
 with one unknown, 7
 polynomial, 82
 quadratic, 22, 76
 system of, 132, 134
 with two unknowns, 39
Euler's number, 115, 238
Event(s), 363
 independent, 388
Expected value, 397
 monetary, 398
Exponential equation, 85, 115, 238, 423
Exponents, 14

F

Factorials, 345
Finite geometric series, 429
Fixed costs, 72
Function, 208, 213
 domain of, 209
 objective, 169
 range of, 209
Fundamental theorem of counting, 341
Fundamental theorem of integral calculus, 295

Future value
 of an annuity, 106, 111
 of money, 92
 net, 107, 110, 111

G

Graphs, 39
Gross national product, 275

I

Identity matrix, 120
Indefinite integral, 289
Independent events, 388
Independent variable, 215
Individual reorder cost, 269
Inequalities, 155
 linear, 157
 strong, 156
 superfluous, 173
 systems of, 161
 weak, 156
Installment loans, 433
 add-on method, 433
 discount method, 434
Instantaneous rate of change, 224
 as a derivative, 236
 as a slope, 224
Integral, 279, 289
 applications of, 299
 definite, 294
 indefinite, 289
Integration. *See* Integral
Interest
 charges, 433
 compound, 87
 effective, 114
 nominal, 90
 rate per conversion period, 90
Intersection, 316
Inventory
 carrying costs, 302
 control, 267
Inverse of a matrix, 149
Inversion, 149

K

Keynesian theory, 274

L

Least squares
 error, 410
 exponential fit, 423

Least squares (*continued*)
  quadratic curve, 418
  straight line, 410
Limit, 115, 229, 287
Linear equation(s), 54, 408
  system of, 133, 144, 153
Linear functional, 168
Linear inequality, 157
  system of, 161
Linear least squares fit, 408
Linear programming, 169, 175, 191
  standard form, 187
Log. *See* Logarithms
Logarithms, 26, 30

M

Main diagonal, 119
Marginal cost, 267, 301
Marginal propensity to consume, 277
Marginal revenue, 267, 299
Matrices, 117
Matrix, 119
  addition, 121
  augmented, 135
  equality, 120
  inversion, 149
  multiplication, 126
  order of, 119
  scalar multiplication, 122
  subtraction, 122
  transposition, 123
  upper triangular, 141
Maximum value, 250
Minimum value, 250
Models, 259
Money
  future value of, 92, 106
  net future value of, 107, 110, 111
  net present value of, 99, 102
  present value of, 94, 99
  time value of, 92
Mortgages, 436
Moving averages, 405
Multiplication
  of matrices, 126
  of signed numbers, 4
Multiplier, 277
Mutually exclusive sets, 375

N

Net future value, 107, 110, 111
Net present value, 99, 102
Nominal interest rate, 90

Nonlinear equations, 133
Null set, 315

O

Objective function, 169
Optimization, 168, 191, 199, 250
Optimum order size, 272
Order of a matrix, 119
Ordinary annuity, 107
Origin, 33
Overhead, 263

P

$P(n, r)$, 346
Parabola, 77
Pascal's triangle, 353
Payment interval, 107
Periodic installments, 107
Permutation, 344
Pivot element, 142, 192
Polynomial equation, 82
Present value
  of an annuity, 111
  of a cash flow, 98
  of money, 92, 94
  net, 99, 102
Probability, 361
  conditional, 381
  definition of, 376
  equal, 362, 364, 367
  rules of, 373
Process, 329
Profit, 155, 261

Q

Quadrant, 36
Quadratic equation, 22, 76, 417
Quadratic formula, 22
  derivation of, 23
Quadratic least squares fit, 417

R

Range, 209
Rate of change, 66, 218
  average, 218
  instantaneous, 224
  of a line, 66
Reduced column, 192
Relative maximum, 252, 259
Relative minimum, 252, 259
Rent, 107
  period, 107

Reorder point, 273
Revenue, 73, 261, 299
  marginal, 267, 299
Roundoff, 12
Row vector, 118
Run, 393

S

$s_{\overline{n}|i}$, 109, 111, 112, 431
Sample space, 329
  equally likely, 362
Scalar multiplication, 122
Scatter diagram, 428
Scientific notation, 24
Second derivative, 246
  test, 259
Sets, 313
  algebra of, 316
Short run, 72
Sigma notation, 47
Signed numbers, 3
Simple annuity, 107
Simplex method, 191, 199
Simultaneous linear equations, 132
  solutions of, 133, 144, 153
Slack variables, 188
Slope
  as an average rate of change, 222
  of a curve, 224
  as an instantaneous rate of change, 224
  of a straight line, 64
Solution
  of an equation, 8, 40
  of a linear inequality, 157
  of a system of equations, 133, 144, 153
  of a system of linear inequalities, 161
Square matrix, 119
Standard form of a linear program, 187
Straight line, 54
  least squares, 410
  properties of, 64
Strong inequality, 156
Subset, 314
Substitution of variables, 306
Subtraction
  of matrices, 122
  of signed numbers, 5
Sum
  of matrices, 121
  of signed numbers, 3

Superfluous inequality, 173
Symmetric difference, 321
Systems
  of linear inequalities, 161
  of simultaneous equations, 132, 144, 153

T

Time diagrams, 99
Time value of money, 92
Total business expenditures, 275
Total carrying cost, 303
Total consumer expenditures, 275
Total inventory cost, 269
Total production cost, 261, 301
Total reordering cost, 269
Total revenue, 73, 261, 299
Total storage cost, 269
Transpose of a matrix, 123
Transposition, 123
Trees, 333

U

Upper triangular form, 141
Union, 316
Unit cost, 262
Unit price, 261
United States rule, 436
Universal set, 319

V

Variable, 58
  continuous, 260
  dependent, 215
  discrete, 260
  independent, 215
Variable cost, 72
Vector
  column, 118
  row, 118
Venn diagram, 321
Void set, 315

W, Y, Z

Weak inequality, 156
Work column, 192
y-intercept, 68
Zero factorial, 345
Zero matrix, 122